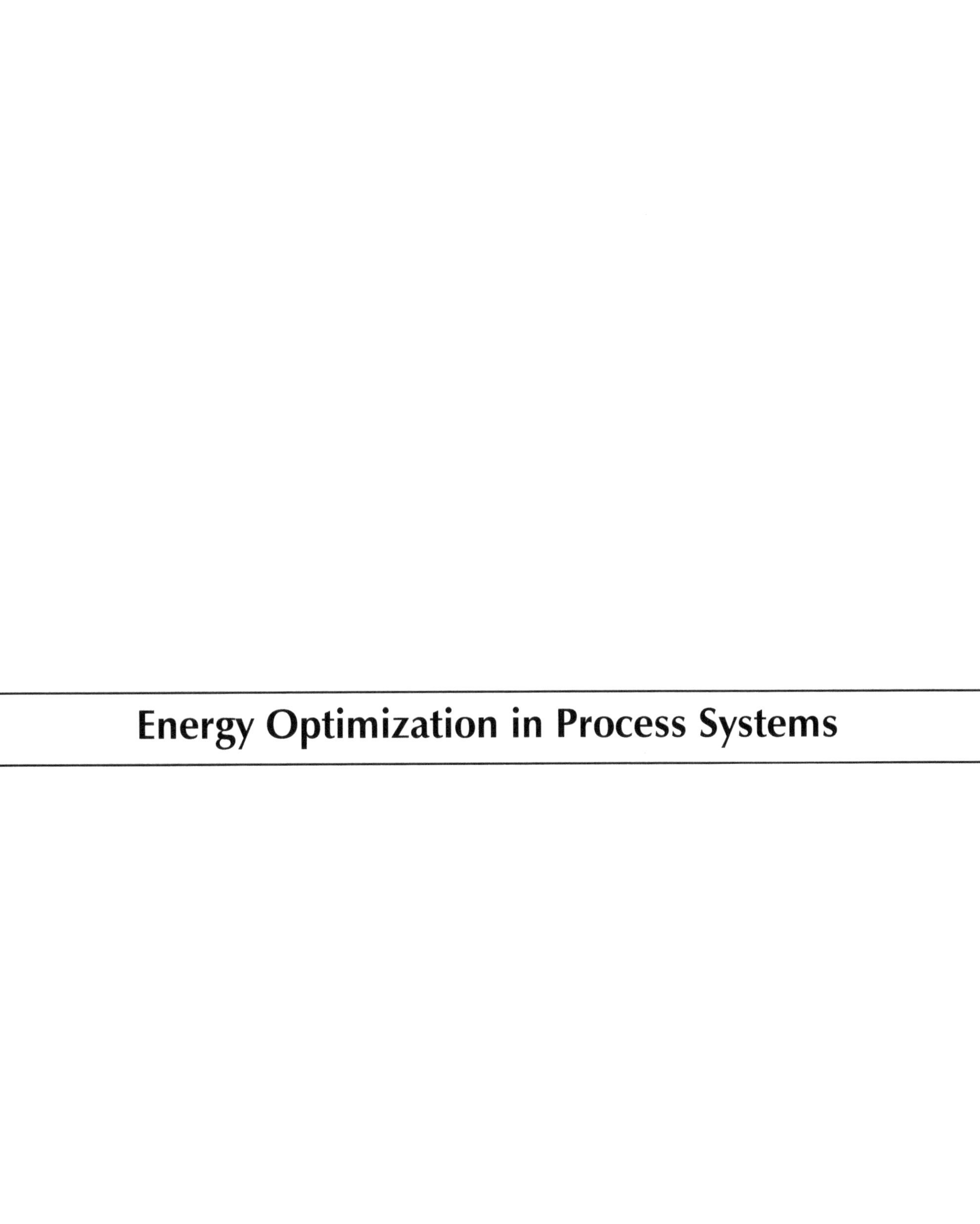

Energy Optimization in Process Systems

Energy Optimization in Process Systems

Editor: Alexia Fisher

www.murphy-moorepublishing.com

MURPHY & MOORE

Cataloging-in-Publication Data

Energy optimization in process systems / edited by Alexia Fisher.
p. cm.
Includes bibliographical references and index.
ISBN 978-1-63987-742-3
1. Power resources--Environmental aspects. 2. Energy facilities--Environmental aspects.
3. Mathematical optimization. I. Fisher, Alexia.
TD195.E49 E54 2023
333.791 4--dc23

Murphy & Moore Publishing
1 Rockefeller Plaza,
New York City,
NY 10020, USA

ISBN 978-1-63987-742-3

Contents

Preface

Energy is the ability to perform any type of work. It can exist in several different forms such as kinetic, chemical, potential, electrical, nuclear and thermal. Constant and reliable supply of energy is the foundation for all production activities. Energy optimization refers to the management of energy in the built environment in order to optimize advantages for the environment and human beings. Energy optimization and system integration plays a crucial role in reducing negative environmental effects of energy production, rising cost of energy and impending energy shortages. The implementation of energy optimization helps in significantly reducing the carbon emissions across different sectors, which can potentially help prevent the worst effects of climate change. Energy optimization can also be done by using process integration techniques that focus on the network of heat exchangers (HEXs) used widely in industries. This book unfolds the innovative aspects of energy optimization in process systems. It consists of contributions made by international experts. This book is a resource guide for experts as well as students.

The researches compiled throughout the book are authentic and of high quality, combining several disciplines and from very diverse regions from around the world. Drawing on the contributions of many researchers from diverse countries, the book's objective is to provide the readers with the latest achievements in the area of research. This book will surely be a source of knowledge to all interested and researching the field.

In the end, I would like to express my deep sense of gratitude to all the authors for meeting the set deadlines in completing and submitting their research chapters. I would also like to thank the publisher for the support offered to us throughout the course of the book. Finally, I extend my sincere thanks to my family for being a constant source of inspiration and encouragement.

Editor

Optimal Energy Management for Microgrids Considering Uncertainties in Renewable Energy Generation and Load Demand

Haotian Wu [1,*], Hang Li [2] and Xueping Gu [1]

[1] School of Electric and Electrical Engineering, North China Electric Power University, Baoding 071003, China; xpgu@ncepu.edu.cn
[2] CEPREI (Beijing) Industrial Technology Research Institute Co., Ltd., Beijing 100041, China; lihang@jimat-tech.com
* Correspondence: dgy@zua.edu.cn

Abstract: This paper proposes an efficient power management approach for the 24 h-ahead optimal maneuver of Mega–scale grid–connected microgrids containing a huge penetration of wind power, dispatchable distributed generation (diesel generator), energy storage system and local loads. The proposed energy management optimization objective aims to minimize the microgrid expenditure for fuel, operation and maintenance and main grid power import. It also aims to maximize the microgrid revenue by exporting energy to the upstream utility grid. The optimization model considers the uncertainties of the wind energy and power consumptions in the microgrids, and appropriate forecasting techniques are implemented to handle the uncertainties. The optimization model is formulated for a day-ahead optimization timeline with one-hour time steps, and it is solved using the ant colony optimization (ACO)-based metaheuristic approach. Actual data and parameters obtained from a practical microgrid platform in Atlanta, GA, USA are employed to formulate and validate the proposed energy management approach. Several simulations considering various operational scenarios are achieved to reveal the efficacy of the devised methodology. The obtained findings show the efficacy of the devised approach in various operational cases of the microgrids. To further confirm the efficacy of the devised approach, the achieved findings are compared to a pattern search (PS) optimization-based energy management approach and demonstrate outperformed performances with respect to solution optimality and computing time.

Keywords: ant colony optimization; energy management; microgrids; optimization; pattern search optimization; renewable energy; wind power; uncertainty

---

1. Introduction

The increasing deployment of distributed generations (DGs), the advantage of renewable energy in reducing carbon emissions, the intermittency of renewable generations, the advent of advanced controllers and the need to have a more reliable and resilient power grid are some of major causes for the ongoing energy transition reforms globally [1–3].

A microgrid (MG) is the assemblage of integrated electricity consumers, distributed generations (DGs) and distributed energy storages (DESs) at a distribution grid voltage level with clear electrical margins. The DGs can be the conventional or fuel-fired dispatchable power resources, such as a diesel generator, microturbine, fuel cell and other related energy sources [4]. The DGs can also contain renewable energy resources (RESs)—for example, wind power generation, photovoltaic (PV) solar generation, biomass and other technologies. While the distributed sources (DSs) can include batteries, flywheels and super-capacitors. Microgrids have a black starting ability and can function either in island mode or in grid-coupled mode [5]. They are not coupled with the upstream utility network when operating in the island mode. However, in the grid-coupled mode, MGs function in connection with the utility grid and exchange (buy or sell) power with the main grid.

Grid-coupled microgrids can export or import electrical energy to/from the larger electricity network. The power exchange with the upstream electricity network has generally been traded with a static, presettled price. Nevertheless, with the recent advent of smart sensor products and technology, it has become possible to precisely measure power generation and demands instantaneously or in real-time. This has created the opportunity for the change to time-of-use dynamic pricing schemes for electricity trading nowadays [6]. Well-established control techniques that can integrate a number of generation resources and storage devices in a microgrid framework are developing to provide electricity users access to obtain sustainable and secure electric power nearby [7,8]. It also paves the way for selling energy during excess power production or at expensive-price hours and buying energy during production shortage or cheap-price hours.

An energy management system (EMS) is a fundamental component of MG control and operational supervision. It takes input information from the generation sources, energy storage devices, load demands and main grid to properly allocate the microgrid energy resources (or decide the amount and duration of energy utilization). If this EMS decision is performed by solving some desired objective function (it can be the minimization of cost or maximization of profit), it is called an optimal EMS. The power-trading advancement inspires microgrid aggregators to adjust their power-exchange engagements with the upstream grid based on time-of-use dynamic pricing schemes in order to reduce power generation costs (fuel expenses), guarantee the enhanced utilization of RESs and DSs and improve the energy-trading profit. To accomplish these objectives, robust and optimal EMS should be implemented and integrated in the microgrid control architecture [6,9–11].

There have been several research works, by various individuals and institutions, on microgrid EMS in the previous couple of years. These works on microgrid EMS have had different objectives, configurations, scenarios and contexts. Some of these works are presented below.

A sensitivity study that augmented EMS, considering the rating of the energy storage system (ESS) and the growing trend of electric loads, was proposed for a microgrid system in Taiwan [12]. It aimed to search the optimal operation points of the microgrid energy resources for maximum profit. The EMS for optimal energy trading between two interconnected microgrid systems was presented in [13]. The objective was to minimize the power generation and transportation costs. In this work, centralized versus decentralized control schemes were investigated, employing iterative approaches and convex optimization techniques.

The work proposed in [14] developed an EMS to find the operating power set points of power sources in an MG. The work was based on an artificial neural network (ANN) and emphases to lower the total power generation expense of a MG that was involved in energy trading with the main electricity grid. Livengood [15] proposed a power management device known as an energy box to manage the electricity consumption of residential communities in a real–time dynamic electricity pricing scenarios. In this work, a stochastic dynamic program was used to solve the EMS objective problem using predictions of the electricity demand, meteorological variables and electricity price. The EMS decisions were the charging/discharging power of the ESS and amount of main grid import/export power.

An EMS with a hierarchical optimization framework was implemented in [16]. It focused on reducing the power generation spending and maximizing the power trading profit of a MG participating in a wholesale electricity-trading platform.

An optimal EMS model implemented employing a mixed-integer-linear-program (MILP) was formulated in [17] for minimizing operation expenditures in community microgrids in a dynamic price electricity market. The microgrid consisted of heating/cooling demands, ESS and dispatchable demands. Malysz [18] described an online power management technique to economically operate ESS devices of grid-coupled microgrids. The technique used a MILP described over a rolling scheduling period, employing forecasted electric loads and renewable power productions.

A genetic algorithm (GA)-based EMS [4] and modified particle swarm optimization (MPSO)-based EMS [5] were proposed for isolated microgrids containing multiple DGs and DSs. These EMS were targeted to reduce the operation costs of the MGs and effectively utilize the renewable and energy storage devices based on predictions of the electricity demand and renewables. References [5,7] proposed optimal EMS configurations depending on the prediction information of the electricity demands and renewable power productions for grid-connected microgrids in a variable electricity price environment. The proposed optimization problems were aimed to maximize profit and solved using the regrouping particle swarm optimization (RegPSO) technique.

Most of the aforementioned research works on microgrid energy management considered an ESS with one storage unit. The possible potential benefit of a microgrid ESS with more than one storage devices has not been investigated. Besides, the microgrids considered did not include multiple and integrated energy resources. Furthermore, the optimization techniques used to solve the EMS optimization problems did not ensure a global optimum solution, which, in turn, obstructed the exploitation of the maximum benefit of the microgrid in the power trading with the upper utility network.

In this paper, we propose an efficient power management technique for the 24 h-ahead optimal operation of mega-scale grid-coupled microgrids that consists of wind energy, a diesel generator, an energy storage system with several units and local (critical and noncritical) loads. The major target of the devised energy management methodology is to reduce the microgrid expenditure for fuel, operation and maintenance and main grid power import. It also targets maximizing the MG profit by exporting electricity to the upstream utility network. The optimization framework considers the stochastic ties of the wind power and electricity consumption in the MG, and pertinent predictions are employed to manage the stochastic ties. The optimization model is formulated for the 24 h-ahead scheduling period with a one-hour resolution, and it is solved using the ant colony optimization (ACO)-based meta-heuristic technique. Actual data and parameters obtained from an operating MG platform in Atlanta, GA, USA are employed to formulate and validate the proposed energy management approach. To assess and compare the efficacy of the devised method, another heuristic technique called a pattern search (PS) was also developed to obtain the EMS solution. The ACO was able to obtain the global best solution of the microgrid EMS problem. In addition, we chose ACO, as it has few parameters to update during the optimization process compared to other AI methods. The main contribution of the paper is the optimization formulation of microgrids considering the uncertainties of the renewable and load demands using integrated forecasting tools. From the optimization point to point, the contribution of the paper lays in implementing the ACO to solve such microgrid energy management optimization problems.

The arrangement of the other sections of the paper is described below. Section 2 outlines the case study microgrid and the proposed EMS optimization model. Section 3 describes the devised EMS framework and the working mechanism of the ACO. The simulation findings and comparative analysis are presented and discussed in Section 4. The study is summarized in Section 5.

2. Case Study Microgrid Framework and Proposed EMS Optimization Model

2.1. Microgrid Framework—Case Study Microgrid

The case study microgrid delivers power to various loads in an industrial park. The schematic illustration of the MG configuration is depicted in Figure 1. It contains wind power, diesel generator and ESS with two storage units. The ESS devices are a vanadium redox battery (VRB) and lithium-ion battery (Li–Ion). The MG is coupled with the upstream electricity network via a 10-kV busbar at the Point of Common Coupling (PCC). The real parametric values of the MG elements depicted in Figure 1 are employed in this work.

The EMS optimization problem, based on the case study microgrid framework, will be formulated in the following subsections. The MG operates in the grid-coupled mode. It can either export electricity to the upstream utility grid or import electricity from the utility power network. The MG EMS targets minimizing the total operating cost, consisting of the fuel expense, operation and maintenance (O&M) expense and grid power purchasing expense. Conversely, the EMS targets to maximize the profit that equals the income due to the power export to the utility grid minus the fuel and O&M costs.

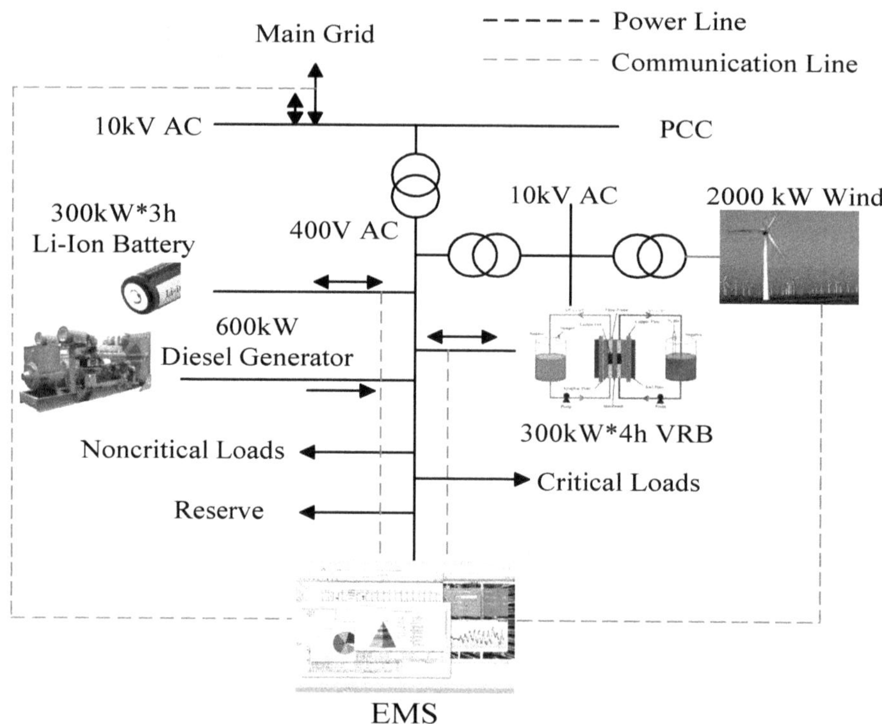

Figure 1. Framework of the case study microgrid. EMS: energy management system.

The decision variables are the charging/discharging powers and state of charges (SOCs) of the ESS devices, the generation from the diesel generator and the amount of energy trading with the upstream utility gird (i.e., grid power).

2.2. *Objective Function*

The optimization problem is formulated based on the following, the input information [19,20]:

- Electricity demand prediction
- Wind power prediction
- Electricity price prediction
- MG system data and component parameters

This information should be known in advance for the EMS to execute the desired optimal decisions.

As described above, the EMS aims to reduce the expense of buying electricity from the upstream utility network (or exploit the profit of selling electricity to the utility network), the fuel cost and O&M costs in the MG. The objective function is formulated as follows:

$$\begin{aligned} Min \sum_{t=1}^{N} \Delta T \Big\{ & c(t)P_g(t) + \sum_{k=1}^{G} \big(F_k(P_k(t)) \cdot \xi_k(t) + SU_k(t) + SD_k(t) + c_{OM,k}(t)P_k(t) \big) \\ & + c_{OM,w}(t)P_{wind}(t) + \sum_{u=1}^{S} c_{OM,u}(t)P_{ESS,u}(t) \Big\} \end{aligned} \tag{1}$$

where N is the scheduling horizon that equals 24 for the day-ahead hourly optimization framework; ΔT is the duration of the time steps that equals one hour for the hourly optimization model; *c(t)* is the electricity price (forecasted or prespecified) at time t, and it equals the purchasing price when the microgrid imports electricity from the upstream utility network and the selling price when the microgrid exports electricity to the main grid; $P_g(t)$ is the electricity of the utility grid—the sign convention here is that $P_g(t)$ is positive when the MG purchases electricity, negative when the MG exports power and zero when no electricity is being transferred between the MG and upper utility network; G is the number of dispatchable DGs in the MG; $P_k(t)$ is yield power of dispatchable DG k and F_k is fuel cost function of dispatchable DG k, and it is defined as follows for the diesel generator [8]:

$$F_k\big(P_k(t)\big) = a_k P_k(t)^2 + b_k P_k(t) + c_k \tag{2}$$

where a_k, b_k, and c_k are the generator cost function parameters; $\xi_k(t)$ is the commitment status of DG k, and its value is 1 if the DG is operating and 0 if the DG is off and $SU_k(t)$ is the startup cost of DG k, and it is expressed below.A

$$SU_k(t) = \begin{cases} SU, & \text{if } \xi_k(t) - \xi_k(t-1) = 1 \\ 0, & \text{otherwise} \end{cases} \tag{3}$$

where SU is a fixed startup cost, and $SD_k(t)$ is the shutdown cost of DG k, and it is described below.

$$SD_k(t) = \begin{cases} SD, & \text{if} \xi_k(t) - \xi_k(t-1) = -1 \\ 0, & \text{otherwise} \end{cases} \tag{4}$$

where SD is a fixed shutdown cost, $c_{OM,k}(t)$ is the O&M expense of DG k, $c_{OM,w}(t)$ is the O&M expense of the wind plant, $P_{wind}(t)$ is the predicted wind power, S is the quantity of the ESS devices, $c_{OM,u}(t)$ is the O&M expense of the ESS device u and $P_{ESS,u}(t)$ is the charge/discharge power of the ESS unit u at time t. The sign convention here is that $P_{ESS,u}(t)$ is positive while the storage devices discharge, negative while they charge and zero when the ESS is not in operation.

2.3. Constraints

The objective problem described above is subjected to constraints formulated in the next subsections.

2.3.1. Power Limits of Dispatchable DGs

$$P_k^{min}(t) \le P_k(t) \le P_k^{max}(t) \tag{5}$$

where $P_k^{min}(t)$ and $P_k^{max}(t)$ are the lower and upper yield power restrictions on dispatchable DG k, respectively. This constraint is the manufacturer power generation capacity of the DGs.

2.3.2. Power Exchange Limits

$$-P_g^{max}(t) \le P_g(t) \le P_g^{max}(t) \tag{6}$$

where $P_g^{max}(t)$ is the maximum permissible electricity exchange between the MG and the main electricity network. This constraint represents the physical limit (capacity) of the switch and transformer connecting the microgrid and the main grid.

2.3.3. Power Balance Constraint

$$\sum_{k=1}^{G} P_k(t) + \sum_{u=1}^{S} P_{ESS,u}(t) + P_{wind}(t) + P_g(t) = P_l(t) \tag{7}$$

where $P_l(t)$ is the predicted electric load of the MG. This constraint is based on the power system stability concept where the sum of all power generations should be equal to the sum of all power demands.

2.3.4. Power Limits of ESS Devices

$$P_{ESS,u}^{min}(t) \leq P_{ESS,u}(t) \leq P_{ESS,u}^{max}(t) \quad (8)$$

where $P_{ESS,u}^{min}(t)$ and $P_{ESS,u}^{max}(t)$ are the lower and upper charging/discharging power of the ESS unit u, respectively. This constraint designates the manufacturer charging and discharging power limits of the energy storage devices.

2.3.5. Dynamic Operation of ESS Units

$$SOC_{ESS,u}(t+1) = SOC_{ESS,u}(t) - \frac{\eta_{ESS,u}(t)P_{ESS,u}(t)}{C_{ESS,u}} \quad (9)$$

$$SOC_{ESS,u}^{min}(t) \leq SOC_{ESS,u}(t+1) \leq SOC_{ESS,u}^{max}(t) \quad (10)$$

where $SOC_{ESS,u}(t)$ is the state of charge of the ESS device u, $\eta_{ESS,u}(t)$ is the charging/discharging efficiency of the ESS device u, $C_{ESS,u}$ is the storage capacity of the ESS unit u and $SOC_{ESS,u}^{min}(t)$ and $SOC_{ESS,u}^{max}(t)$ are the lower and upper SOCs of the ESS unit u at time t. The constraint in Equation (9) expresses the state of charge dynamics of the energy storage device while charging and discharging. The state of charge increases while charging and decreases during discharging. The constraint in Equation (10) denotes the manufacturer-designed storage capacity of the energy storage devices.

Therefore, the decision variables that the EMS optimizer determines are $P_g(t)$, $P_k(t)$, $P_{ESS,u}(t)$ and $SOC_{ESS,u}(t)$ for all t, k and u.

3. Proposed Optimization Solution Approach

The focus of the proposed energy management is to execute safe 24 h-ahead hourly decisions for economic maneuvers of the microgrid. The EMS considers the volatility of renewables, electric loads and electricity prices and uses appropriate forecasts for these values. It also takes into account the fuel expense, O&M costs, various operational and design constraints and system parameters. Figure 2 depicts the schematic representation of the proposed EMS configuration showing the information exchange into and out of the EMS.

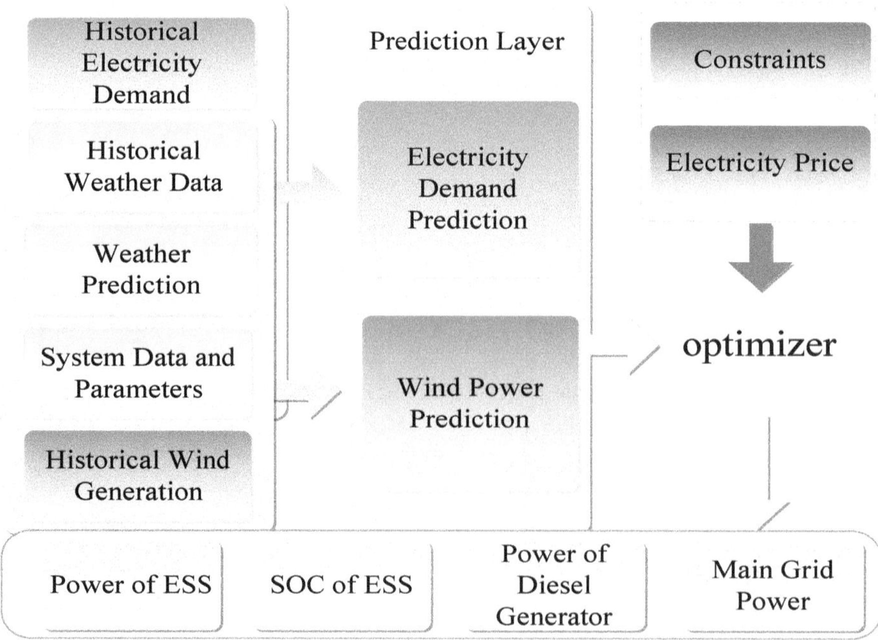

Figure 2. Proposed microgrid energy management system (MG EMS) configuration. SOC: state of charge and ESS: energy storage system.

Ant Colony Optimization (ACO)

The ACO is employed, in this study, for solving the EMS optimization problem formulated in Section 2. The ACO belongs to artificial intelligence (AI)–based modern optimization techniques. It is a likelihood optimization tool to solve functions that are described with graphical search routes. It was stimulated by the activities of ants to search the best routes of food destinations.

The ACO has been broadly used in computer technology and operation researches [21–23]. By mimicking the way real ants communicate each other, the ACO algorithm operates based on the pheromone exchanges among artificial ants [24]. Artificial ants designate population-based modern optimization methods inspired by the behavior of real ants. The combination of artificial ants and searching techniques have come to be one of the solution approaches for plenty of problems consisting of certain sorts of graphs—for example, vehicles and internet routings.

The brief working mechanism of the ACO algorithm is presented as follows. First, the ACO forms a graph through optimization or decision variables, and multiple ants are randomly placed in n nodes (places). The places visited by the ants are recorded by the list rec_a. The rec_a is established for every ant a. The initial pheromone intensity $\zeta_{ij}(0)$ is fixed at zero from all sides. The ants can prefer the next node based on the pheromone intensity in all sides of that node. The likelihood $\rho_{ij}^{a}(t)$ that the ants travel from parameter i to j at the iteration is formulated below.

$$\rho_{ij}^{a}(t) = \begin{cases} \dfrac{\zeta_{ij}^{\alpha}(t)\cdot\eta_{ij}^{\beta}(t)}{\sum_{q\notin rec_a}\zeta_{iq}^{\alpha}(t)\cdot\eta_{iq}^{\beta}(t)} & ,j \notin rec_a \\ 0\,, & \text{otherwise} \end{cases} \tag{11}$$

where η_{ij} is a heuristic memo that is computed as $1/d_{ij}$, where d_{ij} is the Euclidean norm of the space from parameter i to j, $\zeta_{ij}(t)$ is the pheromone intensity of the path from parameter i to j at iteration t and α and β are the memo heuristic weight and expectation heuristic weight applied to assign the coefficients for heuristic information and pheromone intensity. While the ants complete the travels, the memo (information) intensity on each path is updated by the following expression.

$$\zeta_{ij} \leftarrow (1-p)\cdot\zeta_{ij} + p\cdot\sum_{a=1}^{m}\Delta\xi_{ij}^{a} \tag{12}$$

where $p \in (0,1]$ is the weight parameter known pheromone-evaporation ratio, and $\Delta\zeta_{ij}^{a}$ is the pheromone enhancement over the route from parameter i to j during the travels, and it is defined below.

$$\Delta\zeta_{ij}^{a} = \begin{cases} \dfrac{Q}{L_a}\,, & (i,\ j) \in \text{route of } a \\ 0\,, & \text{otherwise} \end{cases} \tag{13}$$

where Q is a fixed term called pheromone strength, and L_a is the path distance of ant a. The ACO iteration terminates when all the ants reach the same solution. Figure 3 illustrates the ACO algorithm flowchart. The parameters of the ACO algorithm used in the study are given in Table 1.

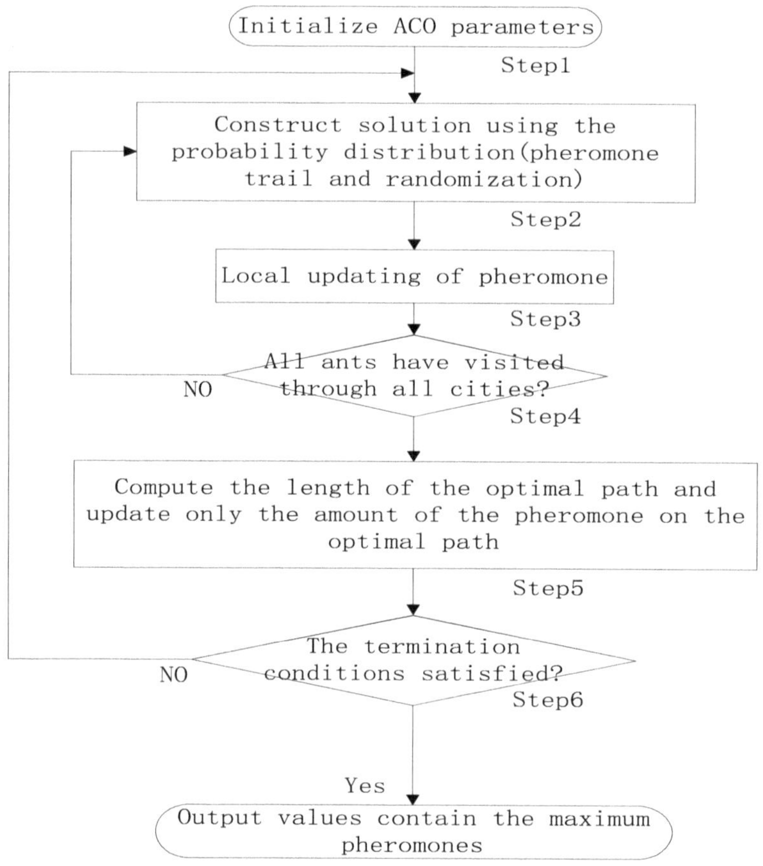

Figure 3. Flowchart of the ant colony optimization (ACO) algorithm.

Table 1. Ant colony optimization (ACO) parameters. EMS: energy management system.

Parameter	Value
# of iterations	100
# of variables	Number of EMS decision variables
Number of ants	20
Initial pheromone	0
Information heuristic coefficient	1.0
Expectation heuristic coefficient	1.0
Pheromone intensity	2
Pheromone evaporation	0.9

4. Simulation Results and Discussions

Case Study

A large-scale grid-connected microgrid containing a 2000-kW wind generator, 600-kW diesel generator and 2100-kW·h ESS (300-kW·4-h VRB and 300-kW·3-h Li–Ion battery) is considered as a case study microgrid platform in this paper. It is a practical microgrid framework in Atlanta, GA, USA, which is designed to supply electricity to industrial park loads with a peak aggregate capacity of 3000 kW.

The lowest and peak SOC limits of the ESS devices are set as 20% and 100%, respectively. The initial SOCs (at 00:00 or 12:00 a.m.) of the ESS devices are assumed to be 20%. The maximum charging/discharging power of the ESS units is taken as 300 kW. The ideal charging and discharging efficiency (100%) is assumed. The diesel generator maximum generation is set as 600 kW, and its parameters are given in Table 2. The maximum grid power exchange is set as 4000 kW, which equals the capacity of the grid-coupling transformer. SU is a fixed startup cost of the diesel generator. SD is a fixed shutdown cost of the diesel generator.

Table 2. Diesel generator parameters.

Parameter	Unit	Value
a	($/kWh)^2	0.00025
b	$/kWh	0.0156
c	$/h	0.3312
SU	$/h	0
SD	$/h	0

SU is a fixed startup cost of the diesel generator. SD is a fixed shutdown cost of the diesel generator.

Several scenarios of the generation and load demands have been investigated to validate the proposed EMS optimization approach based on the information of the case study microgrid. However, for the purpose of illustration and summarizing the findings, the performance of the proposed approach will be discussed next based on a single-day scenario. The 24-ahead predictions of the wind power and electricity demand are depicted in Figures 4 and 5, respectively.

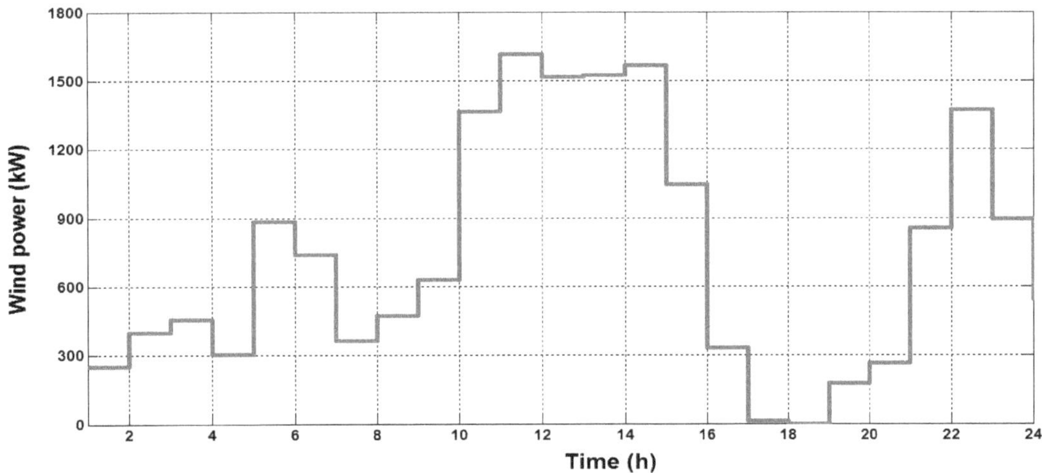

Figure 4. Wind generation prediction.

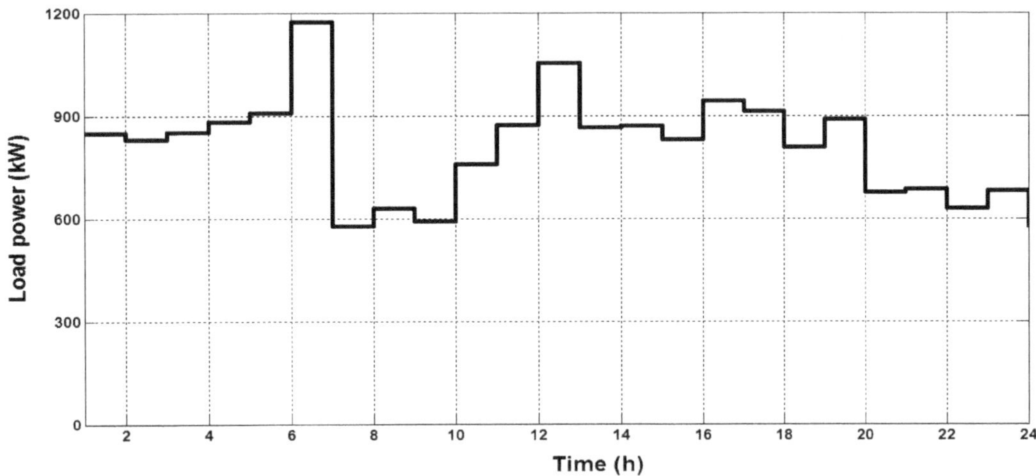

Figure 5. Electric load demand prediction.

The electricity price is shown in Figure 6. The price data is the actual energy cost of industries and big institutions in Atlanta, GA, USA. It is shown that the power-selling bill to the upstream utility network is constant through the operating day, while the electricity purchasing bill from the upstream utility network is dynamic (time-of-use pricing scheme) within the day and has three step-prices within a day (low price period, moderate price period and peak price period).

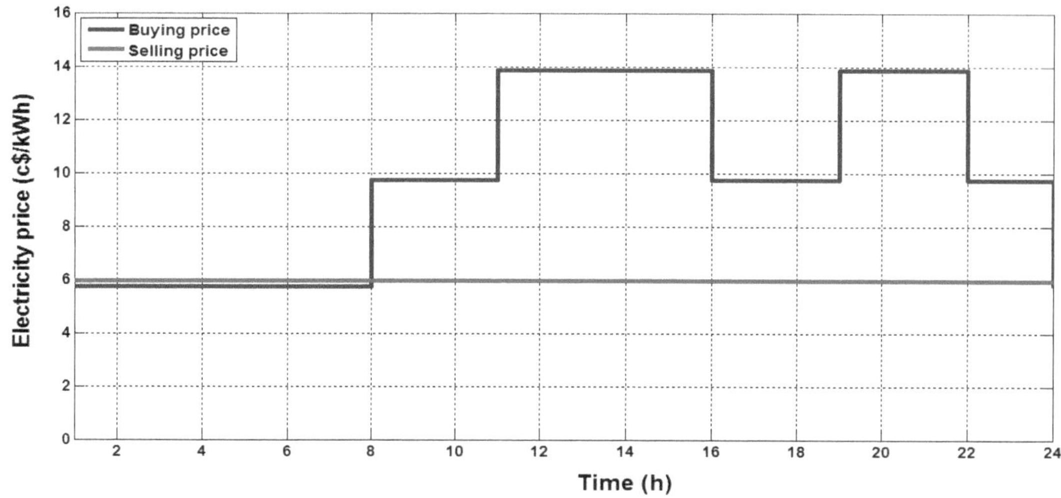

Figure 6. Electricity buying and selling prices.

Table 3 specifies the parameters associated with the O&M expenses for the various components of the microgrid.

Table 3. Operation and maintenance (O&M) cost parameters for various microgrid components. VRB: vanadium redox battery and MG: microgrid.

MG Component	O&M Cost (c$/kWh)
Wind generator	0.3767
Diesel generator	0.5767
VRB	0.003
Li-Ion battery	0.0015

Figure 7 depicts the ACO-based optimal solution for the formulated EMS objective function. The associated SOCs of the ESS devices are also illustrated in Figure 8.

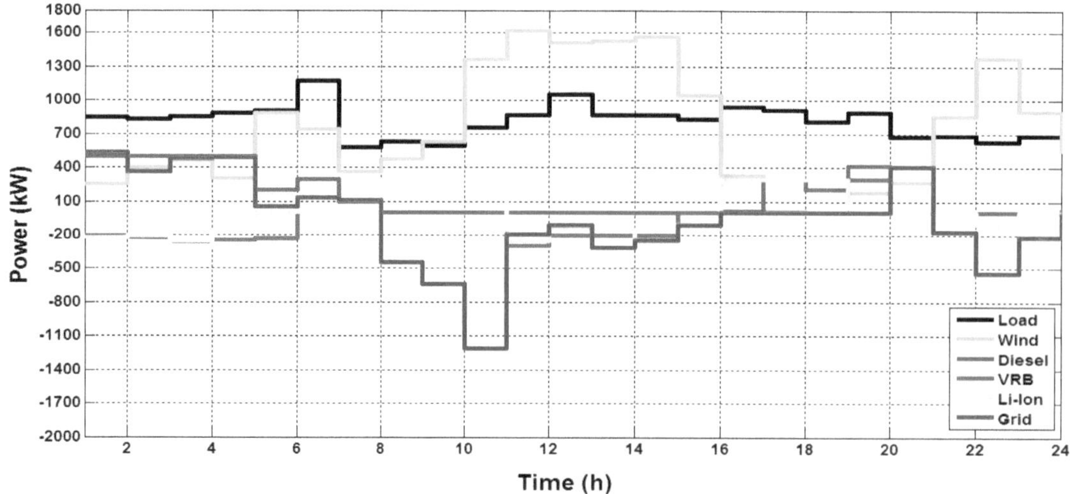

Figure 7. ACO-based optimal EMS solution. VRB: vanadium redox battery.

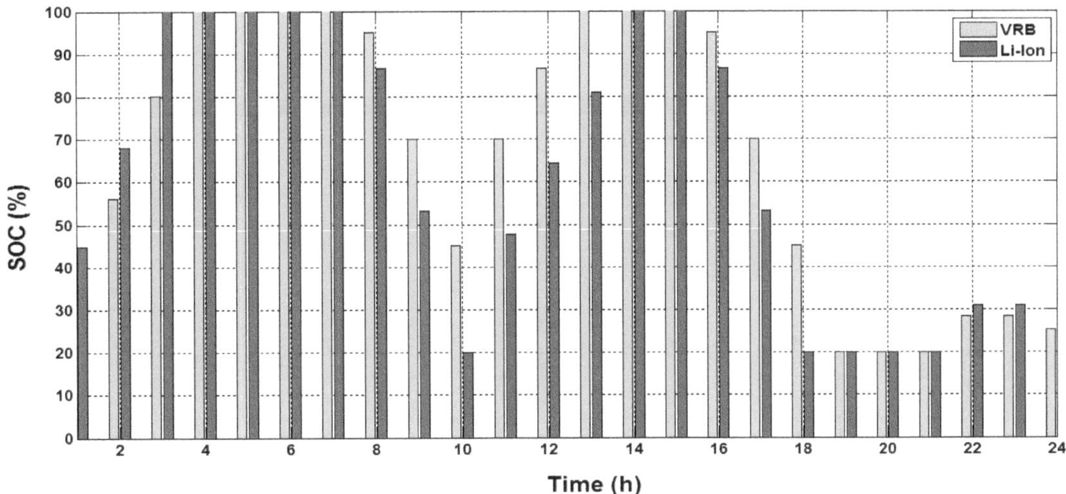

Figure 8. SOCs of the ESS units obtained using ACO.

As clearly observed in Figures 4–8, during the time interval (00:00, 7:00) or (12 a.m., 7 a.m.) the produced wind power cannot deliver the full electricity consumption. However, the utility electricity bill is cheapest in this time interval, and therefore, the MG purchases power from the upstream utility network to support the wind generation to supply the load demand and charge the ESS. The diesel generator also supplies power to support the wind generation in this period. The ESS devices continuously charge and become full in this period.

In the period (7:00, 10:00) or (7 a.m., 10 a.m.), the wind power generation increases and becomes more than the load demand. The ESS units are at their full charge state in this time. Hence, the MG exports the excess energy from the wind power and ESS units in this period and earns profit. The diesel generator stops producing to lower the fuel expense, in this period, as there is excess renewable power production.

During the period (10:00, 15:00) or (10 a.m., 3 p.m.), which is the peak electricity price time interval, the microgrid still has more surplus electricity due to more production by the wind plant. Thus, the MG still keeps selling electricity to the upper utility network and charges the energy storage devices. The diesel generator does not produce power in this period, since there is still excess renewable power production in the MG.

During the period (15:00, 19:00) or (3 p.m., 7 p.m.), the wind power generation decreases and becomes lower than the load demand. The ESS units are at full in this period, and the electricity bill is medium. The microgrid supplies the load demand in this period using the generations from the ESS units and the dispatchable DG, in addition to the wind generation. The MG neither buys nor sells power from/to the utility network in this period.

The wind power generation increases again in the period (20:00, 23:00) and becomes greater than the microgrid electricity consumption. The microgrid exports the surplus generation to the main grid and charges the ESS devices in this period. The diesel generator power is zero in this period to lower the fuel expense, since there is plenty of renewable power production in the MG

Figure 9 shows the PS-based solution of the microgrid EMS optimization problem for the case study microgrid.

The hourly power production fuel cost comparison between the two EMS optimization approaches is shown in Figure 10. It is shown that the ACO-based EMS optimization solution has resulted in lower fuel costs in most of the operating hours of the day.

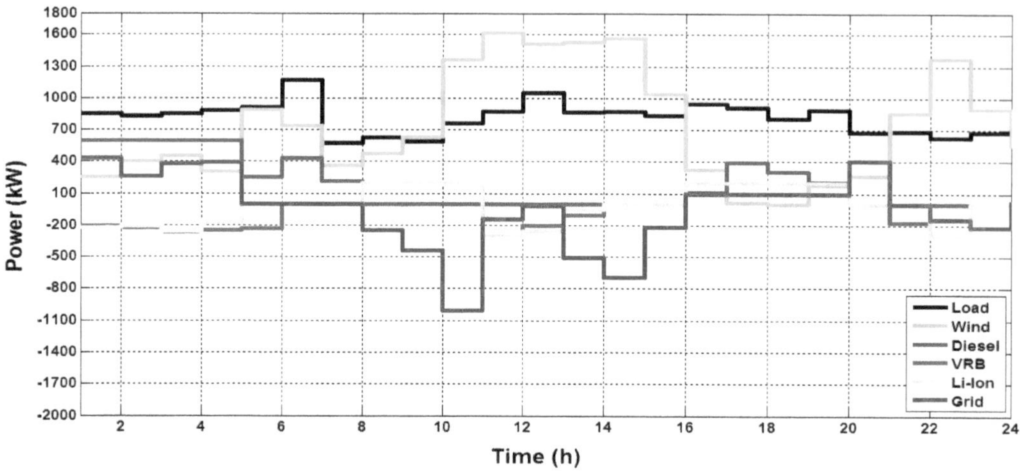

Figure 9. Pattern search PS-based EMS solution.

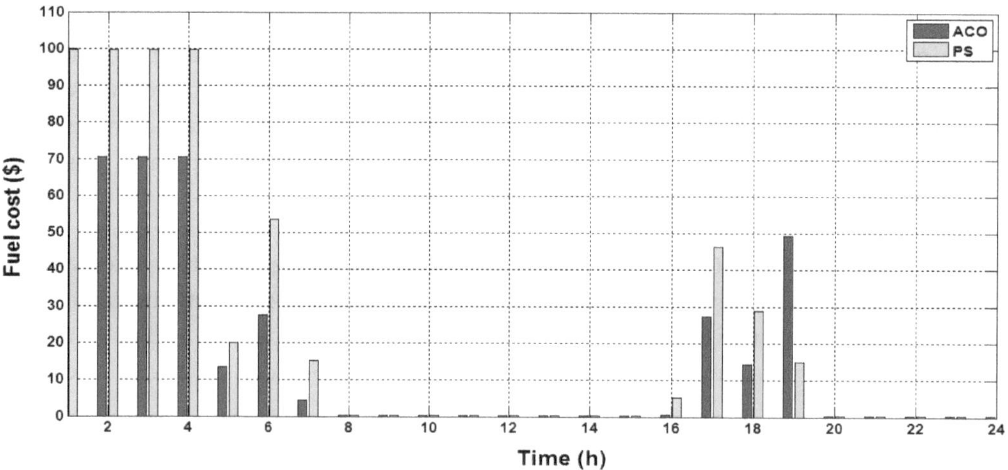

Figure 10. Power production fuel cost comparison.

The hourly main grid power purchasing cost comparison between the two EMS optimization methods is shown in Figure 11. It is observed that the ACO-based EMS optimization solution has given lower purchasing costs in most of the operating hours of the day.

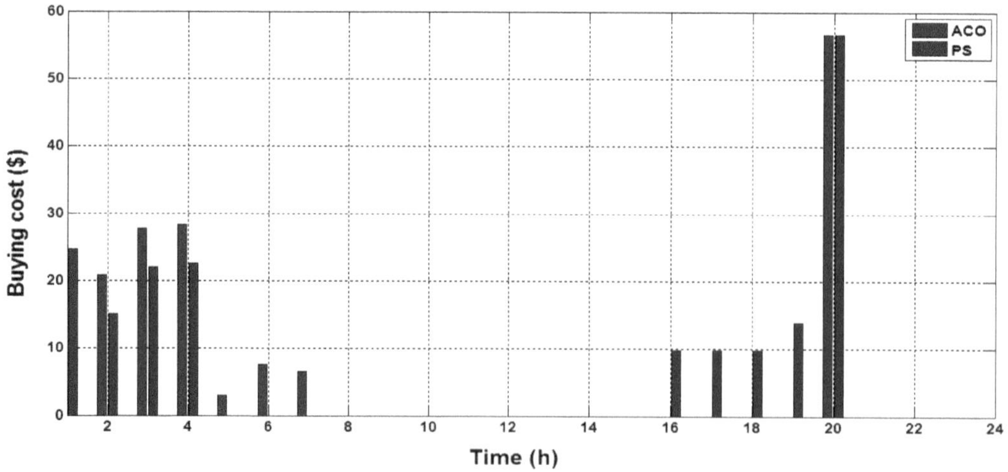

Figure 11. Grid power purchasing cost comparison.

As it is shown in Figures 10 and 11 above, the fuel cost and power purchasing cost are almost zero, as the microgrid has sufficient renewable energy production from the wind during this period.

The comparison of the hourly income of exporting electricity to the upstream utility network between the two EMS optimization techniques is shown in Figure 12. It is clearly shown that the ACO-based EMS optimization solution has achieved a higher income in several hours of the day.

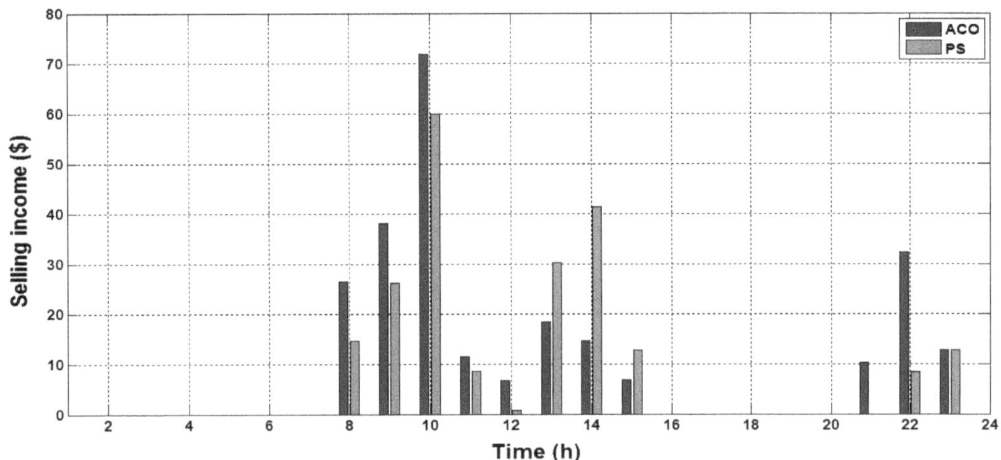

Figure 12. Power selling income comparison.

Table 4 presents the daily power generation fuel cost, grid power purchasing cost and power-selling income by both EMS optimization solutions. It is observed in the table that the proposed ACO-based EMS solution has achieved a lower total fuel cost and grid power purchasing expense and higher total power selling income.

Table 4. Daily cost and income summary. PS: pattern search.

Solution Approach	Total Daily Cost and Income ($)		
	Fuel Cost	Grid Power Purchasing Cost	Grid Power Selling Income
ACO	423.74	180.83	249.56
PS	587.12	186.03	215.58

Table 5 presents the computational time elapsed by both methods to obtain their solutions using the MATLAB/Simulink software (MATLAB2016a, MathWorks.lnc, Natick, MA, USA) platform on a PC with an Intel core i7 CPU, 4.0 GHz processor and 8GB RAM. The ACO-based method has given the optimal solution within a shorter computation time.

Table 5. Computation time comparison.

Solution Approach	Time (s)
ACO	9.52
PS	17.36

5. Conclusions

The optimal energy management approach for a wind-diesel generator ESS microgrid operating in the grid-connected mode was devised based on the ACO algorithm in this paper. The devised approach considers the volatility of wind generation and electricity consumption in the MG, and suitable 24 h-ahead predictions have been used to manage the volatilities. The experimental findings have verified the performance and benefits of the proposed microgrid EMS approach. The approach has achieved lower fuel expenditure and reduced costs of buying electricity from the upstream utility network. It has also resulted in a higher income from the electricity export to the upstream utility grid. The obtained solution has shown effective utilization of the renewable generation and ESS devices. The obtained simulation results have been compared with the PS-based EMS solution and gave outperforming performances with respect to cost, income and computation time. Thus, the obtained numerical results and illustrative demonstrations verify that the proposed EMS approach is effective and robust for the day-ahead power control of MGs with multiple energy resources and storage units. Stochastic energy management with a more robust representation of the uncertainties of the renewable and load demands will be the future research direction and extension of the findings of this paper.

Author Contributions: H.W. and H.L. collected all the microgrid information, developed the optimization model, implemented the optimization algorithm and drafted and revised the paper. X.G. revised the paper format, reviewed the paper and simulated consulting. All authors have read and agreed to the published version of the manuscript.

Acknowledgments: The State Key Lab of Alternative Electrical Power System of North China Electric Power University and the Power Supply Company of Henan Shangqiu supported the research.

References

1. Falcon, P.M.; Lopolito, A.; Sica, E. Instrument mix for energy transition: A method for policy formulation. *Technol. Forecast. Soc. Chang.* **2019**, *148*, 119706. [CrossRef]
2. Hess, D.J.; Renner, M. Conservative political parties and energy transitions in Europe: Opposition to climate mitigation policies. *Renew. Sustain. Energy Rev.* **2019**, *104*, 419–428. [CrossRef]
3. Rosenow, J.; Kern, F.; Rogge, K. The need for comprehensive and well targeted instrument mixes to stimulate energy transitions: The case of energy efficiency policy. *Energy Res. Soc. Sci.* **2017**, *33*, 95–104. [CrossRef]
4. Wang, L.; Zhang, L.; Xu, C.; Eseye, A.T.; Zhang, J.; Zheng, D. Dynamic Economic Scheduling Strategy for a Stand-alone Microgrid System Containing Wind, PV Solar, Diesel Generator, Fuel Cell and Energy Storage: A Case Study. In *IOP Conference Series: Earth and Environmental Science*; IOP Publishing: Bristol, UK, 2018; p. 168.
5. Li, H.; Eseye, A.T.; Zhang, J.; Zheng, D. Optimal Energy 11 Management for Industrial Microgrids with High-Penetration Renewables. *J. Prot. Control Mod. Power Syst.* **2017**, *2*, 12. [CrossRef]
6. Velik, R.; Nicolay, P. Grid-Price-Dependent Energy Management in Microgrids Using a Modified Simulated Annealing Triple-Optimizer. *Appl. Energy* **2014**, *130*, 384–395. [CrossRef]
7. Eseye, A.T.; Zheng, D.; Li, H.; Zhang, J. Grid-price Dependent Optimal Energy Storage Management Strategy for Grid-connected Industrial Microgrids. In *2017 Ninth Annual IEEE Green Technologies Conference (GreenTech)*; IEEE: Piscataway, NJ, USA, 2017.
8. Eseye, A.T.; Jianhua, Z.; Dehua, Z.; Dan, W. Optimal Energy Management Strategy for an Isolated Industrial Microgrid Using a Modified Particle Swarm Optimization. In *2016 IEEE International Conference on Power and Renewable Energy*; IEEE: Piscataway, NJ, USA, 2016.
9. Velik, R. The influence of battery storage size on photovoltaics energy self-consumption for grid-connected residential buildings. *IJARER Int. J. Adv. Renew. Energy Res.* **2013**, *2*, 310–317.
10. Velik, R. Battery storage versus neighbourhood energy exchange to maximize local photovoltaics energy consumption in gridconnected residential neighbourhoods. *IJARER Int. J. Adv. Renew. Energy Res.* **2013**, *2*, 402–411.
11. Velik, R. Renewable energy self-consumption versus financial gain maximization strategies in grid-connected residential buildings in a variable grid price scenario. *IJARER Int. J. Adv. Renew. Energy Res.* **2014**, *3*, 414–422.
12. Chen, Y.; Lu, S.; Chang, Y.; Lee, T.; Hu, M. Economic analysis and optimal energy management models for microgrid systems: A case study in Taiwan. *Appl. Energy* **2013**, *103*, 145–154. [CrossRef]
13. Gregoratti, D.; Matamoros, J. Distributed convex optimization of energy flows: The two-microgrid case. In Proceedings of the 1st International Black Sea Conference on Communication and Networking, Batumi, Georgia, 3–5 July 2013; pp. 201–205.
14. Celli, G.; Pilo, F.; Pisano, G.; Soma, G. Optimal participation of a microgrid to the energy market with an intelligent EMS. In Proceedings of the 7th International Power Engineering Conference, Niigata, Japan, 4–8 April 2005; pp. 663–668.

15. Livengood, D.; Larson, R. The Energy Box: Locally automated optimal control of residential electricity usage. *Serv. Sci.* **2009**, *1*, 1–16. [CrossRef]
16. Mashhour, E.; Moghaddas-Tafreshi, S. Integration of distributed energy resources into low voltage grid: A market-based multiperiod optimization model. *Electr. Power Syst. Res.* **2010**, *80*, 473–480. [CrossRef]
17. Kriett, P.; Salani, M. Optimal control of a residential microgrid. *Energy* **2012**, *42*, 321–330. [CrossRef]
18. Malysz, P.; Sirouspour, S.; Emadi, A. An Optimal Energy Storage Control Strategy for Grid-connected Microgrids. *IEEE Trans. Smart Grid* **2014**, *5*, 1785–1796. [CrossRef]
19. Zhao, Z. Optimal Energy Management for Microgrids. Ph.D. Thesis, Electrical and Computer Engineering, Clemson University, Clemson, SC, USA, January 2012.
20. Borghetti, A.; Bosetti, M.; Grillo, S. Short-term scheduling and control of active distribution systems with high penetration of renewable resources. *IEEE Syst. J.* **2010**, *4*, 313–322. [CrossRef]
21. Ma, Y.; Zhai, M. A Dual-Step Integrated Machine Learning Model for 24 h-Ahead Wind Energy Generation Prediction Based on Actual Measurement Data and Environmental Factors. *Appl. Sci.* **2019**, *9*, 2125. [CrossRef]
22. Dorigo, M.; Gambardella, L.M. Learning Approach to the Traveling Salesman Problem. *IEEE Trans. Evol. Comput.* **1997**, *1*, 214. [CrossRef]
23. Dorigo, M.; Birattari, M. *Ant Colony Optimization*; MIT Press: Cambridge, MA, USA, 2004; ISBN 0-262-04219-3.
24. Nicolas, M.; Frédéric, G.; Patrick, S. *Artificial Ants*; Wiley-ISTE: Hoboken, NJ, USA, 2010; ISBN 978-1-84821-194-0.

2

Multi-Agent Consensus Algorithm-Based Optimal Power Dispatch for Islanded Multi-Microgrids

Xingli Zhai [1] and Ning Wang [2,*]

[1] Jinan Power Supply Company in Shandong Provincial Electric Power Company of State Grid, Jinan 250000, China; 15169199001@163.com
[2] College of Electrical Engineering, Zhejiang University, Hangzhou 310027, China
* Correspondence: 11610045@zju.edu.cn

Abstract: Islanded multi-microgrids formed by interconnections of microgrids will be conducive to the improvement of system economic efficiency and supply reliability. Due to the lack of support from a main grid, the requirement of real-time power balance of the islanded multi-microgrid is relatively high. In order to solve real-time dispatch problems in an island multi-microgrid system, a real-time cooperative power dispatch framework is proposed by using the multi-agent consensus algorithm. On this basis, a regulation cost model for the microgrid is developed. Then a consensus algorithm of power dispatch is designed by selecting the regulation cost of each microgrid as the consensus variable to make all microgrids share the power unbalance, thus reducing the total regulation cost. Simulation results show that the proposed consensus algorithm can effectively solve the real-time power dispatch problem for islanded multi-microgrids.

Keywords: islanded multi-microgrids; real-time power dispatch; multi-agent; consensus algorithm

---

1. Introduction

The emergence of microgrids (MG) provides a new technical means for the comprehensive utilization of renewable energy [1–3]. According to whether there is an electrical connection with the main grid, microgrids feature two typical operation modes, i.e., grid-connected and islanded. In general, microgrids can operate in grid-connected mode to exchange power with the main grid. On the other hand, when being used in a remote area or in an emergency situation, they can also be transferred to islanded mode to guarantee local grid services [4]. Therefore, an islanded microgrid is more suitable for remote areas without grid coverage, which can improve the utilization efficiency of local renewable energy and reduce the cost of power supply in remote areas. Recently, the multi-microgrids (MMGs) system has become an integrated, flexible network that incorporates multiple individual microgrids (MGs) [5,6], which are often geographically close and connected to a distribution bus. If the neighboring islanded microgrid in a remote area can realize the cluster operation through interconnection, the mutual energy between the microgrids not only helps to absorb excess power energy, but also supports each other as a backup power supply [7,8], which is beneficial to improving the overall power supply reliability and economy.

The islanded microgrids plays an important role in renewable energy applications and power sharing among different loads connected to multi-microgrid systems [9,10]. The key issue with islanded microgrids is how to ensure a power balance between generation and demand in a cost-effective way. Hence, problems of microgrid real-time dispatching and operations have received considerable attention in the literature. There are some achievements on the dispatching and operations of individual microgrids that mainly focus on reducing the regulation cost and the coordination of various devices in the microgrid. Due to the intermittency and variability of renewables-based distributed generation (DG), the methods for uncertainty power dispatch of individual microgrid are mainly classified into three categories: stochastic power dispatch [11], robust power dispatch [12,13], and rolling power dispatch [14]. In order to deal with the uncertainty of demand response and renewable energy, the authors of [11] presented a stochastic programming framework for 24-h optimal scheduling of combined heat and power (CHP) systems-based MG. Wang R. and Luo Z. of [12] and [13] adopted a robust optimization approach to accommodate the uncertainties of demand response and renewable energy; it performs better than deterministic optimization in terms of the expected operational costs. For a real renewable-based microgrid in the north of Chile, the authors of [14] proposed a moving horizon optimization strategy to eliminate the forecasting errors caused by renewable energy. The optimal dispatch of an individual microgrid owes more to the coordination of the controllable units in a microgrid. However, for multi-microgrids, the power dispatch between each microgrid is a critical issue, so the optimal dispatch of multi-microgrids is more complex.

Due to the frequent fluctuation of power supply and load on both sides, offline optimal dispatch methods have not been suitable for the multi-microgrids, especially for working in islanded mode, which lacks support from the main grid [15]. Therefore, the power dispatch of the islanded multi-microgrids should focus on real-time optimal dispatch capability in order to maintain a real-time power balance between power generation and load demand, ensure the stable operation of the multi-microgrids system, and take into account the economic operation of the system. As long as the power command calculation process is required for centralized optimization, it will cause a certain computational complexity. In the case of high real-time requirements, the requirements for the generation and delivery speed of dispatching command also increase. With the advantages of distributed control architecture in smart grids, the multi-agent theory [16] and distributed control method [17,18] are applied to the microgrid dispatching model. The authors of [19] developed a new hybrid intelligent algorithm called imperialist competitive algorithm-genetic algorithm (ICA-GA) to determine both the optimal location and operation of an islanded MG. In order to minimize the islanded microgrids' operational losses, the authors of [20] adopted the glow-worm swarm optimization (GSO) algorithm to solve an optimal power flow problem. However, these approaches are a centralized optimization method that collects the whole network's information via a central controller and uses an intelligent optimization algorithm, such as glow-worm swarm optimization, particle swarm optimization, ant colony optimization, etc. Although the centralized optimization method has high regulation accuracy, when the number of network nodes is large, the communication volume is too large, the communication line is required to be high, and the scalability is poor. In addition, the intelligent optimization algorithm is unstable and cannot guarantee convergence to the optimal solution, which could cause a decline in the control performance. A new distributed reinforcement learning approach based on the multi-agent systems algorithm was applied to minimize the power losses under given operational constraints in [21]. The multi-agent system based consensus method [22] provides a new way of solving the real-time power dispatch problem of islanded multi-microgrids. The main issue with a consensus problem is achieving agreement regarding certain quantities of interest associated with agents in multi-agent systems by utilizing a local information exchange [23]. The traditional consensus algorithm is a very simple local coordination rule, which results in agreement at the group level, and no centralized task planner or global information is required by the algorithm. Due to its distributed implementation, robustness, and scalability, multi-agent consensus algorithms have been widely applied in many coordination problems, such as power system economic dispatch [24,25], power allocation [26], optimal control [27,28], etc. Compared with the traditional centralized optimization algorithm, the consensus method only requires each agent to obtain the information on the local and neighboring agents in real time. Hence, it can obtain the ideal convergence value with less transmission information and a shorter optimization time.

Motivated by these works, this paper provides a multi-agent system-based consensus algorithm to solve the real-time power dispatch problem of islanded multi-microgrids, which have a lower communication burden and better dynamic performance. In order to ensure the overall real-time power balance of the islanded multi-microgrids and reduce the power regulation costs, the real-time cooperative power dispatch framework of the islanded multi-microgrids is built by using a multi-agent system consensus algorithm. Simultaneously, the real-time dispatch of power imbalance is optimized to reduce the overall regulation costs of the islanded multi-microgrids, so as to ensure that each microgrid is responsible for the corresponding power regulation tasks according to its own situation.

At the same time, the speed of dispatching command generation and release is accelerated because of avoiding centralized optimization, so the system can better adapt to the dynamic requirements of real-time power dispatch of the islanded multi-microgrids. Based on the consensus method, the real-time power dispatch strategy works in a fully distributed manner without a central coordinator; communication occurs only between the device and its neighbors. The main contributions of this paper are as follows:

(1) A real-time cooperative dispatch framework for islanded multi-microgrids based on multi-agent consensus method is built that can ensure the overall real-time power balance and minimize the power regulation costs.

(2) The consensus method only needs a small amount of information from the local and neighboring microgrids, which reduces the communications burden and increases the reliability compared with the traditional centralized optimization method.

The remainder of this paper is organized as follows. In Section 2, we establish a cooperative power dispatch framework for islanded multi-microgrids based on the consensus algorithm. We then model the regulation cost of each controllable unit in the microgrid to quantify the regulation costs of each controllable unit participating in the real-time control process in Section 3. The power dispatch consensus algorithm is designed for an islanded multi-microgrid in Section 4. Several numerical simulations are conducted and analyzed in Section 5, and Section 6 concludes this paper.

2. Cooperative Power Dispatch Framework of Islanded Multi-Microgrids

2.1. Cooperative Power Dispatch Framework Based on Consensus Algorithm

The microgrid is equipped with a microgrid controller (MGC) according to the requirements of the control. The MGC is responsible for ensuring the stable real-time operation of the microgrid. When many microgrids, through interconnection, constitute a multi-microgrids system, the traditional centralized dispatch framework requires the upper system to gather real-time information on each microgrid. This information is distributed and dispatched to each unit after centralized optimization. Although this method can more fully acquire system information and adapt to various optimization algorithms, it increases the burden on the communication network and the controller. At the same time, it cannot adapt to the requirements of plug and play and real-time control.

The number of controllable units and amount of data in multi-microgrids is large. The multi-microgrids' interconnection makes the operation mode diversified and requires the control mode to be easily extended, which need to satisfies the requirement of "plug and play." Therefore, the architecture and operational characteristics of the islanded multi-microgrids determine that it is suitable to adopt a decentralized control architecture [29]. This paper establishes a real-time cooperative dispatch framework for islanded multi-microgrids based on the multi-agent consensus method, which is shown in Figure 1.

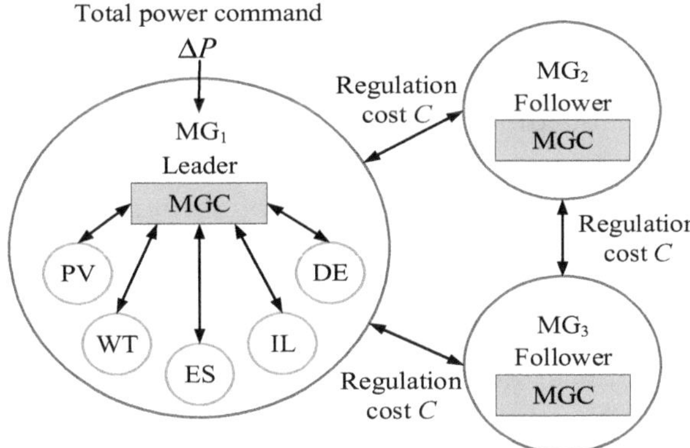

Figure 1. Real-time cooperative dispatch framework for islanded multi-microgrids.

The control objective is to ensure the overall real-time power balance of islanded multi-microgrids and reduce the power regulation costs, which is equivalent to the problem of converting the economic dispatch problem into the consensus of the regulation costs in the power allocation process [30]. Therefore, the role of each MGC is as the agent in the network and the regulation cost of each microgrid is the concerned consensus variable. Under the premise that interconnected multi-microgrids system belongs to the same community of interests, the regulation of each microgrid is guided by the economic signal of the whole system, i.e., the regulation cost signal. Thus the economy of islanded multi-microgrids operation can be improved by reasonably allocating the regulation power. The islanded multi-microgrids adopt the "leader-follower" model [31,32]. The leader MGC is a communication center of all microgrids, which calculates the total power command, communicates and cooperates with other MGs, and balances the power of the entire control area. The follower exchanges the regulation costs with its adjacent MGs. After the communication iteration of each microgrid, the total power command is delivered to each microgrid and the MGC delegates the power commands of each control unit in the microgrid according to the established strategy. Because each MGC does not need to acquire the global information of the system, the information needed by each microgrid in the iteration is only local information, and the regulation cost information transmitted by the neighbors, so the communication burden is small.

2.2. *Power Dispatch Strategy of Microgrid*

After receiving the allocated power command, each microgrid needs to decentralize the power commands of the controllable units according to the power allocation strategy. It is assumed that each microgrid includes the following units: photovoltaic (PV), wind turbine (WT), energy storage device (ES), and controllable micro-power sources, such as diesel engine (DE) or interruptible load (IL).

Considering the regulation costs and dispatching priority of various regulation methods, the decision order of power allocation strategy for positive and negative power command is specified as follows.

(1) Power allocation strategy for positive power command. When the microgrid receives a positive power command, it requires the microgrid to increase the power generation or reduce the power consumption. The energy storage discharge is preferred. If the energy storage reaches the discharge limit but still does not meet the demand, the diesel generator output power is increased. If the output of the diesel generator reaches the limit but still does not meet the demand, then only the load shedding method can be adopted to regulation the interruptible load.

(2) Power allocation strategy for negative power command. When the microgrid receives a negative power command, it requires the microgrid to reduce the power generation. The energy storage charge is preferred. If the energy storage reaches the charge limit but still does not meet the demand, the output of the photovoltaic system or wind turbine will be reduced and part of the power generation will be abandoned.

3. Microgrid Regulation Cost Modeling

In this section, the regulation cost of each controllable unit in the microgrid is modeled to quantify the regulation cost of each controllable unit participating in the real-time control process, thus providing a basis for calculating the consensus power allocation.

Considering that the regulation cost of most controllable units is more suitable to be measured by electricity, the state duration window T_S is introduced here. When calculating the regulation cost of each controllable unit at time t, the cumulated regulation power of each controllable unit in T_S period starting from time t is calculated by using real-time power commands, so that the real-time regulation cost of each microgrid at time t can be quantified according to the amount of electricity. The significance of the regulation cost described in this paper is to calculate the regulation cost of running each microgrid in the state specified by the current power command within a given time window, and to quantify the regulation cost by converting the power into electricity.

3.1. Regulation Cost of Energy Storage Battery

The regulation cost of an energy storage battery includes the damage to the battery life caused by the charge and discharge behavior and the corresponding maintenance cost, which is related to the charge and discharge power and the state of charge (SOC). According to the working requirements of an energy storage battery, four reference limits are selected to divide the battery capacity of an energy storage battery into five areas, as shown in Figure 2.

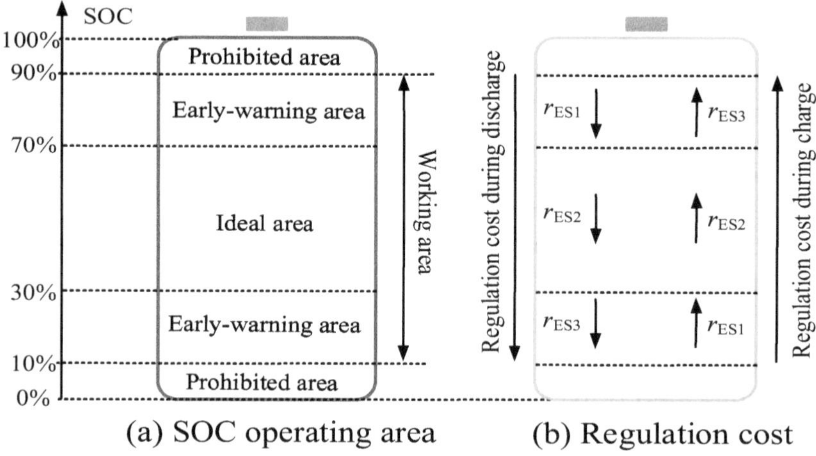

Figure 2. Schematic diagram for energy storage system SOC subarea: (**a**) Description of the SOC operating area; (**b**) Description of the regulation cost.

The four reference limits are SOC_{max}, SOC_{high}, SOC_{low}, and SOC_{min}, which are 90%, 70%, 30%, and 10%, respectively. Between SOC_{min} and SOC_{max} is the working area. It is required that the SOC of an energy storage battery be confined to within this range. The ideal working area (30-70%) is the optimal range of SOC. When the SOC enters the early-warning working area, it should not further approach the prohibited working area. In order to enable the dispatching energy storage battery to conform to the above basic idea, the unit regulation cost of each area is set according to the charge and discharge process in the working area, and it satisfies $r_{ES1} < r_{ES2} < r_{ES3}$. For example, in the early-warning area with high SOC, if the charging continues, the unit regulation cost will be higher. It can be considered that the damage to the battery will be greater, however. Conversely, the discharging will help the SOC approach the ideal working area. It can be considered that this is beneficial to the life of the battery and will lead to lower regulation costs. Therefore, the real-time maximum charge and discharge power of the energy storage system are given as follows:

$$\begin{array}{l} P_{\mathrm{ES}+}^{\max} = \begin{cases} \frac{\eta(SOC(t-\Delta t)-SOC_{\min})V_{\mathrm{ES}}}{\Delta t}, SOC(t) < SOC_{\min} \\ \eta P_{\mathrm{ES}}^{\max},\ SOC(t) \geq SOC_{\min} \end{cases} \\ SOC(t) = SOC(t-\Delta t) - \frac{P_{\mathrm{ES}}^{\max}\Delta t}{V_{\mathrm{ES}}} \end{array} \tag{1}$$

$$\begin{array}{l} P_{\mathrm{ES}-}^{\max} = \begin{cases} P_{\mathrm{ES}}^{\max},\ SOC(t) \leq SOC_{\max} \\ \frac{(SOC_{\max}-SOC(t-\Delta t))V_{\mathrm{ES}}}{\eta\Delta t}, SOC(t) > SOC_{\max} \end{cases} \\ SOC(t) = SOC(t-\Delta t) + \frac{\eta P_{\mathrm{ES}}^{\max}\Delta t}{V_{\mathrm{ES}}} \end{array} \tag{2}$$

where Δt is the time interval of power command; $SOC(t)$ and $SOC(t-\Delta t)$ represent the current SOC calculated value and the actual SOC value of the previous time point, respectively; and V_{ES}, $P_{\mathrm{ES}}^{\max}$, and η are the energy storage system capacity, power limits, and efficiency, respectively.

The real-time regulation cost of the energy storage system can be expressed as follows:

$$\begin{cases} C_{\mathrm{ES}}(\Delta P_{\mathrm{ES}}) = V_{\mathrm{ES}}T_{\mathrm{S}}\sum_{i=1}^{3} r_{\mathrm{ES}i}\Delta S_i \\ \Delta P_{\mathrm{ES}}\Delta P \geq 0 \\ -P_{\mathrm{ES}-}^{\max} \leq \Delta P_{\mathrm{ES}} \leq P_{\mathrm{ES}+}^{\max} \end{cases}, \tag{3}$$

where ΔP_{ES} is the power command of the energy storage device, $C_{\mathrm{ES}}(\Delta P_{\mathrm{ES}})$ is the regulation cost generated when the energy storage system is adjusted to the power command ΔP_{ES} and the regulation cost in the other adjustment modes below is the same, and ΔP is the total power imbalance of the multi-microgrids, that is, the real-time power imbalance. ΔS_i is the SOC change caused by the SOC area with a regulation cost of $r_{\mathrm{ES}i}$ after running the T_{S} period with the current power command. The interval that the SOC may actually traverse during the calculation of the real-time power command is also related to the selection of the state duration window T_{S}. $P_{\mathrm{ES}-}^{\max}$ and $P_{\mathrm{ES}+}^{\max}$ are the maximum charge and discharge power of the energy storage system, respectively.

3.2. Regulation Cost of Diesel Generator

The generation cost of the diesel generator is determined by its fuel consumption coefficient and fuel unit price. In this paper, the power commands of diesel generators in the power allocation process refer to the output value within the range of its adjustable output force, i.e.,

$$\begin{cases} C_{\mathrm{DE}}(\Delta P_{\mathrm{DE}}) = r_{\mathrm{DE}}\Delta P_{\mathrm{DE}}T_{\mathrm{S}} = \lambda r_{\mathrm{oil}}\Delta P_{\mathrm{DE}}T \\ \Delta P_{\mathrm{DE}} = P_{\mathrm{DE}}(t) - P_{\mathrm{DE}}^{\min} \\ \Delta P_{\mathrm{DE}}^{\max} = P_{\mathrm{DE}}^{\max} - P_{\mathrm{DE}}^{\min} \\ P_{\mathrm{DE}}^{\min} = \max\left\{0.3P_{\mathrm{DE}}^{\mathrm{N}}, P_{\mathrm{DE}}(t-\Delta t) - R_{\mathrm{d}}\Delta t\right\} \\ P_{\mathrm{DE}}^{\max} = \min\left\{P_{\mathrm{DE}}^{\mathrm{N}}, P_{\mathrm{DE}}(t-\Delta t) + R_u\Delta t\right\} \\ P_{\mathrm{DE}}^{\min} \leq P_{\mathrm{DE}}(t) \leq P_{\mathrm{DE}}^{\max} \\ 0 \leq \Delta P_{\mathrm{DE}} \leq \Delta P_{\mathrm{DE}}^{\max} \end{cases}, \tag{4}$$

where ΔP_{DE} is the power command of the diesel generator; r_{DE}, λ, and r_{oil} are the unit generation cost, unit fuel coefficient, and fuel unit cost of the diesel generator, respectively. $P_{\mathrm{DE}}^{\mathrm{N}}$, $P_{\mathrm{DE}}^{\max}$, and $P_{\mathrm{DE}}^{\min}$ are the rated power and real-time output upper and lower limit of diesel generators, respectively. $\Delta P_{\mathrm{DE}}^{\max}$ is the upper limit of diesel generator power command. R_u and R_{d} are the rates of increasing and decreasing output of diesel generators, respectively.

Here ΔP_{DE} is the excess part of the current real-time output $P_{\mathrm{DE}}(t)$, relative to the lower limit of output $P_{\mathrm{DE}}^{\min}$ at this time. That is, the $P_{\mathrm{DE}}^{\min}$ output part is regarded as the unadjustable part in the power allocation calculation process, which is recorded as the forced output. The operation cost generated by $P_{\mathrm{DE}}^{\min}$ is recorded as the forced cost, which is not included in the regulation cost of the microgrid.

In this paper, the starting and stopping of the diesel generator are determined by setting the start and stop thresholds, given the start and stop coefficients $k_{\mathrm{DE\_on}}$ and $k_{\mathrm{DE\_off}}$. When the power shortage of the i-th microgrid equipped with diesel generator exceeds $k_{\mathrm{DE\_on}}P_{\mathrm{DE}}^{\mathrm{N}}$ and the diesel generator satisfies the other operational constraints, the diesel generator is started. If the power shortage of the i-th microgrid is lower than $k_{\mathrm{DE\_off}}P_{\mathrm{DE}}^{\mathrm{N}}$, and the diesel generator has satisfied the other operation constraints, the diesel generator is shut down. When the diesel generator enters the start-stop state, it runs according to the increase/decrease output rate until the state transition is completed.

3.3. Regulation Cost of Other Units

For interruptible load and DG, the following assumptions are made in this paper: (1) when a part of the interruptible load needs to be cut off, the current calculation point calculates the current interruptible load amount and the next calculation point restores part of the load under the necessary conditions by default. If the next calculation point is still unable to restore the load, that part will be delayed according to the actual demand; (2) photovoltaic or wind turbines operate in the maximum power point tracking (MPPT) mode. When the output of distributed generation needs to be reduced, the current calculation point calculates the reduction command at the current maximum output, and the next calculation point is still calculated by the maximum output by default, so there is no need to consider the increase of the distributed generation output.

The regulation costs of load shedding and distributed generation abandoned generation are as follows:

$$\begin{cases} C_{\mathrm{IL}}(\Delta P_{\mathrm{IL}}) = r_{\mathrm{IL}}\Delta P_{\mathrm{IL}}T_{\mathrm{S}} \\ 0 \le \Delta P_{\mathrm{IL}} \le P_{\mathrm{IL}}^{\max} \\ P_{\mathrm{IL}}^{\max} = P_{\mathrm{IL}}(t) \end{cases} \tag{5}$$

$$\begin{cases} C_{\mathrm{DG}}(\Delta P_{\mathrm{DG-}}) = r_{\mathrm{DG}}|\Delta P_{\mathrm{DG-}}|T_{\mathrm{S}} \\ -P_{\mathrm{DG-}}^{\max} \le \Delta P_{\mathrm{DG-}} \le 0 \\ P_{\mathrm{DG}}^{\max} = P_{\mathrm{PV}}(t) + P_{\mathrm{WT}}(t) \end{cases}, \tag{6}$$

where ΔP_{IL} and $\Delta P_{\mathrm{DG-}}$ are the power commands of the interruptible load and the distributed generation output reduction, respectively. r_{IL} and r_{DG} are the unit regulation costs of the interruptible load and the distributed generation output reduction, respectively. $P_{\mathrm{IL}}^{\max}$ and $P_{\mathrm{DG-}}^{\max}$ are the upper limit of interruptible load and the distributed generation output reduction, respectively. $P_{\mathrm{IL}}^{\max}$ is the real-time maximum load $P_{\mathrm{IL}}(t)$ of the current interruptible load, i.e., $P_{\mathrm{DG-}}^{\max}$ is the sum of the real-time output of PV ($P_{\mathrm{PV}}(t)$) and the real-time output of ($P_{\mathrm{WT}}(t)$).

3.4. Regulation Cost Function of Microgrid

Combining with the microgrid power allocation strategy and the regulation cost model of each controllable unit, the regulation cost function of each microgrid can be constructed as shown in Figure 3.

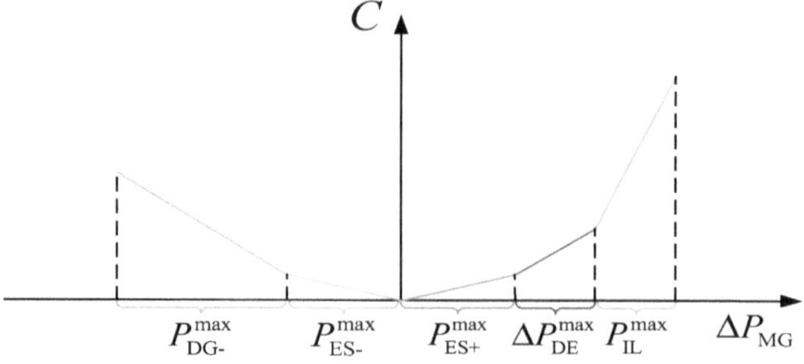

Figure 3. Regulation cost function of microgrids.

From the previous analysis, it can be seen that the regulation cost of each controllable unit can be approximated as a linear or piecewise linear function (e.g., the SOC variation in the charge and discharge process of energy storage system spans two areas of different unit regulation cost). It can be seen from Figure 3 that, when considering the dispatch priority, this paper can combine the regulation cost model of each controllable unit with the segmental linear function according to the power allocation strategy decision order of positive and negative power commands, that is, the microgrid regulation cost function.

This profile has the following salient features:

(1) Because the power allocation strategy determines the dispatching order according to the regulation cost, the unit with low regulation cost takes priority in the task of power regulation. Therefore, the curve along the increasing direction of positive and negative power command is steeper, that is, the slope value increases.
(2) The positive and negative power command parts of the regulation cost function are single-valued functions. In other words, for any positive and negative power command, as long as the microgrid power command ΔP_{MG} is determined, the corresponding power allocation scheme and regulation cost can be uniquely determined. Conversely, a uniquely determined power command can be obtained when the regulation cost is given.
(3) For the positive power command, $P_{\mathrm{ES+}}^{\max}$, $\Delta P_{\mathrm{DE}}^{\max}$, and $P_{\mathrm{IL}}^{\max}$ jointly determine the upper limit of the microgrid power regulation amount, i.e., $\Delta P_{\mathrm{MG}}^{\max} = P_{\mathrm{ES+}}^{\max} + \Delta P_{\mathrm{DE}}^{\max} + P_{\mathrm{IL}}^{\max}$. For the negative power command, $P_{\mathrm{ES-}}^{\max}$ and $P_{\mathrm{DG-}}^{\max}$ jointly determine the lower limit of the microgrid power regulation amount, i.e., $\Delta P_{\mathrm{MG}}^{\min} = P_{\mathrm{ES-}}^{\max} + P_{\mathrm{DG-}}^{\max}$.
(4) Due to the above assumptions in this paper, it is the case that the load-shedding operation and the distributed generation increasing the output after reducing the output can be regarded as the default cost-free adjustment strategy with the highest priority, which is not reflected in the regulation cost function.

4. Consensus Algorithm Design for Power Dispatch

4.1. Cooperative Power Dispatch Model for Islanded Multi-Microgrids

In this paper, a power dispatch consensus algorithm is designed for islanded multi-microgrids, which take the regulation cost of each microgrid as a consensus variable and use the discrete-time first-order consensus algorithm [33] to solve the problem iteratively. Hence, the total regulation cost of the multi-microgrids is chosen as the objective function. The mathematical model is constructed as follows:

$$\begin{cases} \min F = \sum\limits_{i=1}^{n} f_{\mathrm{C}}(\Delta P_{\mathrm{MG}i}) \\ s.t.\ \Delta P = \sum\limits_{i=1}^{n} \Delta P_{\mathrm{MG}i} \\ \qquad = \sum\limits_{i=1}^{n} \left(P_{\mathrm{Load}i} - P_{\mathrm{PV}i} - P_{\mathrm{WT}i} - P_{\mathrm{DE}i}^{\min} \right) \\ \Delta P_{\mathrm{MG}i} = \Delta P_{\mathrm{ES}i} + \Delta P_{\mathrm{DE}i} + \Delta P_{\mathrm{IL}i} + \Delta P_{\mathrm{DG}-i} \\ \Delta P \Delta P_{\mathrm{MG}i} > 0 \\ \Delta P_{\mathrm{MG}i}^{\min} \le \Delta P_{\mathrm{MG}i} \le \Delta P_{\mathrm{MG}i}^{\max} \end{cases}, \tag{7}$$

where the total power command ΔP of the islanded multi-microgrids is the difference between the total load and the total output of the wind power, PV, and the forced output of the energy storage device. $\Delta P_{\mathrm{MG}i}$ is the power command of the i-th microgrid, that is, the power regulation amount of the i-th microgrid. $f_{\mathrm{C}}(\Delta P_{\mathrm{MG}i})$ is the regulation cost of the i-th microgrid, which is a function of the microgrid power regulation. $P_{\mathrm{Load}i}$ is the total load of the i-th microgrid. $\Delta P_{\mathrm{MG}i}^{\max}$ and $\Delta P_{\mathrm{MG}i}^{\min}$ are the upper and lower limits of the power regulation of the i-th microgrid, respectively, which depend on the real-time status of each control unit in the microgrid. n is the number of microgrids.

Therefore, when a power imbalance occurs in the islanded multi-microgrids, it will be jointly undertaken by all the microgrids that can participate in the regulation. Selecting the regulation cost of the microgrid as the consensus variable can enable each microgrid to participate in the regulation according to its own resources. Its essence is to allocate the power regulation amount to each microgrid according to the slope value of each segment of each microgrid regulation cost function, which aims at reducing the overall regulation cost of the multi-microgrids.

4.2. Microgrid Regulation Cost Consensus

The regulation cost of each microgrid is selected as the consensus variable, which is abbreviated as C. Based on the discrete-time first-order consensus algorithm [33], the formula for updating the consensus variables of each agent is as follows:

$$C_i^{(k+1)} = \sum_{j=1}^{n} d_{ij}^{(k)} C_j^{(k)}, \tag{8}$$

where $C_i^{(k)}$ is the regulation cost calculated by the k-th iteration of the i-th microgrid. $d_{ij}^{(k)}$ is the i-th row and j-th column element of the row-stochastic matrix D when iterating at the k-th step. The row-stochastic matrix D is obtained from the Laplacian matrix L of the communication topology and is related to the structure of the communication topology, which is defined by the following:

$$d_{ij} = \frac{|l_{ij}|}{\sum\limits_{j=1}^{n} |l_{ij}|}, \quad i = 1, \cdots, n \text{ with } \begin{cases} l_{ii} = \sum\limits_{i \neq j} a_{ij} \\ l_{ij} = -a_{ij} \end{cases}, \tag{9}$$

where a_{ij} is the (i, j) entry of the adjacency matrix A. In this paper, the adjacency matrix A is

$$A = \begin{bmatrix} 0 & 1 & 1 \\ 1 & 0 & 1 \\ 1 & 1 & 0 \end{bmatrix}. \tag{10}$$

It can be known from Equation (8) that each agent obtains the state information of the previous iteration of the neighboring agent by means of the row stochastic matrix D, which is related to the communication topology to update its state. In order to ensure the power balance, the leader guides the regulation direction and magnitude of the regulation cost. As a leader, the microgrid regulation cost update formula is given as follows:

$$C_i^{(k+1)} = \begin{cases} \sum\limits_{j=1}^{n} d_{ij}^{(k)} C_j^{(k)} + \mu \Delta P_{\text{error}} & \Delta P > 0 \\ \sum\limits_{j=1}^{n} d_{ij}^{(k)} C_j^{(k)} - \mu \Delta P_{\text{error}} & \Delta P < 0 \end{cases}, \tag{11}$$

where μ is the error adjustment step size; ΔP_{error} is the deviation between the total power command and the sum of the microgrid power commands, which ignores the line loss of the islanded multi-microgrids system. The expression of ΔP_{error} is:

$$\Delta P_{\text{error}} = \Delta P - \sum_{i=1}^{n} \Delta P_{\text{MG}i}. \tag{12}$$

The error adjustment step size μ can be artificially given an appropriate parameter or adaptively adjusted by detecting ΔP_{error}. The meaning of Equation (10) can be explained as follows: taking the current total power command $\Delta P > 0$ as an example, if $\Delta P_{\text{error}} > 0$, the sum of the power commands of each microgrid is still insufficient to balance the current power imbalance, and the regulation cost needs to be increased accordingly; if $\Delta P_{\text{error}} < 0$, the regulation cost can be reduced.

In principle, the selection of leaders should be determined by the regulation capability. Selecting the microgrid with the largest regulation capability as the leader can reduce the need to replace leaders. The so-called regulation capability can be reflected by the parameters that reflect the regulatable resource of microgrid, such as energy storage capacity, diesel generator capacity, interruptible load capacity, etc.

In summary, the regulation cost of each microgrid in the iterative process is the weighted average of the regulation cost of the local and neighbors' previous iterations, so the communication burden on the network is small. When some microgrids have reached the limit of the power regulation, they should quit the communication topology and stop updating. The adjacent microgrids should also modify the corresponding row random matrix elements according to the new communication topology. In the convergence process of the consensus algorithm, $|\Delta P_{\text{error}}| \leq \varepsilon$ is taken as the convergence condition, where ε is the convergence error.

When the regulation cost is updated by communication interaction between microgrids, the power command of each unit needs to be inversely solved by the variable of regulation cost and the corresponding regulation cost function. Because the regulation cost function is a piecewise linear function and has a unique solution, the calculation process is not complicated, and each MGC can quickly solve and calculate the microgrid total power command for the next iteration. The flowchart of the proposed consensus algorithm for real-time cooperative power dispatch of islanded multi-microgrids is shown in Figure 4

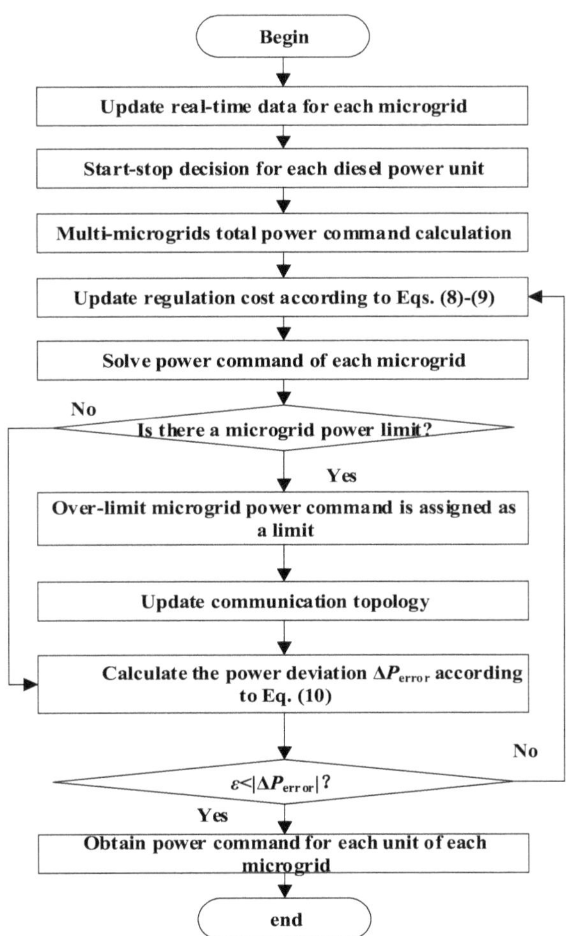

Figure 4. Flowchart of consensus algorithm for real-time cooperative power dispatch of islanded multi-microgrids.

5. Numerical Simulations

5.1. Simulation Parameter Setting

In this paper, an islanded multi-microgrids model formed by three interconnected microgrids is established. The main program is carried out in a Matlab 2015b (USA mathworks company) environment. The communication topology is shown in Figure 1. In the absence of power overrun in microgrids, there is communication between any two microgrids. Considering the regulatable resources of each microgrid, MG1 is the leader, and MG2 and MG3 are followers.

The equipment parameters of the multi-microgrids are given in Table 1. The minimum output of the diesel generator is 30% of the rated output, the minimum running time is 60 min, the minimum stop time is 30 min, and the start and stop coefficients k_{DE_on} and k_{DE_off} are set at 1.5 and 0.5, respectively. The regulation cost parameters are given in Table 2.

Table 1. Equipment parameters of multi-microgrids.

Equipment Parameters	MG1	MG2	MG3
PV capacity/kW	180	100	120
WT capacity/kW	40	80	80
ES capacity/kWh	200	200	300
ES rated power/kW	50	50	100
ES efficiency η	0.9	0.9	0.9
ES initial SOC	0.34	0.25	0.21
DE rated power/kW	50	50	-
DE increase output rate/(kW/min)	2	2	-
DE decrease output rate/(kW/min)	2	2	-
Maximum load/kW	130	180	140

Table 2. Parameters of regulating cost.

Parameters of Regulating Cost ($/kWh)	MG1	MG2	MG3
r_{ES1}	0.05	0.05	0.05
r_{ES2}	0.1	0.1	0.1
r_{ES3}	0.25	0.25	0.25
r_{DE}	1.4	1.4	1.4
r_{DG}	1.6	1.6	1.6
r_{IL}	1.6	1.9	1.8

5.2. Simulation of Multi-Microgrids Power Allocation

The simulation analysis of continuous real-time power allocation is performed at intervals of 1 min. The total duration is 24 h. The shorter interval is also applicable in practical applications. The time window is taken as T_S = 30 min or 0.5 h, the error adjustment step size of consensus algorithm is $\mu = 0.01$, and the convergence error is $\Delta P_{error} = 0.1$ kW.

The load curve of the multi-microgrids, the total output curve of wind power and photovoltaics, and the corresponding power imbalance curve are shown in Figure 5, whose data source is an actual islanded multi-microgrid [34]. It can be seen that the power imbalance mainly occurs at 09:00–22:00. On the one hand, due to the large PV output at noon, there is a situation of excess power, and then a power shortage occurs after the peak load arrives at night.

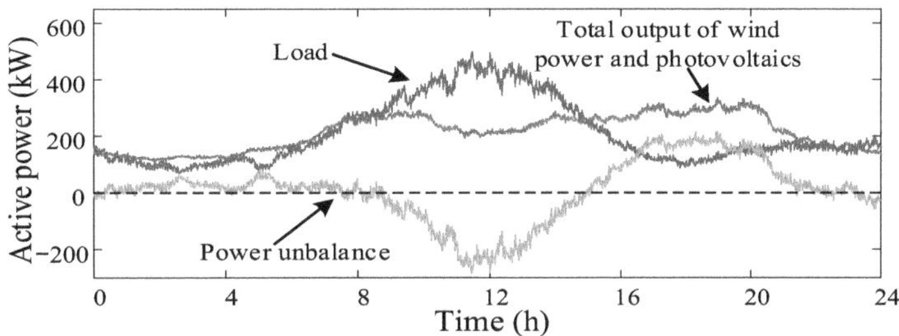

Figure 5. Power unbalance of multi-microgrids.

After simulation, the power command of each microgrid basically follows the trend of the total power command. The regulation power of each unit throughout the day and the corresponding microgrid regulation cost are given in Table 3. The total regulation cost of MG1 and MG2 includes the forced cost corresponding to the unadjustable part of the output of the energy storage device. Since MG1 and MG2 are equipped with energy storage devices, the actual regulation costs of MG1 and MG2 are higher than that of MG3. The power command curves of each microgrid and the output curves of various distributed generations are shown in Figures 6 and 7, respectively.

Table 3. Regulation power and cost.

	MG1	MG2	MG3	Total
Diesel power generation (kWh)	248.07	260.88	0	508.95
Load shedding (kWh)	9.63	8.00	6.18	23.81
Abandoned power generation (kWh)	120.77	116.38	97.09	334.25
Regulation cost ($)	484.06	498.88	140.76	1123.70

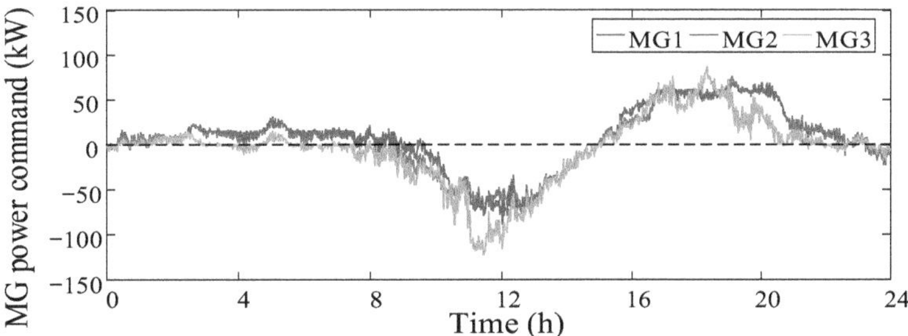

Figure 6. Power command curve of each microgrid.

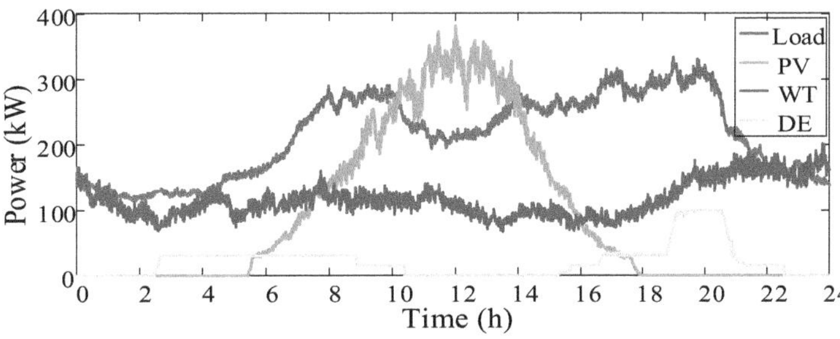

Figure 7. Power curve of DGs.

The results of charge and discharge power control of each microgrid energy storage unit are analyzed. The real-time power curve and SOC curve of each microgrid energy storage unit are shown in Figures 8 and 9. It can be seen that the energy storage units of each microgrid are charged during the period of excessive photovoltaic output during the daytime, and discharge is completed during the peak load period, which is in accordance with the scheduling rules. The SOC change trend of each energy storage is basically the same, the capacity can be effectively utilized, and the power command allocation is reasonable.

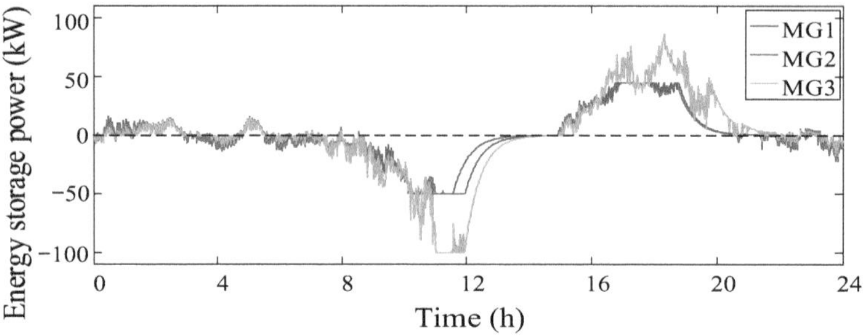

Figure 8. Power curve of each ES.

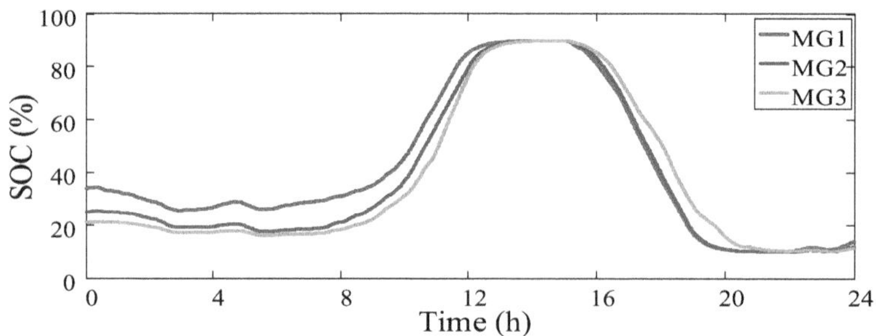

Figure 9. SOC curve of each ES.

The consensus convergence process of the two sections in the continuous simulation process is selected. The power command is $\Delta P = -205$ kW, $\Delta P = +210$ kW, respectively. The convergence process curves are shown in Figures 10 and 11.

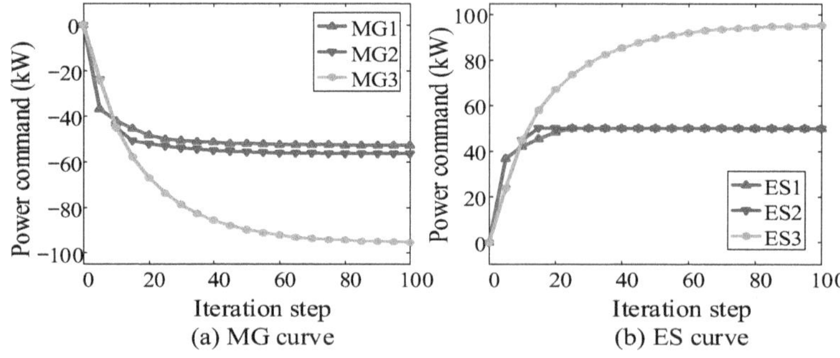

Figure 10. Consensus convergence process ($\Delta P = -205$ kW): (**a**) Description of the three MG iteration process; (**b**) Description of the three ES iteration process.

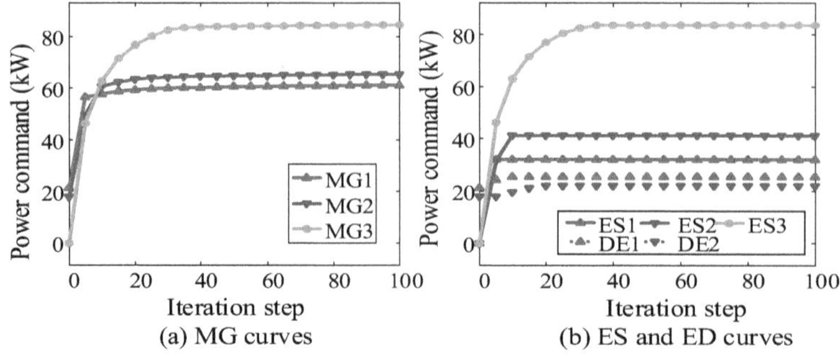

Figure 11. Consensus convergence process ($\Delta P = +210$ kW): (**a**) Description of the three MG iteration processes; (**b**) Description of the three ES and two DE iteration processed.

As can be seen from Figure 10, the current total power command is negative; the energy storage device in each microgrid charges to balance the corresponding unbalance amount, of which MG3 bears the largest regulation power. The reason is that the energy storage device of MG3 has the maximum capacity and rated power. When the energy storage devices of MG1 and MG2 reach the charge limit in the iteration process, it is necessary to adopt a regulation approach with a higher regulation cost. The energy storage device bears the low cost of power regulation. When each microgrid consumes a consensus regulation cost, MG3 would bear more power regulation.

Similarly, as shown in Figure 11, when the current total power command is positive, each microgrid should increase its power generation to balance the corresponding unbalance. Each energy storage device discharges and two diesel generators have been turned on. When the output of ES1 and ES2 reaches the limit, the corresponding DE1 and DE2 gradually increase their output in the iteration process to promote the consensus regulation cost of each microgrid.

5.3. *Comparative Analysis under Different Operation Modes*

In order to reflect the advantages of the interconnected operation of microgrid and consensus power allocation model, the different operation modes of microgrid are compared and analyzed in this section, including the following three cases.

Mode 1: each microgrid operates independently [35].

Mode 2: the multi-microgrids are interconnected to form a microgrid cluster and each microgrid gives priority to autonomous operation. When its own power commands are different from the total power command of the microgrid cluster, it actively exchanges power with the microgrid cluster system [36].

Mode 3: the multi-microgrids are interconnected to form a microgrid cluster and the power allocation is based on the real-time cooperative power dispatch model established in this paper.

The three models are simulated continuously throughout the day and the regulation costs of each microgrid and microgrid cluster are given in Table 4. From the perspective of total regulation costs, the order is mode 1 > mode 2 > mode 3. Under mode 1, each microgrid operates independently and can only rely on its own supply for balance. When the generation and load are not balanced, there will be more abandoned power generation or load shedding, which would lead to the regulation costs being higher. The interconnection of the microgrids in mode 2 will reduce the regulation costs, which reflects the advantages of cluster operation of adjacent microgrids to a certain extent. However, the benefits of clustering operations have not been fully explored due to the priority of autonomous balance of microgrids. Mode 3 adopts the cooperative power dispatch strategy in which three microgrids in total are dispatched, and the energy transfer between each microgrid is mutually beneficial. Each microgrid allocates the regulation power reasonably according to the signal of regulation costs. Overall, the regulation capability of the microgrid cluster system has been optimized and the economy has been improved.

Table 4. Regulation costs under the three modes.

Microgrid	Regulation Cost ($)		
	Mode 1	**Mode 2**	**Mode 3**
MG1	597.86	560.29	484.06
MG2	694.19	644.66	498.88
MG3	317.22	338.23	140.76
Total	1609.27	1543.18	1123.70

Figure 12 shows the MG1 power command under the three modes; the PV configuration of MG1 is the largest. It can be seen that the negative power command of MG1 at noon is smaller than that of the other two modes. Modes 1 and 2 decrease obviously at noon due to the bias towards autonomy.

Under mode 3, due to the interconnection of each microgrid, the surplus output of MG1 can partly support the other two microgrids, thus reducing the negative power command in this period.

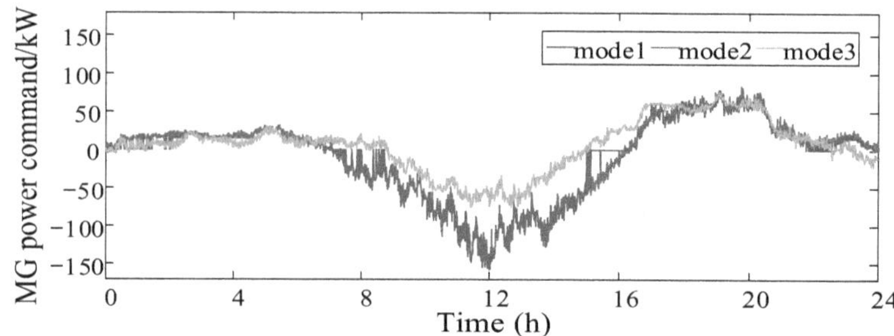

Figure 12. Power command of MG1 under the three modes.

5.4. Comparison with Centralized Optimization

From the perspective of engineering applications, this paper compares the optimization of performance with the centralized optimization method. Aiming at the time section of power command P = +210 kW, a multiple population genetic algorithm (MPGA) [37] is selected for centralized optimization, in which the MPGA runs 10,000 times and the results are compared with those of the proposed consensus algorithm. The CPU (central processing unit) used in the simulation is 3.2 GHz, with 2 GB memory. A performance comparison of the two optimization approaches is given in Table 5.

Table 5. Comparison of optimization performance.

	MPGA		Consensus Algorithmic
Regulation cost ($)	optimal	20.35	21.16
	mean	23.44	
Optimization time (s)	mean	1.66	0.02
Optimization approach	centralized, static		distributed, dynamic

From Table 5, it can be seen that the consensus algorithm belongs to distributed optimization technology, and its convergence time is greatly reduced compared with centralized optimization. As the scale of the microgrid clusters further increases, its advantages will become more obvious. MPGA is optimized 10,000 times and the optimal regulation cost is better than that of the consensus algorithm. It can be seen that the optimization result of the consensus algorithm is not the global optimal solution. The mean value of 10,000 optimization results obtained by MPGA is slightly lower than that of the consensus algorithm. According to statistics, the probability that the MPGA optimization result will be better than the consensus algorithm is about 30.69%. For the comparison of the optimization results shown in Figure 13, most of the optimization algorithms belong to static optimization, and it is difficult to solve the real-time dynamic problem.

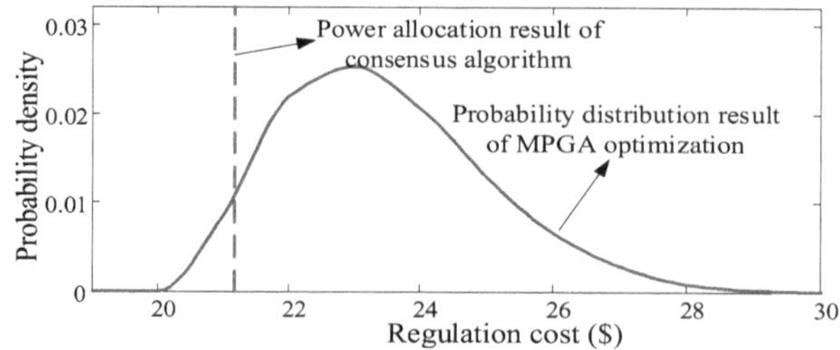

Figure 13. Comparison of optimization results.

5.5. Impact Analysis of Different T_S

The value of time window T_S will affect the results, as reflected in the power control of the energy storage. The reason is that the SOC electricity calculation of the energy storage is related to the length of the time window. This section compares the impact of different T_S.

Taking ES3 as an example, the power curve and SOC curve of ES3 are shown in Figures 14 and 15, respectively. Figure 14a shows the overall situation throughout the day and Figure 14b,c shows that the time period has a significant impact under different T_S. Between 11:00 and 14:00, because of the high photovoltaic output at noon, the charge of the energy storage device is close to the SOC upper limit. When T_S is small (such as T_S = 10/60 h), the power command of the energy storage device will quickly drop to 0 and stop charging until the SOC reaches the upper limit. If T_S is large (such as T_S = 90/60 h), the energy storage device will start earlier to reduce the charge power. This proves that the control of the TS energy storage device tends to be conservative when T_S is large, i.e., a larger T_S can control the SOC of the energy storage device to avoid approaching the limit value.

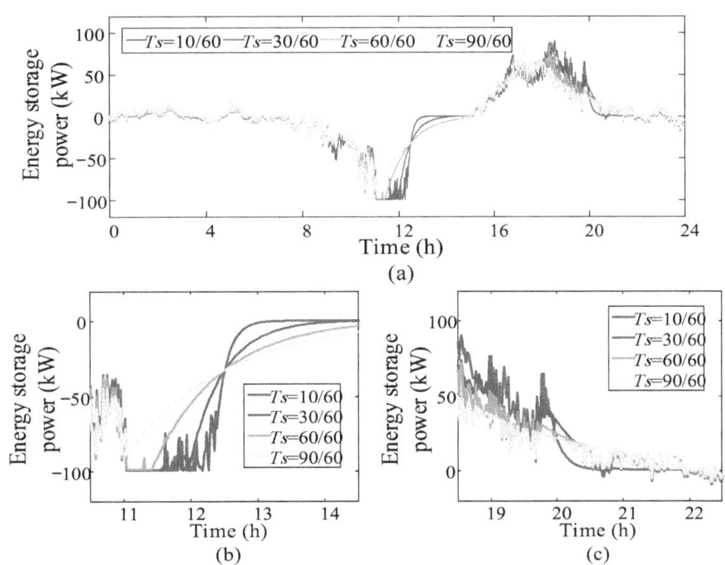

Figure 14. Power curve of ES3 with different *T*s: (**a**) Description of the Power curve of ES3 with different *T*s from 0:00 to 24: 00; (**b**) Description of the Power curve of ES3 with different *T*s from 10:30 to 14: 20; (**c**) Description of the Power curve of ES3 with different *T*s from 18:40 to 22: 20.

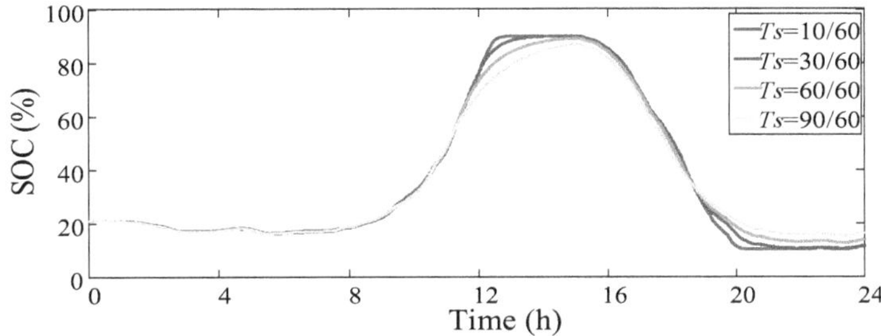

Figure 15. SOC curve of ES3 with different *T*s.

However, the larger T_S will increase the regulation cost of other controllable units, that is, the power regulation capability of the energy storage device is restricted in some time periods and the corresponding parts need to be borne by other controllable units. Table 6 gives the regulation cost under different values. Generally speaking, T_S should not be too large or too small. The value of T_S can be determined by reference to the type of energy storage battery and the consideration of the life of the energy storage battery, as well as the actual application requirements. The recommended value is 0.5–1.5 h.

Table 6. Regulation cost with different Ts.

T_S (h)	MG1 Regulation Cost ($)	MG2 Regulation Cost ($)	MG3 Regulation Cost ($)	Total ($)
10/60	472.14	493.30	146.29	1111.73
30/60	484.06	498.88	140.76	1123.70
60/60	493.24	512.18	141.82	1147.24
90/60	515.62	532.72	147.83	1196.17

6. Conclusions

In recent years, the development of microgrids has attracted interest to its related topics. This paper studies the real-time power allocation problem of islanded multi-microgrids. Considering the architecture and operational characteristics of islanded multi-microgrids, it is suitable to adopt a decentralized control architecture. A real-time cooperative power dispatch framework for islanded multi-microgrids is proposed. The microgrid regulation cost is modeled and a consensus algorithm is introduced to realize the power allocation. The conclusions are as follows:

(1) The consensus power allocation algorithm proposed in this paper can allocate power reasonably and ensure the real-time power balance of the islanded multi-microgrids.

(2) Each microgrid is guided by the regulation cost, and rationally optimizes the regulation power allocated by each microgrid, thereby improving the overall economics of the microgrid cluster system.

(3) The communication burden of the control architecture is small. Each microgrid only needs a small amount of information from the local and neighboring microgrids, which can satisfy the real-time dynamic power allocation requirements of islanded multi-microgrids.

Author Contributions: All the authors have contributed to this paper in different aspects. X.Z. proposed the original concept and wrote the original draft. N.W. acted as supervisor and gave suggestions on paper improvement.

Abbreviations and Nomenclature

MG	microgrid
MMGs	multi-microgrids
CHP	combined heat and power
MGC	microgrid controller
ICA-GA	imperialist competitive algorithm-genetic algorithm
GSO	glowworm swarm optimization
PV	photovoltaic
WT	wind turbine
ES	energy storage device
DE	diesel engine
IL	interruptible load
SOC	state of charge
DG	distributed generation
MPPT	maximum power point tracking
MPGA	multiple population genetic algorithm
Δt	time interval of power command
$SOC(t)$	current SOC calculated value
$SOC(t-\Delta t)$	previous actual SOC value
V_{ES}	energy storage system capacity
$P_{\mathrm{ES}}^{\max}$	energy storage system power limits
η	energy storage system efficiency

T_S	state duration window
ΔP_{ES}	power command of the energy storage device
$C_{ES}(\Delta P_{ES})$	regulation cost generated by power command ΔP_{ES}
ΔP	total power imbalance
ΔS_i	SOC change caused by the SOC area with the
r_{ESi}	unit regulation cost of ES
P_{ES-}^{max}	maximum charge power
P_{ES+}^{max}	maximum discharge power
ΔP_{DE}	power command of the diesel generator
r_{DE}	unit generation cost of the diesel generator
λ	unit fuel coefficient of the diesel generator
r_{oil}	fuel unit cost of the diesel generator
P_{DE}^{N}	rated power of diesel generators
P_{DE}^{max}	upper limit of diesel generators output
P_{DE}^{min}	lower limit of diesel generators output
ΔP_{DE}^{max}	upper limit of diesel generator power command
R_u	rates of decreasing output of diesel generators
R_d	rates of increasing output of diesel generators
ΔP_{DE}	excess part of the current real-time output
k_{DE_on}	start coefficients of the diesel generator
k_{DE_off}	stop coefficients of the diesel generator
ΔP_{IL}	power commands of interruptible load
ΔP_{DG-}	power command of distributed generation output reduction
r_{IL}	unit regulation cost of interruptible load
r_{DG}	unit regulation cost of distributed generation output reduction
P_{IL}^{max}	upper limit of interruptible load
P_{DG-}^{max}	upper limit of distributed generation output reduction
ΔP_{MG}	microgrid power command
ΔP_{MGi}	power command of the i-th microgrid
$f_C(\Delta P_{MGi})$	regulation cost of the i-th microgrid
P_{Loadi}	total load of the i-th microgrid
ΔP_{MGi}^{max}	upper limit of the power regulation of the i-th microgrid
ΔP_{MGi}^{min}	lower limit of the power regulation of the i-th microgrid
n	number of microgrid
μ	error adjustment step size
ΔP_{error}	power deviation commands
ε	convergence error

References

1. Hatziargyriou, N.; Asano, H.; Iravani, R.; Marnay, C. Microgrids. *IEEE Power Energy Mag.* **2007**, *5*, 78–94. [CrossRef]
2. Katiraei, F.; Iravani, R.; Hatziargyriou, N.; Dimeas, A. Microgrids management. *IEEE Power Energy Mag.* **2008**, *6*, 54–65. [CrossRef]
3. Che, L. Microgrids and distributed generation systems: Control, operation, coordination and planning. *Diss. Theses-Gradworks* **2015**, *63*, 33–39.
4. Katiraei, F.; Iravani, M.R.; Lehn, P.W. Micro-grid autonomous operation during and subsequent to islanding process. *IEEE Trans. Power Deliv.* **2005**, *20*, 248–257. [CrossRef]

5. Amoateng, D.O.; Al Hosani, M.; El Moursi, M.S.; Turitsyn, K.; Kirtley, J.L. Adaptive voltage and frequency control of islanded multi-microgrids. *IEEE Trans. Power Syst.* **2018**, *33*, 4454–4465. [CrossRef]
6. Hualei, Z.; Shiwen, M.; Yu, W.; Zhang, F.; Chen, X.; Cheng, L. A survey of energy management in interconnected multi-microgrids. *IEEE Access* **2019**, *7*, 72158–72169.
7. John, T.; Ping Lam, S. Voltage and frequency control during microgrid islanding in a multi-area multi-microgrid system. *IET Gener. Transm. Distrib.* **2017**, *11*, 1502–1512. [CrossRef]
8. Farrokhabadi, M.; Canizares, C.A.; Bhattacharya, K. Frequency control in isolated/islanded microgrids through voltage regulation. *IEEE Trans. Smart Grid* **2017**, *8*, 1185–1194. [CrossRef]
9. Chowdhury, D.; Khalid Hasan, A.; Rahman Khan, M.Z. Scalable DC microgrid architecture with phase shifted full bridge converter based power management unit. In Proceedings of the 2018 10th International Conference on Electrical and Computer Engineering (ICECE), Dhaka, Bangladesh, 20–22 December 2018; pp. 22–25.
10. Khalid Hasan, A.; Chowdhury, D.; Rahman Khan, M. Scalable DC microgrid architecture with a one-way communication based control interface. In Proceedings of the 2018 10th International Conference on Electrical and Computer Engineering (ICECE), Dhaka, Bangladesh, 20–22 December 2018; pp. 265–268.
11. Alipour, M.; Mohammadi-Ivatloo, B.; Zare, K. Stochastic scheduling of renewable and CHP based microgrids. *IEEE Trans. Ind. Inform.* **2015**, *11*, 1049–1058. [CrossRef]
12. Wang, R.; Wang, P.; Xiao, G. A robust optimization approach for energy generation scheduling in microgrids. *Energy Convers. Manag.* **2015**, *106*, 597–607. [CrossRef]
13. Luo, Z.; Wei, G.U.; Zhi, W.U.; Wang, Z.; Tan, Y. A robust optimization method for energy management of CCHP microgrid. *J. Mod. Power Syst. Clean Energy* **2018**, *6*, 132–144. [CrossRef]
14. Palma-Behnke, R.; Benavides, C.; Lanas, F.; Severino, B.; Reyes, L.; Llanos, J.; Sáez, D. A microgrid energy management system based on the rolling horizon strategy. *IEEE Trans. Smart Grid* **2013**, *4*, 996–1006. [CrossRef]
15. Solanki, B.V.; Canizares, C.A.; Kankar, B. Practical energy management systems for isolated microgrids. *IEEE Trans. Smart Grid* **2019**, *10*, 4762–4775. [CrossRef]
16. Dou, C.X.; An, X.G.; Yue, D. Multi-agent system based energy management strategies for microgrid by using renewable energy source and load forecasting. *Electr. Power Compon. Syst.* **2016**, *44*, 2059–2072. [CrossRef]
17. Wang, Z.; Feng, L.; Low, S.H.; Zhao, C.; Mei, S. Distributed frequency control with operational constraints, Part I: Per-node power balance. *IEEE Trans. Smart Grid* **2019**, *10*, 40–52. [CrossRef]
18. Wang, Z.; Feng, L.; Low, S.H.; Zhao, C.; Mei, S. Distributed frequency control with operational constraints, Part II: Network power balance. *IEEE Trans. Smart Grid* **2019**, *10*, 53–64. [CrossRef]
19. Hamed, M.G.; Mohammad, K. A novel optimal control method for islanded microgrids based on droop control using the ICA-GA algorithm. *Energies* **2017**, *10*, 485.
20. Quynh, T.T.T.; Maria, L.D.S.; Riva Sanseverino, E.; Zizzo, G.; Pham, T. Driven primary regulation for minimum power losses operation in islanded microgrids. *Energies* **2018**, *11*, 2890.
21. Sanseverino, E.R.; Silvestre, M.L.D.; Mineo, L.; Favuzza, S.; Nguyen, N.Q.; Tran, Q.T.T. A multi-agent system reinforcement learning based optimal power flow for islanded microgrids. In Proceedings of the 2016 IEEE 16th International Conference on Environment and Electrical Engineering (EEEIC), Florence, Italy, 7–10 June 2016; pp. 1–6.
22. Li, Q.; Gao, D.W.; Zhang, H.; Wu, Z.; Wang, F.Y. Consensus-based distributed economic dispatch control method in power systems. *IEEE Trans. Smart Grid* **2019**, *10*, 941–954. [CrossRef]
23. Yang, S.; Tan, S.; Xu, J. Consensus based approach for economic dispatch problem in a smart grid. *IEEE Trans. Power Syst.* **2013**, *28*, 4416–4426. [CrossRef]
24. Pourbabak, H.; Luo, J.; Chen, T.; Su, W. A novel consensus-based distributed algorithm for economic dispatch based on local estimation of power mismatch. *IEEE Trans. Smart Grid* **2018**, *9*, 5930–5942. [CrossRef]
25. Tang, Z.; Hill, D.J.; Liu, T. A novel consensus-based economic dispatch for microgrids. *IEEE Trans. Smart Grid* **2018**, *9*, 3920–3922. [CrossRef]
26. Zhang, X.; Yu, T.; Yang, B.; Li, L. Virtual generation tribe based robust collaborative consensus algorithm for dynamic generation command dispatch optimization of smart grid. *Energy* **2016**, *101*, 34–51. [CrossRef]
27. Lu, L.; Chu, C. Consensus-based droop control of isolated micro-grids by ADMM implementations. *IEEE Trans. Smart Grid* **2018**, *9*, 5101–5112. [CrossRef]

28. Liu, Y.; Li, Y.; Xin, H.; Gooi, H.B.; Pan, J. Distributed optimal tie-line power flow control for multiple interconnected AC microgrids. *IEEE Trans. Power Syst.* **2019**, *34*, 1869–1880. [CrossRef]
29. Zhao, B.; Wang, X.; Lin, D.; Calvin, M.M.; Morgan, J.C.; Qin, R.; Wang, C. Energy management of multiple-microgrids based on a system of systems architecture. *IEEE Trans. Power Syst.* **2018**, *33*, 6410–6421. [CrossRef]
30. Bui, V.H.; Hussain, A.; Kim, H.M. A multiagent-based hierarchical energy management strategy for multi-microgrids considering adjustable power and demand response. *IEEE Trans. Smart Grid* **2018**, *9*, 1323–1333. [CrossRef]
31. Asimakopoulou, G.E.; Dimeas, A.L.; Hatziargyriou, N.D. Leader-follower strategies for energy management of multi-microgrids. *IEEE Trans. Smart Grid* **2013**, *4*, 1909–1916. [CrossRef]
32. Xu, Y.; Li, Z. Distributed optimal resource management based on the consensus algorithm in a microgrid. *IEEE Trans. Ind. Electron.* **2015**, *62*, 2584–2592. [CrossRef]
33. Zhang, Z.; Chow, M.Y. Convergence analysis of the incremental cost consensus algorithm under different communication network topologies in a smart grid. *IEEE Trans. Power Syst.* **2012**, *27*, 1761–1768. [CrossRef]
34. Alegria, E.; Brown, T.; Minear, E.; Lasseter, R.H. CERTS microgrid demonstration with large-scale energy storage and renewable generation. *IEEE Trans. Smart Grid* **2014**, *5*, 937–943. [CrossRef]
35. Olivares, D.E.; Canizares, C.A.; Kazerani, M. A centralized energy management system for isolated microgrids. *IEEE Trans. Smart Grid* **2014**, *5*, 1864–1875. [CrossRef]
36. Fang, X.; Yang, Q.; Wang, J.; Yan, W. Coordinated dispatch in multiple cooperative autonomous islanded microgrids. *Appl. Energy* **2016**, *162*, 40–48. [CrossRef]
37. Wang, B.; Li, I. Load balancing task scheduling based on multi-population genetic algorithm in cloud computing. In Proceedings of the 2016 35th IEEE Chinese Control Conference (CCC), Chengdu, China, 27–29 July 2016; pp. 5261–5266.

3

The Direct Speed Control of Pmsm Based on Terminal Sliding Mode and Finite Time Observer

Yao Wang *, HaiTao Yu *, Zhiyuan Che, Yuchen Wang and Yulei Liu

School of Electrical Engineering, Southeast University, Nanjing 210096, China; zhiyuanche@foxmail.com (Z.C.); 220172743@seu.edu.cn (Y.W.); wangyhbcd@126.com (Y.L.)

* Correspondence: 230179191@seu.edu.cn (Y.W.); htyu@seu.edu.cn (H.Y.)

Abstract: A non-singular terminal sliding mode control based on finite time observer is designed to achieve speed direct control for the permanent magnet synchronous motor (PMSM) drive system. Speed and current are regulated in one loop under the non-cascade structure, taking place of the cascade structure control method in the vector control of PMSM. Based on the second-order speed function of the PMSM, the disturbance and parameters uncertainties are estimated by the designed finite time observer (FTO), and compensate to the drive system. The estimated value of the finite time observer will converge to the actual disturbance value in a finite time. A second-order non-singular terminal sliding mode controller is proposed to realize the speed and current single-loop, which can track the reference speed and reference current in a finite time. Rigorous stability analysis is established. Comparative results verified that the proposed method has faster speed tracking performance and disturbance rejection property.

Keywords: non-singular terminal sliding mode control (NTSMC); finite-time observer (FTO); mismatched/matched disturbance/uncertainties; permanent magnet synchronous motor (PMSM)

---

1. Introduction

By reason of the high-power density, torque-to-inertia ratio and high efficiency, the permanent magnet synchronous motor (PMSM) are widely used in industrial areas, such as, aerospace, servo control, numerical control machine and robot [1–5]. In these applications, the dynamic response performance and disturbance rejection property of PMSM are very important.

In recent years, with the progress of technology, the control periods between the speed loop and current loop of PMSM gradually decreased, or even vanished [6]; making it possible to realize the speed-current single-loop of PMSM drive system under the non-cascade structure. Generally speaking, in the traditional cascade control method for PMSM, the control period of the speed loop is 5–10 times that of the current loop, reducing the real-time control performance of the speed [7–9]. When the same control algorithm is adopted, different from the cascade control structure, the number of adjustable parameters is reduced and the speed can be directly controlled. These are the virtues of the non-cascade control structure [10,11]. Despite its advantages, there is little research on non-cascade control structures for the PMSM system in recent years. A non-cascade structure control based on model predictive control is proposed in [12], in which the dynamic performance of the system is improved and the computational complexity is reduced, compared with the traditional cascade predictive control method. In [13], under the non-cascade structure, the speed and current are adjusted in one proportion integration differentiation (PID) controller. Rigorous theoretical derivation and experimental analysis verified that the proposed method has better dynamic performance and disturbance rejection ability. Considering the influence of various disturbances on the PMSM system, a new non-cascade structure controller is established in [14], which can directly control the speed of PMSM. PMSM speed and current are adjusted in one loop based on terminal sliding mode and nonlinear disturbance observer under non-cascade structure control in [15]. However, when without the nonlinear disturbance observer, the proposed method has a poor ability to deal with the load sudden change. A direct speed control method based on radial basis function (RBF) is designed in [16], which avoided the control of current, simplified the control structure and improved the control performance. A model predictive direct speed control method based on voltage vector control is proposed in [17]. In this method, the voltage vector does not need to be measured, the computational burden of the system is reduced, and the output current is constrained within a certain range. A model predictive direct speed controller is proposed in [18], which overcomes the shortcoming of cascade linear controller in high-speed control, and the results show that the proposed method has better stability performance. Based on the state-dependent Riccati equation (SDRE) and Convex constrained optimization, a direct speed controller was proposed in [19], which can make the PMSM control system achieve high dynamic and accurate stability performance, and the input voltage and stator current can be constrained.

Due to the nonlinear and strong coupling characteristics of the PMSM drive system, ideal control results can hardly be achieved in traditional PI controller [20,21]. Many nonlinear control methods have been applied in PMSM drive systems, such as sliding mode control, model predictive control, auto-disturbance rejection control, finite time control, etc., [22–26]. Among these methods, it can converge in finite time and has a better disturbance rejection performance. The terminal sliding mode is widely used in control systems. In [27], a new terminal sliding mode controller is designed to adjust the speed of the PMSM servo system, which can make the system reach the reference speed in a finite time, ensuring a fast convergence performance and a better tracking accuracy of the system. In [28], a non-singular terminal sliding mode control based on state observer is investigated to realize the pressure control. In the proposed method, the pressure tracking error can converge to the equilibrium point in finite time and the chattering of the sliding mode is weakened. In [29] according to euler discrete technology, a new discrete time fast terminal sliding mode method is proposed and applied to the control of permanent magnet synchronous linear motor (PMLSM), and the reference position of PMLSM can be quickly tracked. In [30], a fractional-order terminal sliding mode controller based on fractional-order disturbance observer is proposed, under which the speed can converge to the reference speed in a finite time. In [31], a higher speed tracking accuracy can be achieved by a continuous fast terminal sliding mode control, and the robustness of the PMSM system can be improved when the disturbance is feedforward to the system by the extended state observer. In [32], a nonsingular terminal sliding mode based on improved extended state observer is investigated to realize the direct voltage control for the stand-alone doubly-fed induction generator (DFIG) system, which can achieve a balanced stator voltage.

Load change, parameters uncertainty and unmodeled dynamics are considered to be important factors affected the control performance. At present, in order to improve the robustness, the disturbance will be estimated by state observer and feed forward to the system before it affects the system. In [33], to improve the robustness in surface permanent magnet synchronous motor, the lumped disturbance consisted of the external disturbance and mismatched parameters can be estimated by a Luneburg observer, and compensate to the PMSM system. In [34] the parameters uncertainties and disturbances in DC-DC converters are considered as lumped disturbance, estimated by a reduced order generalized proportional integral observer and fed forward to the system, which improves the dynamic performance of the system. In [35], the lumped disturbance in air-breathing hypersonic vehicles is calculated by a disturbance observer, and the accuracy of speed and position control is improved when the disturbance feedforward to the system. In [36], a high-gain generalized proportional integral observer is designed, to estimate the load change and parameters uncertainties in PMSM. In [37], the disturbance is estimated and compensated by a nonlinear disturbance observer to improve the disturbance rejection property of the system. Then, a nonlinear controller is used to control the system, and the semi-global stability of the designed nonlinear controller and nonlinear disturbance observer is proved. In [38], a robust nonlinear observer is proposed for the Lipschitz nonlinear system. On the one hand, the new observer does not need to be added to small Lipschitz constants; on the other hand, the state estimation error of the system can quickly approach zero in the face of large additional disturbances. Disturbance also exists in the PMSM drive system under the non-cascade structure. In order to improve the anti-disturbance ability, it is necessary to estimate and compensate the disturbance to the system.

2. Preliminaries

2.1. The Mathematical Model of Pmsm

The ideal model of a surface mounted PMSM in the d-q frame can be expressed as follows.

$$\begin{cases} \frac{di_d}{dt} = \frac{-Ri_d + n_p\omega L i_q}{L} + \frac{1}{L}u_d \\ \frac{di_q}{dt} = \frac{-Ri_q - n_p\omega L i_d - n_p\omega\psi_f}{L} + \frac{1}{L}u_q \\ \frac{d\omega}{dt} = -\frac{B\omega}{J} + \frac{n_p\psi_f}{J}i_q - \frac{T_L}{J} \end{cases} \tag{1}$$

where, i_d, i_q are the d-axis and q-axis stator currents, respectively; u_d, u_q are the d-axis and q-axis stator voltages, respectively; L is the inductor; R is stator resistance; n_p is the number of pole pairs; ω is angular velocity; ψ_f is rotor flux linkage; T_L is load torque; B is viscous frictional coefficient; J is rotor inertia.

2.2. The Mathematic Model of Speed-Current Single-Loop

Let $x_1 = \omega_{ref} - \omega$, and its derivative can be expressed as

$$\dot{x}_1 = x_2 = \dot{\omega}_{ref} - \dot{\omega} = \dot{\omega}_{ref} + a_1\omega - a_2 i_q + a_3 T_L. \tag{2}$$

where, $a_1 = \frac{B}{J}, a_2 = \frac{n_p\psi_f}{J}, a_3 = \frac{1}{J}$.

When the parameters uncertainties are considered, the following expression can be obtained.

$$\dot{x}_1 = x_2 + d_1 = \dot{\omega}_{ref} + a_1\omega - a_2 i_q + a_3 T_L + d_1 \tag{3}$$

where, $d_1 = \Delta a_1\omega - \Delta a_2 i_q + \Delta a_3 T_L$ is considered as mismatched uncertainties. And a_1, a_2, a_3 are nominal parameter values, a_{t1}, a_{t2}, a_{t3} are the actual parameter values. $\Delta a_1 = a_{t1} - a_1, \Delta a_2 = a_{t2} - a_2$, $\Delta a_3 = a_{t3} - a_3$.

The second order differential equation of speed error can be expressed as follows

$$\dot{x}_2 = \ddot{\omega}_{ref} - \ddot{\omega} = \ddot{\omega}_{ref} + a_1\dot{\omega} - a_2\dot{i}_q + a_3\dot{T}_L. \tag{4}$$

Considering system (1), the (4) can be rewritten as

$$\dot{x}_2 = \ddot{\omega}_{ref} + a_1\dot{\omega} + a_2 b_1 i_q + a_2 b_2\omega i_d + a_2 b_3\omega - a_2 b_4 u_q + a_3\dot{T}_L. \tag{5}$$

where $b_1 = \frac{R}{L}, b_2 = n_p, b_3 = \frac{n_p\psi_f}{L}$ and $b_4 = \frac{1}{L}$.

Taking the parameters uncertainties and disturbance into consideration, system (5) can be expressed as follows

$$\dot{x}_2 = -a_1 x_2 - a_2 b_3 x_1 - a_2 b_4 u_q + d_2. \tag{6}$$

where, $$\begin{aligned} d_2 = {} & \ddot{\omega}_{ref} + a_1\dot{\omega}_{ref} + a_2 b_3\omega_{ref} + a_2 b_1 i_q + a_2 b_2\omega i_d + a_3\dot{T}_L - \Delta a_1 x_2 - \Delta a_2\Delta b_3 x_1 + \\ & \Delta a_1\dot{\omega}_{ref} + \Delta a_2\Delta b_3\omega_{ref} + \Delta a_2\Delta b_1 i_q + \Delta a_2\Delta b_2\omega i_d - \Delta a_2\Delta b_4 u_q + \Delta a_3\dot{T}_L \end{aligned}.$$

b_1, b_2, b_3, b_4 are the nominal parameter values, b_{t1}, b_{t2}, b_{t3}, b_{t4} are the actual parameter values, $\Delta b_1 = b_{t1} - b_1, \Delta b_2 = b_{t2} - b_2, \Delta b_3 = b_{t3} - b_3, \Delta b_4 = b_{t4} - b_4$.

The second-order speed regulation system can be expressed as follows

$$\begin{cases} \dot{x}_1 = x_2 + d_1 \\ \dot{x}_2 = -a_1 x_2 - a_2 b_3 x_1 - a_2 b_4 u_q + d_2 \end{cases} \tag{7}$$

3. Control Design

Based on the method designed in this paper, the PMSM control structure block diagram is shown in Figure 1.

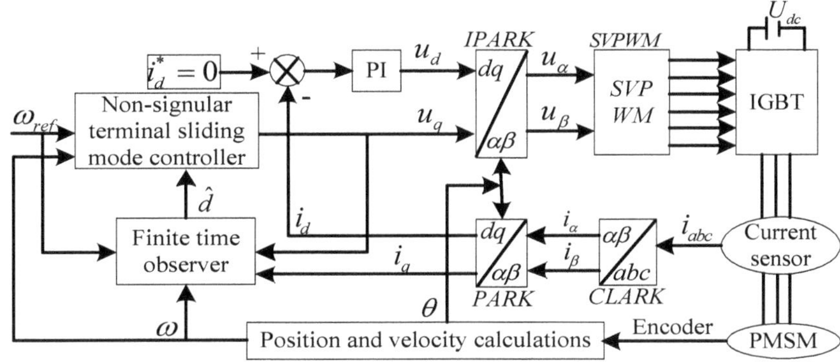

Figure 1. Permanent magnet synchronous motor (PMSM) control system based on the method designed in this paper.

3.1. Finite Time Observer

The disturbance is estimated and feedforward to the system based on the finite time observation method. The stability of the finite time control system is defined as follows.

Lemma 1. *[39] the following system is considered*

$$\dot{x} = f(x), x \in U \subseteq R^n, f(0) = 0 \tag{8}$$

where, $f : U \to R^n$ is a continuous function of x in domain of definition. For the equilibrium solution of the system, $x = 0$ is defined as finite time stability, which requires the system to be both stable and convergent in finite time. Finite time convergence means there are $\forall x_0 \in U_0 \subset R^n$ and a continuous function $T(x)$: $U_0 \backslash \{0\} \to (0, +\infty)$, making the solution $x(t, x_0)$ of the system (3) satisfied the following conditions: when $t \in [0, T(x_0)), x(t, x_0) \in U_0 \backslash \{0\}$ and $\lim_{x \to T(x_0)} x(t, x_0) = 0$ are true; When $t > T(x_0)$, $x(t, x_0) = 0$ is always true. If $U = U_0 = R^n$ existed, the system is considered globally finite time stable.

Notation 1. *For writing convenience, denote $sig^\beta(x) = \mathrm{sgn}(x)|x|^\beta$, where $x, \beta \in \mathbb{R}$, and $\mathrm{sgn}(\cdot)$ is the sign function.*

Lemma 2. *[40] Consider the following system*

$$\dot{x} = f(x) + \hat{f}(x), f(0) = 0, x \in \mathbb{R}^n \tag{9}$$

where, $f(x)$ is a continuous homogeneous vector field, and $f(x)$ has negative homogeneous degree k with respect to expansion vector $(r_1, r_2, \ldots, r_n)$. $\hat{f}(x)$ is the estimated disturbance of the system, which satisfies $\hat{f}(0) = 0$. $x \in \mathbb{R}^n$ refers that x is belonged to the n-dimensional vector. Suppose that the asymptotically stable equilibrium point of system $\dot{x} = f(x)$ is $x = 0$ and satisfies the following conditions, $\forall x \neq 0$

$$\lim_{\varepsilon \to 0} \frac{\hat{f}_i(\varepsilon^{r_1} x_1, \ldots, \varepsilon^{r_n} x_n)}{\varepsilon^{k + r_i}} = 0, i = 1, 2, \ldots, n \tag{10}$$

Then $x = 0$ is a locally finite time equilibrium point of system (9).

Lemma 3. *[41–43] Consider the nonlinear system*

$$\begin{cases} \dot{x}_i = x_{i+1}, i = 1, \ldots, n-1 \\ \dot{x}_n = u \\ y = x_1 \end{cases} \tag{11}$$

where, $x = (x_1, x_2, \ldots, x_n)^T \in R^n$ are the state variables of the system; $u \in R$ and $y \in R$ are the input and output of the system, respectively.

For system (11), the estimated state values $(\hat{x}_1, \hat{x}_2, \ldots, \hat{x}_n)$ can converge to the real states $(x_1, x_2, \ldots, x_n)$ of system (11) in a finite time by the following global finite time observer.

$$\begin{cases} \dot{\hat{x}}_i = \hat{x}_{i+1} + \lambda_i sig^{\beta_i}(x_1 - \hat{x}_1), i = 1, 2, \ldots, n-1 \\ \dot{\hat{x}}_n = u + \lambda_n sig^{\beta_n}(x_1 - \hat{x}_1) \end{cases} \tag{12}$$

where, $\beta_i > 0$, $i = 1, 2, \ldots, n$, it's a Hurwitz polynomial $s^n + \beta_1 s^{n-1} + \cdots + \beta_{n-1}s + \beta_n$.

$$\beta_i = i\beta - (i-1), i = 1, \ldots, n, \beta \in (1 - \frac{1}{n}, 1] \tag{13}$$

Lemma 4. *[44,45] A second-order system can be expressed as follows*

$$\begin{cases} \dot{x}_1 = x_2 - \lambda sig^{\frac{1}{2}}(x_1) \\ \dot{x}_2 = -v sign(x_1) + F(t) \end{cases} \tag{14}$$

If there is a positive real number f^+, $|F(t)| < f^+$ is true, and v, λ satisfies the following description.

$$v > f^+, \lambda > \sqrt{\frac{2}{v - f^+} \frac{(v + f^+)(1 + \mu)}{(1 - \mu)}} \tag{15}$$

where, μ is a constant, $0 < \mu < 1$. Then the state (x_1, x_2) of the system (14) will be converged to the equilibrium point 0 in a finite time, and the system (14) is globally stable in finite time.

Assumption 1. *The disturbance d_1,d_2 in system (7) are second-order and first-order differentiable, respectively.*

Let $\overline{x}_1 = x_1, \overline{x}_2 = x_2 + d_1, d = \dot{d}_1 + d_2 + a_1 d_1$, the following system can be derived from system (7)

$$\begin{cases} \dot{\overline{x}}_1 = \overline{x}_2 \\ \dot{\overline{x}}_2 = -a_1\overline{x}_2 - a_2 b_3 \overline{x}_1 - a_2 b_4 u_q + d \\ y = \overline{x}_1 \end{cases} \tag{16}$$

Finite time state observers are designed for the state variables of system (7) and system (14) according to Lemma 3.

$$\begin{cases} \dot{\hat{\overline{x}}}_1 = x_2 + \hat{d}_1 - \lambda_1 sig^{\beta_1}(\hat{\overline{x}}_1 - x_1) \\ \dot{\hat{d}}_1 = -\lambda_2 sign(\hat{\overline{x}}_1 - x_1) \end{cases} \tag{17}$$

$$\begin{cases} \dot{\hat{\overline{x}}}_2 = -a_1\hat{\overline{x}}_{22} - a_2 b_3 \hat{\overline{x}}_{11} - a_2 b_4 u_q + \hat{d}_2 + a_1 \hat{d}_1 - \overline{\lambda}_1 sig^{\beta_2}(\hat{\overline{x}}_2 - \overline{x}_2) \\ \dot{\hat{d}}_2 = -\overline{\lambda}_2 sign(\hat{\overline{x}}_2 - \overline{x}_2) \end{cases} \tag{18}$$

where, $\lambda_1, \lambda_2, \overline{\lambda}_1, \overline{\lambda}_2$ are the observation gain of the finite time observer, β_1, β_2 are the fractional power of the finite time observer, $\hat{\overline{x}}_1, \hat{\overline{x}}_2, \hat{d}_1, \hat{d}_2$ are the estimated values of $\overline{x}_1, \overline{x}_2, d_1, d_2$. The following function can be acquired $\hat{d} = a_1\hat{d}_1 + \hat{d}_2$, where, $\hat{d}$ is the estimated value of d.

Proof: The estimated errors are defined as $\widetilde{x_1} = \hat{\overline{x}}_1 - \overline{x}_1, \widetilde{d_1} = \hat{d}_1 - d_1$. The error equation obtained by system (7) and system (17) can be expressed as

$$\begin{cases} \dot{\widetilde{x_1}} = \widetilde{d_1} - \lambda_1 sig^{\beta_1}(\widetilde{x_1}) \\ \dot{\widetilde{d_1}} = -\lambda_2 sig^{\beta_2}(\widetilde{x_1}) - \dot{d}_1 \end{cases} \tag{19}$$

From Assumption 1, $-L \le \dot{d}_1 \le L$ can be obtained. When the gain meets (15), the error system can be reached stability within a finite time. Namely, $\hat{d}_1$ can converge to the true value d_1 in finite time. After this moment, $\hat{d}_1 \equiv d_1, \hat{\overline{x}}_1 \equiv \overline{x}_1$ are always true. Then $\overline{x}_2 \equiv x_2 + \hat{d}_1$ is true. The proof of the finite time stability of the error system for the system (17) is the same as above.

In conclusion, the observation state $(\hat{\overline{x}}_1, \hat{\overline{x}}_2, \hat{d}_1, \hat{d}_2)$ estimated by the finite time observer will converge to the actual values $(\overline{x}_1, \overline{x}_2, d_1, d_2)$ of system (7) and system (18) within a finite time.

3.2. Non-Singular Terminal Sliding Mode Control

Consider the following second-order system

$$\begin{cases} \dot{e}_1 = e_2 \\ \dot{e}_2 = f(e) + u + d(t) \end{cases} \tag{20}$$

where, e_1,e_2 are the state variables, $d(t)$ is the disturbance, $|d(t)| \le D$.

The non-singular terminal sliding mode surface is selected as follows

$$s = e_1 + \frac{1}{\eta} e_2^{p/q} \tag{21}$$

where, $\eta > 0$, and $p > q > 0$ are odd.

In order to make the system state converge to the actual value in finite time, the control law can be designed as

$$u = -(D + \varepsilon)sign(s) - f(e) - \eta \frac{q}{p} e_2^{2-p/q} \tag{22}$$

where, ε is the robustness coefficient, $\varepsilon > 0, 1 < p/q < 2$.

In order to prove the stability of the designed system, the Lyapunov function is selected as

$$V = \frac{1}{2}s^2 \tag{23}$$

The derivative of V is as following

$$\begin{aligned} \dot{V} &= s\dot{s} \\ &= s(\dot{e}_1 + \tfrac{1}{\eta}\tfrac{p}{q}e_2^{p/q-1}\dot{e}_2) \\ &= s[e_2 + \tfrac{1}{\eta}\tfrac{p}{q}e_2^{p/q-1}(f(e) + u + d(t))] \\ &= s[e_2 + \tfrac{1}{\eta}\tfrac{p}{q}e_2^{p/q-1}(f(e) + u + d(t))] \\ &= s[e_2 + \tfrac{1}{\eta}\tfrac{p}{q}e_2^{p/q-1}(-(D+\varepsilon)sign(s) - \eta\tfrac{q}{p}e_2^{2-p/q} + d(t))] \\ &= s[\tfrac{1}{\eta}\tfrac{p}{q}e_2^{p/q-1}(-(D+\varepsilon)sign(s) + d(t))] \\ &\le \tfrac{1}{\eta}\tfrac{p}{q}e_2^{p/q-1}(-\varepsilon)|s| \end{aligned} \tag{24}$$

where p, q are positive odd integers and $1 < p/q < 2$, thus $e_2^{p/q-1} > 0$. Then $\dot{V} < 0$ is always true.

According to the above analysis, the control law (22) designed for system (20) can ensure the system convergence.

Assume that the system state reaches the sliding mode surface at t_r, that is to say $s(t_r) = 0$, then

$$\begin{cases} e_1 + \frac{1}{\eta} e_2^{p/q} = 0 \\ \dot{e}_1 = \eta e_1^{q/p} \end{cases} \tag{25}$$

The time it takes for the system to stabilize to the equilibrium point can be expressed as

$$t_s = \frac{p}{\eta(p-q)} \left| e_1(t_r) \right|^{1-q/p} \tag{26}$$

As can be seen from the time function (26), the larger η is, the smaller t_s is to the stable state; However, if η is too large, the effect of switching item will be strengthened due to the change of s symbol, and the control output will be weakened.

For system (16) (17) (18), the non-singular terminal sliding mode surface function is selected as

$$s = \overline{x}_1 + \frac{1}{\eta} \hat{x}_2^{p/q} \tag{27}$$

The control law is designed as

$$u_q = \frac{1}{a_2 b_4} [-a_1 \hat{\overline{x}}_2 - a_2 b_3 \overline{x}_1 + \hat{d} + \eta \frac{q}{p} \hat{x}_2^{2-p/q} + (D + \varepsilon) sign(s)] \tag{28}$$

Choose the Lyapunov function as

$$V = \frac{1}{2} s^2 \tag{29}$$

Derivation of (29)

$$\begin{aligned} \dot{V} &= s\dot{s} \\ &= s(\dot{\overline{x}}_1 + \frac{1}{\eta} \frac{p}{q} \hat{x}_2^{p/q-1} \dot{\hat{x}}_2) \\ &= s[\hat{\overline{x}}_2 + \frac{1}{\eta} \frac{p}{q} \hat{x}_2^{p/q-1} (-a_1 \hat{\overline{x}}_2 - a_2 b_3 \overline{x}_1 - a_2 b_4 u_q + d)] \end{aligned} \tag{30}$$

Consider the control law (28)

$$\begin{aligned} \dot{V} &= s[\frac{1}{\eta} \frac{p}{q} \hat{x}_2^{p/q-1} (-(D + \varepsilon) sign(s) - \hat{d} + d)] \\ &= s[\frac{1}{\eta} \frac{p}{q} \hat{x}_2^{p/q-1} (-\varepsilon sign(s) + d - \hat{d} - D sign(s))] \\ &\leq \frac{1}{\eta} \frac{p}{q} \hat{x}_2^{p/q-1} (-\varepsilon) |s| \\ &\leq 0 \end{aligned} \tag{31}$$

In (31), p, q are positive odd integers and $1 < p/q < 2$, thus $\hat{x}_2^{p/q-1} > 0$ is true, and $\eta > 0$, so $\frac{1}{\eta} \frac{p}{q} |s| > 0$ is true, because $\varepsilon > 0$, then $\frac{1}{\eta} \frac{p}{q} \hat{x}_2^{p/q-1} (-\varepsilon) |s| < 0$ can be proved.

It can be known from (31) that the second-order PMSM system can reach a stable state in a finite time based on the composite strategy of finite time observer and non-singular fast terminal sliding mode.

4. Simulation and Analysis

In order to verify the effectiveness of the proposed method, comparative simulations are built on the traditional cascade PID, cascade sliding mode, and the proposed method this paper. The simulations are based on Asus notebook FX503VD, Intel(R)Core i7 7700HQ, CPU@2.80GHz, RAM 7.88GB (Hynix

DDR4 2400MHz), SanDiskSD8SN8U128G1002(128GB/solid state disk), Nvidia GeForce GTX 1050 (4GB/Asus), 64-bit operating system, matlab 2017b (ASUSTek Computer Inc., Taiwan, China). In order to ensure the fairness of the comparison, the bus voltage is set to 36 V. The reference speed of PMSM is set at 1000 r/min. The PMSM parameters used for simulation are shown in Table 1. The parameters of the traditional cascade PID, the traditional cascade sliding mode control and NTSMC-FTO proposed in this paper are shown in Tables 2–4, respectively. In cascade SMC controller, SMC and PID are used for speed loop and current loop, respectively. $s = cx_1 + x_2$ is taken as the sliding mode surface of SMC, and $i_q^* = \frac{2J}{3n_p\psi_f}\int_0^t [c(x_2) + Mu \times sign(s) + \kappa s]dt$ as the expression of output.

Table 1. Rated parameters of the permanent magnet synchronous motor (PMSM).

Rated Power	P_N	200	W
line resistance	R	0.33	Ω
line inductance	L	9×10^{-4}	H
magnetic poles	n_p	4	pairs
torque constant	K_t	0.087	N·m/A
rated power	U_N	36	VAC
rated current	I_N	7.5	A
rotor inertia	J	1.89×10^{-5}	kg·m^2
rated speed	n_N	3000	r/min

Table 2. The cascade PID controller.

Description	Parameter	Value
speed loop proportional gain	K_1	0.01
speed loop integral gain	I_1	0.95
speed loop proportional gain	K_2	50
speed loop integral gain	I_2	100,000
current loop Id proportional	K_{p1}	2000
current loop Id integral gain	K_{I1}	100,000

Table 3. The cascade sliding mode control (SMC) controller.

Description	Parameter	Value
error gain of SMC	c	10.8
switch gain of SMC	Mu	100
sliding mode surface gain of SMC	κ	12
speed loop proportional gain	K_2	50
speed loop integral gain	I_2	100,000
current loop Id proportional	K_{p1}	2000
current loop Id integral gain	K_{I1}	10,000

Table 4. The proposed controller this paper.

Description	Parameter	Value
the power of NTSMC	p	37
the power of NTSMC	q	35
proportional gain of NTSMC	η	5100
switch gain of NTSMC	ε	200,000,000,000
the gain of observer1	λ_1	1,000,000
the gain of observer1	λ_2	10
the gain of observer2	$\overline{\lambda}_1$	50,000,000
the gain of observer2	$\overline{\lambda}_2$	500
current loop i_d proportional	K_{p1}	2000
current loop i_d integral gain	K_{I1}	10,000

There are two groups of comparative simulations, one is the response curve at the phase of startup, and the other is the response curve when the load torque suddenly changes at a constant speed stage. It can be found from the comparison results that the NTSMC-FTO proposed in this paper, which regulate the speed and current of PMSM in one loop, has a better dynamic performance and disturbance rejection property than the traditional PID and SMC.

Case I: Phase of start. The reference speed of PMSM is set at 1000 r/min, and the motor starts without load torque. Figure 2a–c are ω, i_q, i_d response curves of startup, respectively. The solid (blue) line is NTSMC-FTO controller, the dotted (pink) line is PID controller, and the dotted (black) line is SMC controller. It can be summarized that when the motor starts without speed overshoot, When the motor starts without speed overshoot, it takes 0.0028 s for NTSMC-FTO to reach the steady state, compared with the cascade SMC and PID 0.045 s is needed. The cost to reach steady state is reduced by 0.0422 s. The d-axis and q-axis currents chattering of NTSMC-FTO are smaller than the cascade SMC and PID controller. The comparative simulation results of startup can be seen in Table 5.

Table 5. The comparative simulation results of start up.

Method	Reference Speed	Time to Reach Steady State
NTSMC-FTO	1000 r/min	0.0028 s
the cascade SMC	1000 r/min	0.045 s
the cascade PID	1000 r/min	0.045 s

Case II: Load torque is changed suddenly. The load torque has a sudden change from $T_L = 0\ \mathrm{N \cdot m}$ to $T_L = 0.1\ \mathrm{N \cdot m}$ at $t = 0.1$ s. Figure 3a–c, are ω, i_q, i_d response curves of load torque sudden change, respectively. When the load torque changed suddenly, the speed of NTSMC-FTO is decreased by 2.5 r/min (0.25%), while SMC and PID are 87 r/min (8.7%) and 74 r/min (7.4%), respectively. The recovery time of NTSMC-FTO, SMC and PID to 1000 r/min are 0.0004s, 0.06s and 0.06 s, respectively. The comparative simulation results of load changed suddenly can be seen in Table 6.

Table 6. The comparative simulation results of load changed suddenly.

Method	Reference Speed	Decreased Value of Speed	The Recover Time of Steady State
NTSMC-FTO	1000 r/min	2.5 r/min	0.0004 s
the cascade SMC	1000 r/min	87 r/min	0.06 s
the cascade PID	1000 r/min	74 r/min	0.06 s

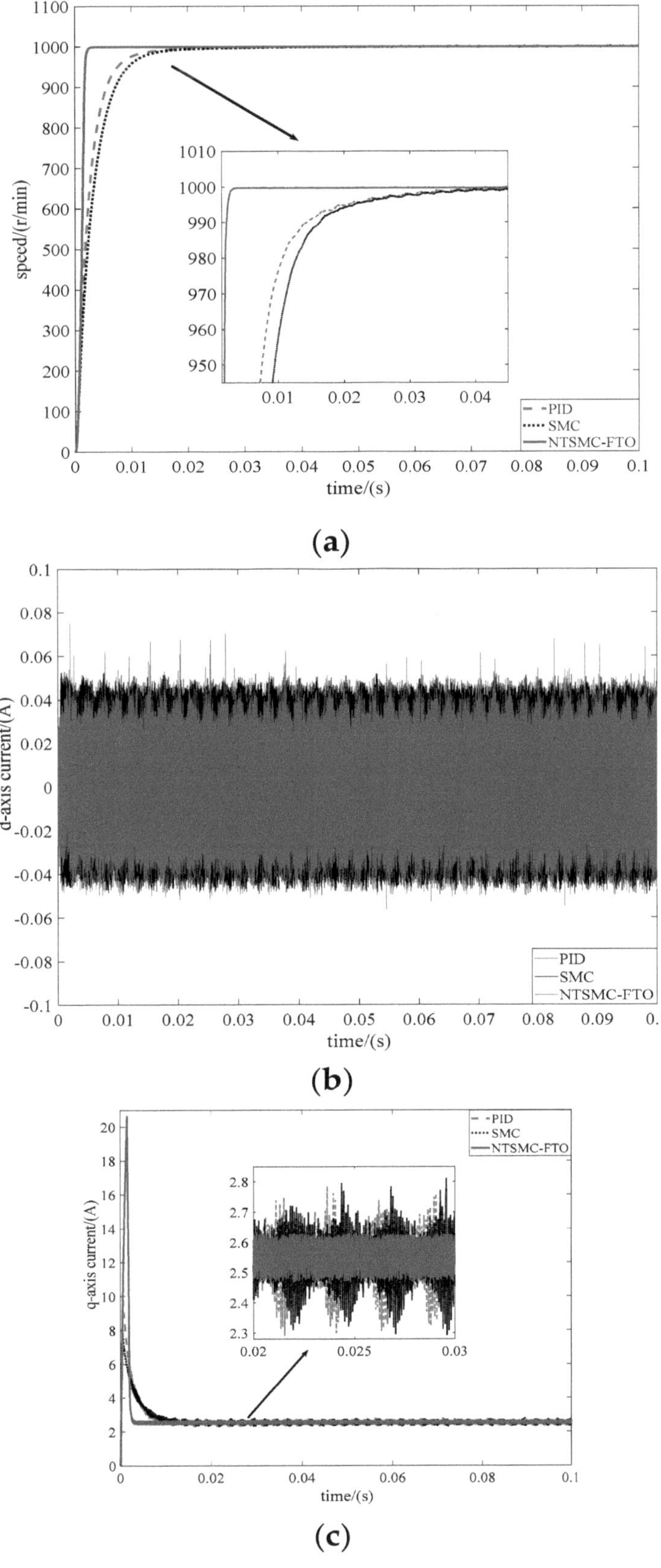

Figure 2. Performance comparisons under the PID, permanent magnet synchronous motor (SMC) and NTSMC-FTO at the phase of startup. (**a**) Speed response curves. (**b**) d-axis current curves. (**c**) q-axis current curves.

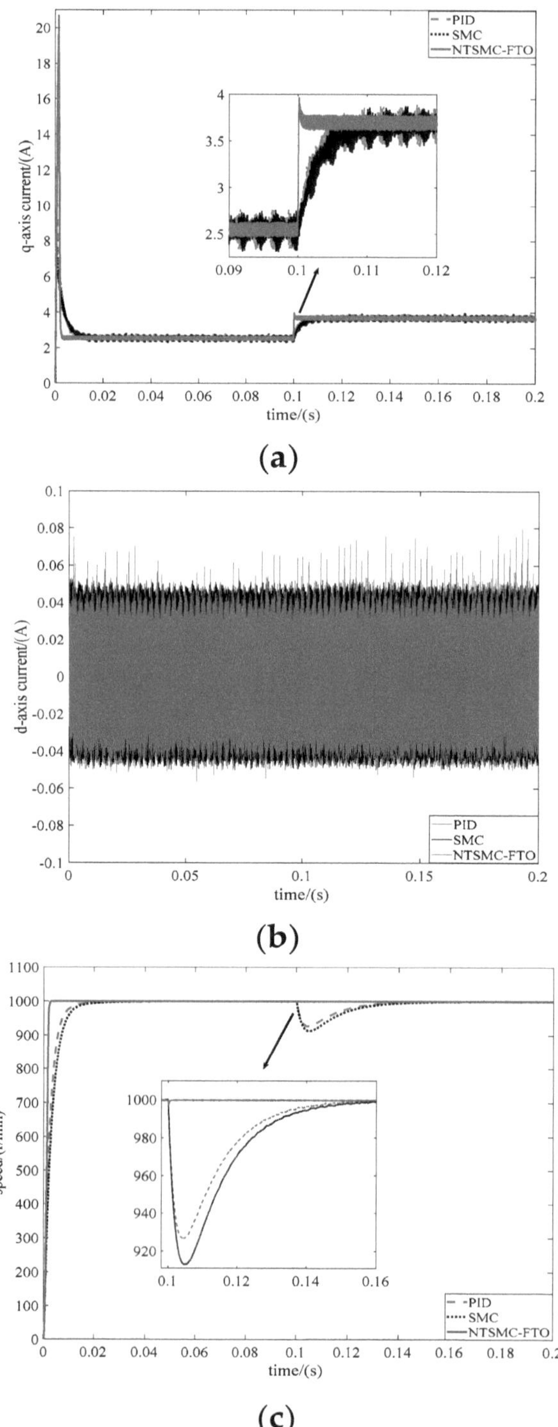

Figure 3. Performance comparisons under the PID, SMC and NTSMC-FTO with sudden load torque change. (**a**) Speed response curves. (**b**) d-axis current curves. (**c**) q-axis current curves.

Figure 4a,b are disturbance d_1 and disturbance d_2 curves estimated by the finite time observers, respectively.

It can be concluded that compared with the traditional SMC and PID, the NTSMC-FTO proposed in this paper, which put the speed and current in one loop to regulate, has a faster tracking speed and a better disturbance rejection performance, demonstrating that the proposed method in this paper has strong robustness.

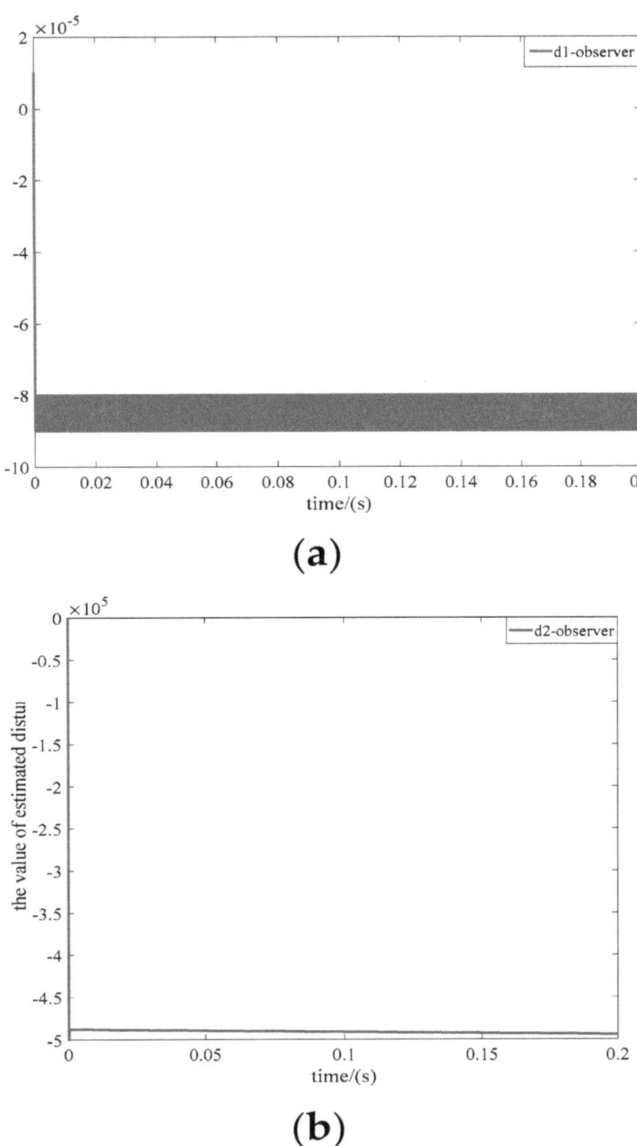

Figure 4. The estimated value based on the finite-time observer. (**a**)The estimated value of d_1 based on the finite-time observer. (**b**)The estimated value of d based on the finite-time observer.

5. Conclusions

In this paper, a novel speed-current single-loop controller for the PMSM drive system has been proposed. Simulations have verified that compared with the cascade PID and cascade SMC method, the proposed method has a faster start up response and a better disturbance rejection performance. The disturbance can be accurately estimated and compensated by the proposed disturbance observer. Future research can be carried out from the state constraint of the proposed method to reduce the q-axis transient current and improve the safety of the system.

Author Contributions: This is a joint work and the authors were in charge of their expertise and capability: Y.W. (Yao Wang) for investigation and analysis; H.Y. for funding support; Z.C., Y.W. (Yuchen Wang), and Y.L. for manuscript revision.

Acknowledgments: The authors would like to express their gratitude to all those who helped them during the writing of this paper. And the authors would like to thank the reviewers for their valuable comments and suggestions.

References

1. Lara, J.; Xu, J.; Chandra, A. Effects of Rotor Position Error in the Performance of Field Oriented Controlled PMSM Drives for Electric Vehicle Traction Applications. *IEEE Trans. Ind. Electron.* **2016**, *63*, 1. [CrossRef]
2. Kommuri, S.K.; Defoort, M.; Karimi, H.R.; Veluvolu, K.C. A Robust Observer-Based Sensor Fault-Tolerant Control for PMSM in Electric Vehicles. *IEEE Trans. Ind. Electron.* **2016**, *63*, 7671–7681. [CrossRef]
3. Liu, H.X.; Li, S.H. Speed control for PMSM servo system using predictive functional control and extended state observer. *IEEE Trans. Ind. Electron.* **2012**, *59*, 1171–1183. [CrossRef]
4. Chaoui, H.; Khayamy, M.; Aljarboua, A.A. Adaptive Interval Type-2 Fuzzy Logic Control for PMSM Drives with a Modified Reference Frame. *IEEE Trans. Ind. Electron.* **2017**, *64*, 3786–3797. [CrossRef]
5. Valente, G.; Formentini, A.; Papini, L.; Gerada, C.; Zanchetta, P. Performance improvement of bearingless multisector PMSM with optimal robust position control. *IEEE Trans. Power Electron.* **2019**, *34*, 3575–3585. [CrossRef]
6. Sun, Z.X.; Li, S.H.; Wang, J.G.; Zhang, X.H.; Mo, X.H. Adaptive composite control method of permanent magnet synchronous motor systems. *Trans. Inst. Meas. Control* **2018**, *11*, 3345–3357. [CrossRef]
7. Yu, J.; Shi, P.; Zhao, L. Finite-time command filtered backstepping control for a class of nonlinear systems. *Automatica* **2018**, *92*, 173–180. [CrossRef]
8. Liang, W.; Fei, W.; Luk, P.C.-K. An Improved Sideband Current Harmonic Model of Interior PMSM Drive by Considering Magnetic Saturation and Cross-Coupling Effects. *IEEE Trans. Ind. Electron.* **2016**, *63*, 4097–4104. [CrossRef]
9. Ren, H.P.; Liu, D. Nonlinear feedback control of chaos in permanent magnet synchronous motor. *IEEE Trans. Circuits Syst. II Express Briefs* **2016**, *53*, 45–50.
10. Formentini, A.; Trentin, A.; Marchesoni, M.; Zanchetta, P.; Wheeler, P. Speed Finite Control Set Model Predictive Control of a PMSM Fed by Matrix Converter. *IEEE Trans. Ind. Electron.* **2015**, *62*, 6786–6796. [CrossRef]
11. Cheema, M.A.M.; Fletcher, J.E.; Farshadnia, M.; Xiao, D.; Rahman, M.F. Combined Speed and Direct Thrust Force Control of Linear Permanent-Magnet Synchronous Motors with Sensorless Speed Estimation Using a Sliding-Mode Control with Integral Action. *IEEE Trans. Ind. Electron.* **2017**, *64*, 3489–3501. [CrossRef]
12. Lang, X.Y.; Yang, M.; Xu, H.D.; Long, J.; Xu, D.G. A non-cascade predictive speed and current controller with PWM modulation for PMSM. In Proceedings of the IECON 2016-42nd Annual Conference of the IEEE Industrial Electronics Society, Florence, Italy, 23–26 October 2016.
13. Guo, T.L.; Sun, Z.X.; Wang, X.Y.; Li, S.H.; Zhang, K.J. A simple current-constrained controller for permanent-magnet synchronous motor. *IEEE Trans. Ind. Inform.* **2019**, *15*, 1486–1495. [CrossRef]
14. Yan, Y.; Yang, J.; Sun, Z.; Zhang, C.; Li, S.; Yu, H. Robust Speed Regulation for PMSM Servo System with Multiple Sources of Disturbances via an Augmented Disturbance Observer. *IEEE/ASME Trans. Mechatron.* **2018**, *23*, 769–780. [CrossRef]
15. Liu, X.; Yu, H.; Yu, J.; Zhao, L. Combined Speed and Current Terminal Sliding Mode Control with Nonlinear Disturbance Observer for PMSM Drive. *IEEE Access* **2018**, *6*, 29594–29601. [CrossRef]
16. Chaoui, H.; Khayamy, M.; Okoye, O. Adaptive RBF network based direct voltage control for interior PMSM based vehicles. *IEEE Trans. Veh. Technol.* **2018**, *67*, 5740–5749. [CrossRef]
17. Zhang, X.G.; He, Y.K. Direct voltage-selection based model predictive direct speed control for PMSM drives without weighting factor. *IEEE Trans. Power Electron.* **2019**, *34*, 7838–7851. [CrossRef]
18. Preindl, M.; Bolognani, S. Model predictive direct speed control with finite control set of PMSM drive systems. *IEEE Trans. Power Electron.* **2013**, *28*, 1007–1015. [CrossRef]
19. Smidl, V.; Janous, S.; Adam, L.; Peroutka, Z. Direct Speed Control of a PMSM Drive Using SDRE and Convex Constrained Optimization. *IEEE Trans. Ind. Electron.* **2018**, *65*, 532–542. [CrossRef]
20. Liu, B.; Zhou, B.; Ni, T.H. Principle and stability analysis of an improved self-sensing control strategy for surface-mounted PMSM drives using second-order generalized integrators. *IEEE Trans. Energy Convers.* **2018**, *33*, 126–136. [CrossRef]
21. Mynar, Z.; Veselý, L.; Vaclavek, P. PMSM Model Predictive Control with Field-Weakening Implementation. *IEEE Trans. Ind. Electron.* **2016**, *63*, 5156–5166. [CrossRef]
22. Liu, J.; Li, H.W.; Deng, Y.T. Torque ripple minimization of PMSM based on robust ILC via adaptive sliding mode control. *IEEE Trans. Power Electron.* **2018**, *33*, 3655–3671. [CrossRef]

23. Tarczewski, T.; Grzesiak, L.M. Constrained state feedback speed control of PMSM based on model predictive approach. *IEEE Trans. Ind. Electron.* **2016**, *63*, 3867–3875. [CrossRef]
24. Wang, W.-C.; Liu, T.-H.; Syaifudin, Y. Model Predictive Controller for a Micro-PMSM-Based Five-Finger Control System. *IEEE Trans. Ind. Electron.* **2016**, *63*, 3666–3676. [CrossRef]
25. Xia, C.; Li, S.; Shi, Y.; Zhang, X.; Sun, Z.; Yin, W. A Non-Smooth Composite Control Approach for Direct Torque Control of Permanent Magnet Synchronous Machines. *IEEE Access* **2019**, *7*, 45313–45321. [CrossRef]
26. Zhang, G.Q.; Wang, G.L.; Yuan, B.H. Active disturbance rejection control strategy for signal injection-based sensorless IPMSM drives. *IEEE Trans. Transp. Electrif.* **2018**, *1*, 330–339. [CrossRef]
27. Li, S.H.; Zhou, M.M.; Yu, X.H. Design and Implementation of terminal sliding mode control method for PMSM speed regulation system. *IEEE Trans. Ind. Inform.* **2013**, *9*, 1879–1891. [CrossRef]
28. Li, S.H.; Wu, C.; Sun, Z.X. Design and implementation of clutch control for automotive transmissions using terminal-sliding-mode control and uncertainty observer. *IEEE Trans. Veh. Technol.* **2016**, *65*, 1890–1898. [CrossRef]
29. Du, H.B.; Chen, X.P.; Wen, G.H.; Yu, X.H.; Lü, J.H. Discrete-time fast terminal sliding mode control for permanent magnet linear motor. *IEEE Trans. Ind. Electron.* **2018**, *65*, 9916–9927. [CrossRef]
30. Wu, F.; Li, P.; Wang, J. FO improved fast terminal sliding mode control method for permanent-magnet synchronous motor with FO disturbance observer. *IET Control. Theory Appl.* **2019**, *13*, 1425–1434. [CrossRef]
31. Xu, W.; Junejo, A.K.; Liu, Y.; Islam, M.R. Improved Continuous Fast Terminal Sliding Mode Control with Extended State Observer for Speed Regulation of PMSM Drive System. *IEEE Trans. Veh. Technol.* **2019**, in press. [CrossRef]
32. Guo, L.; Wang, D.; Diao, L.; Peng, Z. Direct voltage control of stand-alone DFIG under asymmetric loads based on non-singular terminal sliding mode control and improved extended state observer. *IET Electr. Power Appl.* **2019**, *13*, 958–968. [CrossRef]
33. He, L.; Wang, F.; Wang, J.; Rodriguez, J. Zynq Implemented Lunenberger Disturbance Observer Based Predictive Control Scheme for PMSM Drives. *IEEE Trans. Power Electron.* **2019**, in press. [CrossRef]
34. Yang, J.; Cui, H.Y.; Li, S.H.; Zolotas, A. Optimized active disturbance rejection control for DC-DC buck converters with uncertainties using a reduced-order GPI observer. *IEEE Trans. Circuits Syst. I Regul. Pap.* **2018**, *65*, 832–841. [CrossRef]
35. An, H.; Liu, J.X.; Wang, C.H.; Wu, L.G. Disturbance observer-based anti-windup control for air-breathing hypersonic vehicles. *IEEE Trans. Ind. Electron.* **2016**, *63*, 3038–3049. [CrossRef]
36. Hebertt, S.R.; Jesús, L.F.; Carlos, G.R.; Marco, A.C.O. On the control of the permanent magnet synchronous motor: An active disturbance rejection control approach. *IEEE Trans. Control Syst. Technol.* **2014**, *22*, 2056–2063.
37. Chen, W.-H. Disturbance Observer Based Control for Nonlinear Systems. *IEEE/ASME Trans. Mechatron.* **2004**, *9*, 706–710. [CrossRef]
38. Chen, M.-S.; Chen, C.-C. Robust Nonlinear Observer for Lipschitz Nonlinear Systems Subject to Disturbances. *IEEE Trans. Autom. Control.* **2007**, *52*, 2365–2369. [CrossRef]
39. Khalil, H. *Nonlinear Systems*, 2nd ed.; Prentice-Hall: Upper Saddle River, NJ, USA, 1996.
40. Hong, Y.; Huang, J.; Xu, Y. On an output feedback finite-time stabilization problem. *IEEE Trans. Autom. Control* **2001**, *46*, 305–309. [CrossRef]
41. Shen, Y.J.; Huang, Y.H. Uniformly observable and globally Lipschitzian nonlinear systems admit global finite-time observers. *IEEE Trans. Autom. Control* **2009**, *54*, 2621–2625. [CrossRef]
42. Perruquetti, W.; Floquet, T.; Moulay, E. Finite-Time Observers: Application to Secure Communication. *IEEE Trans. Autom. Control.* **2008**, *53*, 356–360. [CrossRef]
43. Du, H.B.; Qian, C.J.; Yang, S.Z.; Li, S.H. Recursive design of finite time convergent observers for a class of time varying nonlinear systems. *Automatica* **2013**, *49*, 601–609. [CrossRef]
44. Davila, J.; Fridman, L.; Levant, A. Second-order sliding-mode observer for mechanical systems. *IEEE Trans. Autom. Control.* **2005**, *50*, 1785–1789. [CrossRef]
45. Lin, C.-K. Nonsingular Terminal Sliding Mode Control of Robot Manipulators Using Fuzzy Wavelet Networks. *IEEE Trans. Fuzzy Syst.* **2006**, *14*, 849–859. [CrossRef]

4

A Modular Framework for Optimal Load Scheduling under Price-Based Demand Response Scheme in Smart Grid

Ghulam Hafeez [1,2,*], Noor Islam [3], Ammar Ali [1], Salman Ahmad [1,4], Muhammad Usman [2] and Khurram Saleem Alimgeer [1]

1 Department of Electrical and Computer Engineering, COMSATS University Islamabad, Islamabad 44000, Pakistan
2 Department of Electrical Engineering, University of Engineering and Technology, Mardan 23200, Pakistan
3 Department of Electrical Engineering, CECOS University of IT & Emerging Sciences, Peshawar 25124, Pakistan
4 Department of Electrical Engineering, Wah Engineering College, University of Wah, Wah Cantt 47070, Pakistan
* Correspondence: ghulamhafeez393@gmail.com

Abstract: With the emergence of the smart grid (SG), real-time interaction is favorable for both residents and power companies in optimal load scheduling to alleviate electricity cost and peaks in demand. In this paper, a modular framework is introduced for efficient load scheduling. The proposed framework is comprised of four modules: power company module, forecaster module, home energy management controller (HEMC) module, and resident module. The forecaster module receives a demand response (DR), information (real-time pricing scheme (RTPS) and critical peak pricing scheme (CPPS)), and load from the power company module to forecast pricing signals and load. The HEMC module is based on our proposed hybrid gray wolf-modified enhanced differential evolutionary (HGWmEDE) algorithm using the output of the forecaster module to schedule the household load. Each appliance of the resident module receives the schedule from the HEMC module. In a smart home, all the appliances operate according to the schedule to reduce electricity cost and peaks in demand with the affordable waiting time. The simulation results validated that the proposed framework handled the uncertainties in load and supply and provided optimal load scheduling, which facilitates both residents and power companies.

Keywords: smart grid; demand response; load scheduling; home energy management; enhanced differential evolution; hybrid gray wolf-modified enhanced differential evolutionary algorithm

1. Introduction

With the emergence of information and communication technology (ICT), smart grid (SG) can make a robust and reliable system for the energy management of residential homes. ICT and sensors have moved the world towards automation. Thus, excessive use of electricity for every activity has increased demand-side energy consumption. The high demand for electricity and limited fossils fuels lead to increased penetration of renewable energy resources (RERs) [1]. Electricity production from RERs is not a part of this discussion. However, through scheduling and coordination of appliances, this high energy consumption can be managed. In [2], the authors reported that 38% increase in electricity consumption of power sector and 16% increase in electricity consumption of both residential and commercial sectors are expected by the year 2020.

Considering this repaid energy consumption growth, there is a need for a system to manage the resident demand according to generation in such a manner to alleviate the gap between demand

and supply [3]. In this regard, the traditional grid is renovated by SG with the integration of ICT. Advanced metering infrastructure (AMI) is responsible for bi-directional communication between the power company and the resident [4]. In a SG, the power company organizes the consumer demand using particular set of programs. These programs are known as demand response (DR) programs [5]. Various DR incentives schemes are introduced by the power companies for the encouragement of the residents to efficiently use available resources as explained in [6]. Price-based DR schemes such as real-time pricing scheme (RTPS), time of use pricing scheme (TOUPS), critical peak pricing scheme (CPPS), flat-rate pricing scheme (FRPS), a day-ahead pricing scheme (DAPS), and inclined block rate scheme (IBRPS) are widely used for load scheduling. The HEM controller (HEMC) receives the pricing signal from the electric power company and electric load profile from the resident to schedule the household load. The HEMC schedule the household load using a pricing signal and the load of the residents. The home appliances are synchronized with the schedule through infrared, ZigBee, Z-Wave, and Wi-Fi [7].

The main focus of research and development (R&D) is on load shifting from ON-peak timeslots to OFF-peak timeslots using demand-side management (DSM) strategies such as peak clipping, strategic conservation, peak shifting, and valley filling. Load shifting helps in two ways: minimize electricity cost by shifting the load to low-price timeslots and minimize peaks in demand by building load in OFF-peak timeslots [8]. However, load shifting reduces electricity cost at the expense of increase user frustration in terms of waiting time. To reduce electricity cost and peaks in demand with affordable waiting time, heuristic techniques are mostly adopted because they are fast converging and simple.

To overcome this rapidly increasing electricity demand of the residential sector, a hybrid gray wolf-modified differential evolution (HGWmEDE) algorithm is proposed to resolve this problem and enhance the sustainability of the electric grid. The proposed algorithm under the price-based DR encourages resident to take part in DSM via load scheduling. In this work, the main focus is on optimal load scheduling based on HGWmEDE under price-incentive-based DR schemes in smart homes. The main contribution and distinguish features of this paper are as follows:

- A modular framework is introduced for optimal load scheduling, which has four modules: power company module, restricted Boltzmann machine (RBM)-based forecaster module, HGWmEDE-based HEMC module, and resident module. Furthermore, smart home appliances are classified into three categories based on power rating and behavior: schedulable, non-schedulable, and controllable. Moreover, each appliance has a different length of operational time (LOT) and each home has different operation time interval (OTI). Four parameters, i.e., energy consumption, electricity cost, peaks in demand, and waiting time are taken into account.
- A deep neural network technique, i.e., RBM is adopted to forecast the pricing signals of price-based DR scheme for optimal load scheduling.
- Finally, the HGWmEDE algorithm is proposed, which is a hybrid of gray wolf optimization and enhanced differential evolutionary algorithms. The proposed algorithm has global powerful search capability and generalization. The proposed algorithm optimizes the performance by fine-tuning the control parameters.

To analyze the proposed scheme in terms of electricity expense, peaks in demand, and discomfort, simulations are conducted in MATLAB 2016. Moreover, the convergence rate and performance trade-off are also evaluated.

The organization of the paper is as follows: In Section 2, recent and relevant work is demonstrated. The proposed modular framework is presented in Section 3. Section 4 describes the proposed scheme. In Section 5, simulations results and discussion are described. The paper is concluded along with future research directions in Section 6.

2. Recent and Relevant Work

In the last few years, a lot of research has been conducted in the area of HEM based on optimization algorithms in the SG to economically use electrical energy. Some recent and relevant research work is presented, in this section.

A heuristic algorithm (genetic algorithm (GA) and bacterial foraging algorithm (BFA))-based HEM model is proposed in [9]. The performance evaluation of these algorithms is conducted using three price-based DR schemes, i.e., RTPS, TOUS, and CPPS. The focus of the authors is to shift the load from ON-peak timeslots to OFF-peak timeslots to minimize electricity cost and smooth out the demand curve.

In [10], three heuristic algorithms, i.e., differential evolution algorithm (DEA), GA, and binary-particle swarm optimization algorithm (BPSOA) were implemented for load scheduling to minimize electricity cost and PAR. On the other hand, carbon emission was alleviated using RERs. DAPS was chosen as a DR scheme. The primary aim was not load scheduling, but also to prioritize the operation of appliances according to resident demand. The grid sustainability was maintained by keeping a balance between the demand and supply side. However, the balance is maintained at the expense of user discomfort.

In [11], authors presented DR program under the corporate sector. The purpose is to perform HEM and maintain using two different pricing schemes, i.e., DAPS and RTPS. In [12], a GA-based optimization model was proposed for electric load scheduling for 24 h time horizon. The energy consumption and load pattern are calculated using power rating and status of appliances for overall time horizon. However, the user-comfort is compromised, and the convergence rate is reduced.

An intelligent decision support system (IDSS) was used for resolving certain HEM problems [13]. Moreover, IDSS was integrated to AMI for bi-directional communication between the power company and residents. Wind-driven optimization (WDO) algorithm with knapsack (K-WDO) was implemented for electricity cost minimization and user-comfort maximization in [14]. The minimum-maximum constraints of K-WDO were defined. The smart home appliances were classified based on consumer behavior and name-plat power rating. TOUPS was used to shift load from ON-peak hours to OFF-peak hours according to user preference and priority; however, peaks in demand emerged at the expense of increased system complexity.

An ant colony optimization algorithm (ACOA)-based model was proposed by [15] for optimal power flow (OPF). The OPF objective was to determine the load to satisfy the end user by providing a continuous energy supply. In the literature, some statistical methods, i.e., newton method, linear programming (LP), non-LP (NLP), and the interior point method were used to solve such problems.

In [16], load balancing via load scheduling was the main focus of the authors. Thus, a multi-agent system was proposed for load balancing, in this system each consumer act as an independent agent and the consumer electric load was divided into time frames for each agent. Power was supplied at a particular time frame for each agent. In this paper, three sectors of demand-side were considered, i.e., residential, commercial, and industrial. In [17,18], the basic concepts of DEA and enhanced version of DEA (EDEA) with five trial vectors were discussed. The mutant vector and trial vectors were created to update population. Moreover, DE-based scheduling model for electricity cost reduction was also presented.

The multi-objective optimization problem was discussed in [19]. The optimization problem was tested on pareto sets (PS) using DEA. The proposed model was also named as multi-objective evolutionary algorithm (MOEA) and was capable in complex PS shapes mapping. A hybrid evolutionary approach-based forecasting model was proposed to cater varying electricity prices by [20]. The model forecasts the day-ahead and week-ahead price profiles. The hybrid evolutionary approach was a combination of PSO and a neuro-fuzzy logic network. This hybrid approach was used to handle uncertainty in the pricing rates of the electricity market. In [21], price and load correlation were developed to modify the energy consumption pattern of ON-peak timeslots and OFF-peak timeslots. Both generalized mutual information (GMI) and wavelet packet transform (WPT) was adopted to formulate the multiple inputs and multiple output model. Electricity price was forecasted to analyze variation in their pattern. The ACOA was applied for optimization purposes.

Teaching and learning-based optimization algorithm (TLBOA) and shuffled-frog leaping (SFL)-based energy management was presented in [22]. The proposed framework was validated using different tariff schemes such as TOUPS, RTPS, CPPS, and without pricing scheme. The household load scheduling was conducted for varying time interval and pricing schemes. Power storage is incorporated in the system model to ensure continuous operation of the sensitive load [23]. A day-ahead of schedule is generated by virtual power play for load and energy consumption. The increased energy demand encourage electricity market participator generation from distributed generation. The intensive demand of residents was catered using a vehicle to grid station (V2GS) strategy. Load balancing among multiple distribution units was performed using BPSO along with MILP. A complex mathematical model was formulated for day-ahead electricity price forecasting. For experimental evaluation 1000, electric vehicle stations and 180 distributed units were used.

Energy consumption is a crucial parameter in electricity bill and peaks in demand reduction. Taking into account this fact, the resident load was scheduled using a hybrid of GA and artificial neural network (ANN-GA) scheme [24]. The load was scheduled on a weekly basis for a single home with four bedrooms. Obtained results show 25%, 40%, and 10% reduction in grid electricity consumption. However, dynamic and different OTI were not taken into account while all homes have not same OTI.

In [25], the authors focused on residential sector DSM. Multiple homes considered were smart homes and have bi-directional communication between the power company and residents. The GA, BPSO, WDO, and BFOA-based HEMC was installed for home load scheduling. The proposed model was evaluated in terms of electricity bill, user-comfort, and peaks in demand. However, the trade-off effect of conflicting parameters was ignored.

A distributed algorithm was used in [26] for energy management of 2560 households. The purpose was to reduce electricity bill with reasonable appliances waiting time. The authors in [27] proposed a harmony search algorithm (HSA)-based model for load scheduling. However, in [28], authors focused on electricity bill reduction. Game theory-based framework in [29,30] was proposed to reduce PAR by load scheduling and DR program.

In the aforementioned recent and relevant literature, the authors and R&D did not completely use the key features of SG. Some authors minimized peaks in demand, electricity cost, and waiting time. On the other hand, some authors focused on user-comfort and user discomfort in terms of waiting time. However, the conflicting parameters were not catered simultaneously in R&D by any of the authors. Furthermore, dynamic and different OTI were not catered while all homes in a city have not same OTI and energy consumption. In this work, the electricity cost reduction and peaks in demand reduction with affordable discomfort are catered simultaneously. The objective is to cope with the increasing demand of residents with the generation of the power company and reduce the burden on both parties. The comprehension of recent and relevant work is listed in Table 1.

Table 1. Summary of recent and relevant work.

Methodologies	Features	Targets Achieved	Limitations and Remarks
MILP	Optimal domestic load scheduling [5]	Electricity cost reduction	The cost was reduced at the expense of user discomfort
Greedy algorithm	Heuristic optimization base generic model [13]	Both electricity bill and user frustration are reduced	The PAR was compromised and complexity of the system is increased
Multi-agent system	DSM via load shifting and DR programs in SG [16]	Electricity cost reduction via load scheduling	The user-comfort is compromised due to the trade-off between electricity cost and user-comfort

Table 1. *Cont.*

Methodologies	Features	Targets Achieved	Limitations and Remarks
BPSO and neuro-fuzzy logic	Hybrid evolutionary-adaptive-based price forecasting model [21]	Improved forecast accuracy	The convergence rate and complexity increased while improving forecast accuracy
MIMO	Future price and load forecasting in SG [22]	Improved load and price profile forecast accuracy	The convergence and complexity were ignored which have a direct influence on the forecast accuracy
TLBOA and SFL	DR programs optimization in SG [23]	Cost reduction	The cost can further be reduced with RERs integration
PSO and MILP	A multi-objective model was used for resource and load scheduling [24]	Scheduling of virtual power play	Requirements to improve reliable power grids was ignored
MIMO	A hybrid optimization algorithm was used for both price and load forecasting [25]	Accuracy improvement	Computational time and execution time was impractical
GWO	Economic dispatch optimization under the GWO technique [31]	Optimization of dispatch problems	The problem arborized of handling the constraints
ANN-GA	Smart energy management using ANN-GA [32]	Efficiency improvement	The model was limited for a small number of appliances
Game theory algorithm (GTA)	Game theory-based household load scheduling under DR [33]	Electricity cost reduction	RERs integration was ignored
MILP and heuristic algorithms	Household load scheduling [34]	Household load balancing	Electricity cost reduction is ignored
MINLP	Efficient household appliances scheduling under DR [35]	Electricity cost reduction	PAR was compromised
GWO and ILP	GWO-based economic load dispatch [36]	Electric load dispatching in low-price timeslots	The electric load is economically dispatched

3. Proposed Modular Framework

The main objective of home energy management in this work is to minimize electricity cost and peaks in demand under price-based DR scheme by scheduling the smart home appliances. The overall proposed modular framework is demonstrated in Figure 1. The proposed framework has four modules: power company module, forecaster module, HEMC module, and resident module. Different electricity pricing signals (RTPS, TOUS, CPPS, DAPS, IBR, and variable time pricing) are defined by the power company for residents to take part in the price-based DR. Timeslots in which consumer demand reaches to the maximum value is known as peak timeslots. Electricity tariffs are usually high in these peak timeslots. However, in this paper, the power company module provides price-based DR information (RTPS and CPPS) and load pattern to the forecaster module. The forecaster module is based on RBM. The primary goal of this module is to devise a framework which is enabled through learning to forecast future load and pricing signals (RTPS and CPPS). The data for training RBM is collected from [37,38]. The predicted pricing signals must be accurate to obtain optimal load scheduling. The forecasted profile of load and pricing signals (RTPS and CPPS) is illustrated in Figure 2a–c. It is obvious that RBM-based forecast closely follows the real curve. This observation in terms of the numerical value is 0.4% for load, 0.5% for RTPS, and 0.2% for CPPS, respectively. The reason for this accurate performance is the adaption of deep learning technique, i.e., RBM. The forecaster module provides forecasted load pattern and pricing signals to the HEMC module, which is based on the proposed HGWmEDE algorithm. The HEMC based on HGWmEDE schedule the household load under the pricing signals provided by the forecaster module. The schedule developed by HEMC module is forwarded to the resident module. The resident module comprised of a smart home with 17 appliances [39]. Each appliance has its own power rating and behavior. These appliances are scheduled to according to HEMC schedule to achieve the objective function.

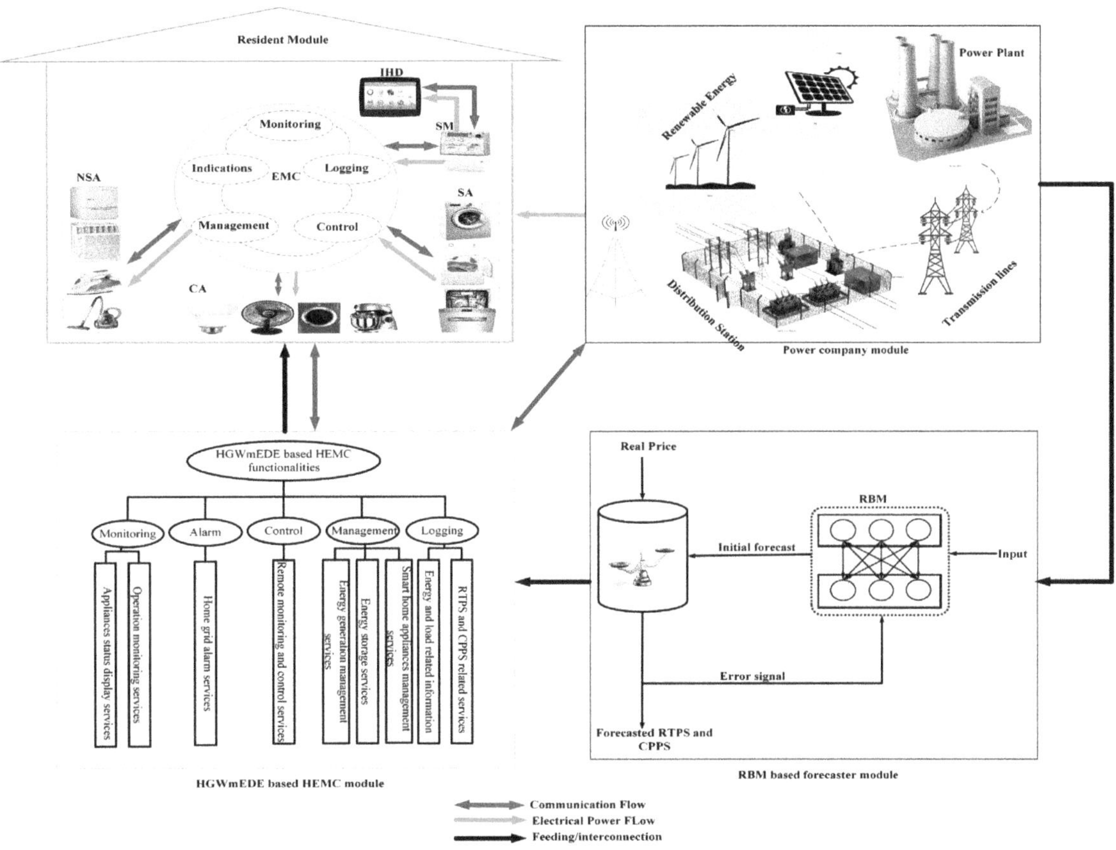

Figure 1. Proposed modular framework.

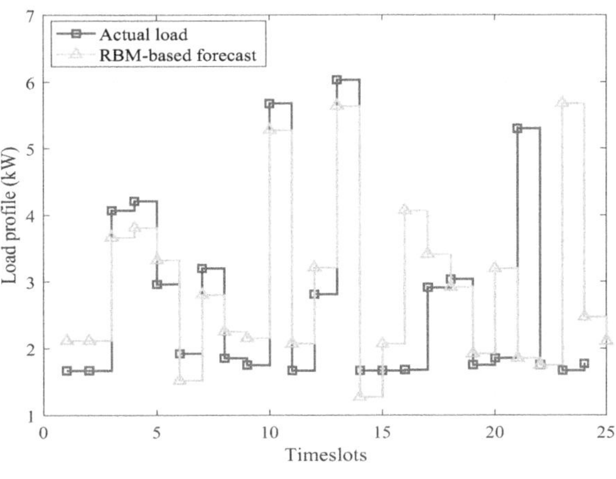

(a) RBM-based load forecast

Figure 2. *Cont.*

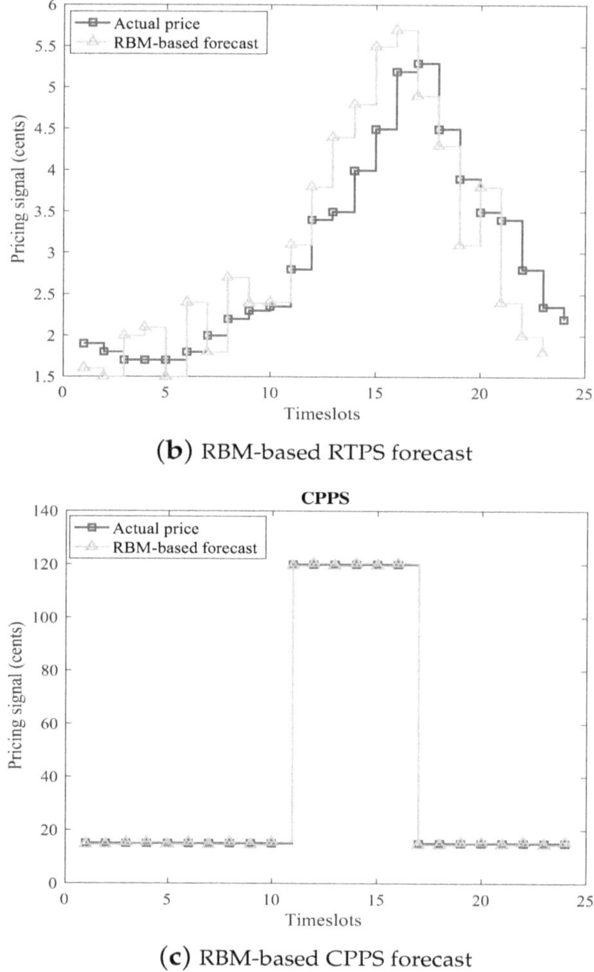

(**b**) RBM-based RTPS forecast

(**c**) RBM-based CPPS forecast

Figure 2. Price-based DR schemes.

In the proposed modular framework bi-directional communication exists between the power company and residents via AMI. Power company sent price-based DR information to the forecaster module, the forecaster module forecast load and pricing signals of RTPS and CPPS. The HEMC module based on HGWmEDE algorithm receives forecasted results to schedule the household load on the basis of pricing signals such that high power-rating appliances cannot be switched on in peak timeslots. The HEMC module dispatches the load schedule to the power company, and ultimately, the power company sent demanded power of residents. The HEMC can perform the functionalities of logging, management, control, monitoring, and alarm. The purpose is to optimally schedule household load and reduce the frustration of both power company and residents. The profiles of forecasted RTPS and CPPS are illustrated in Figure 2a,b. These two pricing schemes are generated using the data of [37,38].

3.1. Smart Home Appliances Categorization

Smart home appliances are classified into three categories based on operating behavior and energy consumption pattern. The smart home total number of appliances is denoted by a set A_n^t which includes shiftable appliances, non-shiftable appliances and controllable appliances and is defined by Equation (1).

$$A_n^t = \{A_s^s,\ A_n^s,\ A_c^s\}, \tag{1}$$

where A_s^s denotes shiftable appliances, A_n^s represents non-shiftable appliances, and A_c^s denotes controllable appliances. The detailed description of the classification is as follows.

3.1.1. Shiftable Appliances

In shiftable appliances, shifting to any timeslot is allowed, but interruption during operational time is prohibited. Such type of appliances are also known as deferrable appliances. Once operation of such type appliances is started, it cannot be stopped/interrupted until to finish the assigned task [17]. These appliances are the subset of total appliances and are defined by Equation (2).

$$A_s^s = \{WM,\ DW,\ HS,\ HD,\ MW,\ TP,\ CP,\ OV, CK,\ IR,\ TR,\ EK,\ PR\}, \tag{2}$$

where *WM* denotes washing machine, *DM* represents dishwasher, *HS* indicates hair straightener, *HD* denotes hair dryer, *MW* represents microwave, *TP* denotes telephone, *CP* represents computer, *OV* represents oven, *CK* denotes cooker, *IR* represents iron, *TR* represents toaster, *EK*, represents electric kettle, and *PR* represents printer. Each appliance power rating and status at a particular timeslot t is denoted by p_r^a and X_A^S, respectively. Equation (3) indicates the ON and OFF status of an appliance.

$$X_A^S = \begin{cases} 1 & ON \\ 0 & \text{otherwise} \end{cases} \tag{3}$$

3.1.2. Controllable Appliances

The controllable appliances have constant operational time and cannot be changed; for example, heating system, lightning, and air conditioning. Such type of appliances are also known as interruptible appliances. The controllable appliances are given by Equation (4).

$$A_c^s = \{AC,\ LT,\ HT\}, \tag{4}$$

where *AC* represents air conditioner, *LT* denotes lighting, and *HT* represents heater.

3.1.3. Non-Shiftable Appliances

Non-shiftable appliances are also known as base appliances. Such type of appliances are uncontrollable, and their operational behavior and energy consumption cannot be altered. Televisions and refrigerators are kept in this category due to resembles. The set of non-shiftable appliances is defined by Equation (5).

$$A_n^s = \{TV,\ RG\}, \tag{5}$$

where *TV* represents television and *RG* denotes refrigerator. The overall categorization of smart home appliances is presented in Table 2.

Table 2. Smart home appliances classification and parameters.

Categories	Type	Power Rating (kW)	Start Time (hours)	End Time (hours)	LOT (hours)
Shiftable appliances	Washing machine	1.4	6	10	1–3
	Dish washer	1.32	15	20	1–3
	Hair straightener	0.055	18	8	1–2
	Hair dryer	1.8	18	8	1–2
	Microwave	1.2	18	8	3–5
	Telephone	0.005	9	17	1–24
	Computer	0.15	18	24	6–12
	Oven	2.4	6	10	1–3
	Cooker	0.225	18	24	2–4
	Iron	2.4	18	24	3–5
	Toaster	0.8	9	17	1–2
	Electric Kettle	2	18	8	1–2
	Printer	0.011	18	24	1–2
Non-shiftable appliances	TV	0.095	8	16	6–14
	Refrigerator	1.75	0	23	0–23
Controllable appliances	Air conditioner	1.14	16	23	6–8
	Lightning	0.1	0	23	12–20

4. Problem Description and Formulation

In DSM, optimal household load scheduling and alignment of residents' random demand under the generation of the power company is a challenging task. In the literature, various models have been developed to address energy management via the resident load scheduling. For example, the scheduling algorithm is proposed for the energy management of smart homes to reduce the electricity bill, reverse power flow, and peak load shaving [40]. A GWDO algorithm is proposed to schedule the household load under RTPS + IBRS to increase the revenue of residents and reduce the peak to average ratio [41]. Load scheduling is performed using GA under RTPS to alleviate cost and PAR. However, peaks in demand may emerge in OFF-peak hours and user-comfort may compromise while reducing electricity bill because these parameters are conflicting parameters. Thus, a modular framework is proposed, which is based on our proposed algorithm HGWmEDE, for a load scheduling of smart homes with three types of appliances: shiftable appliances, non-shiftable appliances, and controllable appliances, in order to alleviate electricity cost and peaks in demand with affordable appliances waiting time. The purpose is to facilitate both residents and power companies by reducing burden (electricity bill and generation, respectively) on both parties. The peaks in demand reduction are favorable for both residents and power companies because it alleviates the need for peak power plants, which power companies operate when peaks in demand emerged and charged more cost from the residents. To perform effective load scheduling dynamic OTI, RTPS, and CPPS are used. The formulation for energy consumption, cost, peaks in demand, and load scheduling are demonstrated as follows.

4.1. Smart Home Appliances Energy Consumption

The HEMC schedule the smart home appliances under forecasted RTPS and CPPS over a 24-hours time horizon. These smart home appliances when operating according to the schedule consume electrical energy, which can be defined as the electrical energy used by an appliance in unit time and can be measured in the *kWh* unit. The electrical energy consumed by an appliance can be calculated by Equation (10);

$$E_c^a(t) = p_r^a \times X_s^a, \tag{6}$$

where $E_c^a(t)$ is the energy consumption of an appliance a at timeslot t, p_r^a and X_s^a is the power rating and status of an appliance, respectively. The aggregated electrical energy consumption of the smart home is calculated from the following formula:

$$E_T^a = \sum_{t=1}^{T} \left(\sum_{a=1}^{n} E_c^a(t) \right) \tag{7}$$

where E_T^a represents the aggregated energy consumption of the smart home appliances.

4.2. Smart Home Appliances Electricity Cost

The electricity cost is defined as the bill deposited by the residents to the power company for the used energy per unit time and per unit price. It is measured in the units of cents. For electricity cost determination, the power company provides various pricing schemes such as RTPS, CPPS, TOUPS, DAPS, and FPS; however, we adopted RTPS and CPPS for the proposed scheme. The RTPS and CPPS are the Midwest independent system operator (MISO) daily electricity pricing schemes taken from federal energy regulatory commission (FERC) [41,42]. The cost for the energy used by the residents under forecasted RTPS and CPPS is as follows:

- electricity cost of resident energy consumption is determined using forecasted RTPS as:

$$C_R^a = \sum_{t=1}^{T} \left(\sum_{a=1}^{n} E_c^a(t) \times p_s^r(t) \right), \tag{8}$$

where C_R^a is the total electricity cost by the resident to power company under RTPS denoted by $p_s^r(t)$.

- electricity cost paid by the resident to power company under forecasted CPPS is determined as:

$$C_p^a = \sum_{t=1}^{T}\left(\sum_{a=1}^{n} E_c^a(t) \times p_s^c(t)\right), \tag{9}$$

where C_p^a is the total electricity cost paid by the resident to power company under CPPS denoted by $p_s^c(t)$.

4.3. Peaks in Demand

Peaks in demand is defined as the highest demand emerged over a specified horizon of time such as daily, weekly, monthly, annually, and seasonally, or the highest points of resident electricity consumption. The electric power company cost high from the resident during the peak demand periods because they supplied continues power to load by bringing peak power plants online. It is measure in the units of power (Watts). The proposed framework tries to smooth out the demand curve by reducing the peaks in demand to avoid blackout situation. The peaks in demand can be determined as:

$$D_d^p(t) = \max\left(E_c^a(t)\right), \tag{10}$$

where $D_d^p(t)$ represents the highest possible peaks in the demand over a specified horizon of time.

4.4. Smart Home Load Scheduling Formulation

The smart home load scheduling problem is formulated as a minimization problem because the main objectives of this work are to alleviate peaks in demand and electricity cost with affordable appliances waiting time.

$$\min\left(\sum_{t=1}^{T}\left(\sum_{a=1}^{n} E_c^a(t) \times p_s^r(t)\right)\right) + \min\left(D_d^p(t)\right) \tag{11}$$

subjected to:

$$E_T^a \leqslant capacity \tag{12a}$$

$$\sum_{a \in A_n^t} E_T^{a,\, unsch} = \sum_{a \in A_n^t} E_T^{a,\, sch} \tag{12b}$$

$$\sum_{a \in A_n^t} T_o^{a,\, unsch} = \sum_{a \in A_n^t} T_o^{a,\, sch} \tag{12c}$$

$$X_a^{s,\, unsch} \neq X_a^{s,\, sch} \tag{12d}$$

The constraint (12a) ensures that total energy consumption of residents must be under the capacity of power company. The constraint (12b) and (12c) ensure that the total energy consumption of residents before and after scheduling must be equal subjected to fair comparison. The scheduling of smart appliance is conformed from constraint (12d).

5. Description of Adapted and Proposed Algorithms

In this section, the adapted and proposed heuristic algorithms for smart home load scheduling are discussed. Electricity is consumed in the three demand-side sectors, i.e., residential, commercial, and industrial sectors. However, the main focus is to perform DSM via optimal residential load scheduling. For this purpose, HEMC based on GWO, mEDE, and the proposed HGWmEDE algorithms schedule the smart home appliances to reduce peaks in demand and electricity cost with affordable user discomfort. In the literature, various optimization schemes have been proposed for load scheduling.

Some of these techniques outperform in cost reduction and others well perform either in PAR or user-discomfort minimization. In this regard, an optimization technique is proposed, i.e., HGWmEDE, which is a hybrid of mEDE and GWO algorithm. The proposed scheme simultaneously caters the minimization objectives of electricity cost, peaks in demand, and waiting time. The existing and proposed techniques are implemented in MATLAB and their detail description is as follows.

5.1. mEDE

The mEDE is a modified enhanced version of DE. The DE at very first time proposed by Storn in 1995 and enhanced and modified by [43]. It is a meta-heuristic population-based algorithm, which includes four main steps, i.e., population creation, crossover, mutation, and selection [44]. Initially, random population is generated by Equation (13) as follows:

$$P_{k,n} = B_n^l + (rand \times (B_n^u - B_j^l)). \tag{13}$$

To form a mutant vector, a random function is generated to create three vectors, i.e., v_{r1}, v_{r2}, and v_{r3}. First vector is the target vector and mutant vectors are generated using Equation (14) as given below:

$$m_{k,G+1} = v_{r1,G} + S(v_{r2,G} - v_{r3,G}), \tag{14}$$

where S is a scaling factor. Mutant vector is generated, then, first three trial vectors are generated by Equations (15)–(17). Then, the best trail vector is selected by comparing with the target vector to update the population with the best trial vectors.

$$B^u_{n,k,G+1} = \begin{cases} m_{n,k,G+1} & if \;\; randb(n) \leq 0.30 \\ v_{n,k,G} & Otherwise \end{cases} \tag{15}$$

$$B^u_{n,k,G+1} = \begin{cases} m_{n,k,G+1} & if \;\; randb(n) \leq 0.60 \\ v_{n,k,G} & Otherwise \end{cases} \tag{16}$$

$$B^u_{n,k,G+1} = \begin{cases} m_{n,k,G+1} & if \;\; randb(n) \leq 0.90 \\ v_{n,k,G} & Otherwise \end{cases} \tag{17}$$

Then, 4th and 5th trial vectors are generated by Equations (18) and (19), respectively, as given below:

$$B^u_{n,k,G+1} = randb(n) \cdot v_{n,k,G} \tag{18}$$

$$B^u_{n,k,G+1} = randb(n) \cdot m_{n,k,G} + (1 - randb(n)) \cdot v_{n,k,G} \tag{19}$$

The mEDE pseudocode is depicted in Algorithm 1. The maximum iterations are denoted by $Mx.itr$; the total population is represented by $POPLAT$, it shows the number of possible solutions. The crossover ratio is denoted by R_c, which is taken as 0.30, 0.60, and 0.90. The mutant, trial, and target vectors are represented by m, μ, and v, respectively.

Algorithm 1: mEDE

```
Parameters initialization Mx.itr, R_c, POPLAT, andhour;
Initially, population is randomly generated by Equation (13) ;
for a = 1:T do
    Compute mutant vector by Equation (14);
    for itr= 1:Mx.itr do
        Compute 1st trial vector with crossover rate of 0.30;
        if rand(.) ≤ 0.30 then
            μ_n = m_n
            else
            μ_n = v_n
        end
        Compute 2nd trial vector with crossover rate of 0.60;
        if rand(.) ≤ 0.60 then
            μ_n = m_n
            else
            μ_n = v_n
        end
        Compute 3rd trial vector with crossover rate 0.90;
        if rand(.) ≤ 0.90 then
            μ_n = m_n
            else
            μ_n = v_n
        end
        Create 4th and 5th trial vector using Equations (18) and (19);
        Find out trial vector which is best ;
        X_new ← best of μ_n ;
        Compare trial vector and target vector;
        if (P_new) < (P_n) then
            P_n = P_new
        end
    end
end
```

5.2. GWO

It is a heuristic technique, motivated by the wolves hunting and leadership nature [44]. For leadership 4 levels are defined: α, β, δ, and γ. The α is the most intuitive leader among the group, which provides guidance on hunting strategies to other wolves. The β and δ come after α in the chronological order, and γ is the feebler member among the group. Thus, γ has a lack of leadership qualities and cannot be considered. In HEMC, α is taken as the fittest member to schedule the smart home load to reduce cost and peaks in demand. Initially, the population is randomly generated by Equation (20):

$$P(k,n) = rand(POPLAT, A_n^t), \tag{20}$$

where $POPLAT$ represents the population of gray wolves and A_n^t is the overall appliances in the smart home, which is used in the proposed framework. The objective function of each search agent can be evaluated using co-efficient D and E.

5.2.1. Encircling Prey

Before hunting the gray wolves encircle a prey. The encircling behavior of gray wolves is mathematically modeled using Equations (21) and (22). These Equations taken from [45].

$$P(t+1) = P_p(t) - D \times A_n^t, \tag{21}$$

$$A_n^t = |E \times P_{p(t)} - P(t)|, \tag{22}$$

where the position of prey is represented by P_p, while P is gray wolf position at t_{th} epoch, which is calculated using Equation (21). The co-efficient vectors D and E are determined according to Equation (23) and Equation (24), respectively:

$$\overrightarrow{D} = 2\overrightarrow{b} \times \overrightarrow{r}_1 - \overrightarrow{b} \tag{23}$$

$$\overrightarrow{E} = 2 \times \overrightarrow{r_2} \tag{24}$$

where $\overrightarrow{r_1}$ and $\overrightarrow{r_2}$ are vectors with random values between 0 and 1. Value of D after multiple epochs is reduced from 2 to 0 while the value of E is randomly taken between 0 and 2. This value of E defines the weight of attractiveness for prey.

5.2.2. Hunting

The α provides guidance for hunting, while the β, and δ are secondary participants. The secondary participants follow α, due to best knowledge about the prey position. The 3 best solutions are achieved and the other participants such as γ update its position according to the best solution. The wolves' position is updated using Equation (25).

$$\overrightarrow{P_{t+1}} = \frac{\overrightarrow{v_1} + \overrightarrow{v_2} + \overrightarrow{v_3}}{3} \tag{25}$$

where $\overrightarrow{v_1}$, $\overrightarrow{v_2}$ and $\overrightarrow{v_3}$ are determined by Equations (26)–(28).

$$\overrightarrow{v_1} = \overrightarrow{v_\alpha} - \overrightarrow{D_1} \times (\overrightarrow{d_\alpha}) \tag{26}$$

$$\overrightarrow{v_2} = \overrightarrow{v_\beta} - \overrightarrow{D_2} \times (\overrightarrow{d_\beta}) \tag{27}$$

$$\overrightarrow{v_3} = \overrightarrow{v_\delta} - \overrightarrow{D_3} \times (\overrightarrow{d_\delta}) \tag{28}$$

where $\overrightarrow{v_\alpha}$, $\overrightarrow{v_\beta}$ and $\overrightarrow{v_\delta}$ are the best solutions obtained at the t_{th} iteration; $\overrightarrow{D_1}$, $\overrightarrow{D_2}$, $\overrightarrow{D_3}$ are determined using Equation (23), while $\overrightarrow{P_\alpha}$, $\overrightarrow{P_\beta}$, $\overrightarrow{P_\delta}$ are determined using Equations (29)–(31):

$$\overrightarrow{P_\alpha} = \overrightarrow{E_1} \times \overrightarrow{v_\alpha} - \overrightarrow{v} \tag{29}$$

$$\overrightarrow{P_\beta} = \overrightarrow{E_2} \times \overrightarrow{v_\beta} - \overrightarrow{v} \tag{30}$$

$$\overrightarrow{P_\delta} = |\overrightarrow{E_3} \times \overrightarrow{v_\delta} - \overrightarrow{v}, \tag{31}$$

where E_1, E_2, and E_3 are calculated using Equation (24). The gradation of variable g is conducted in the last step; exploration and exploitation trade-off is controlled by considering value between 0 to 2 in each epoch as depicted in Equation (32).

$$g = 2 - t\frac{2}{Mx.itr} \tag{32}$$

Mathematical modeling of objective function is shown in Equation (33), which use the power rating and status of an appliance.

$$FitnessF = p_r^a \times X_a^S(t) \tag{33}$$

From Algorithm 2, the maximum iterations are represented by $Mx.itr$, the total population is denoted by $POPLAT$, the total number of smart home appliances is A_n^t and $FitnessF$ is the fitness function. α is best among the group participants, which provides a primary optimal solution in the hunting behavior, while β and δ come after α in the group and provide secondary optimal solutions.

Algorithm 2: GWO

```
Parameters initialization Mx.itr, POPLAT, A_n^t, α, β, and δ;
Initially, gray wolves population is generated Pk(k = 1,2,3,...,Pn);
P(k,n) = rand(POPLAT, A_n^t);
while itr < Mx.itr do
    for k = 1:POPLAT do
        Compute fitness by Equation (33);
        if fitness < α_score then
            α_score = fitness;
            α_Pos = P(k,:);
        end
        if fitness > α_score and fitness< β_score then
            β_score = fitness;
            β_Pos = P(k,:);
        end
        if fitness > α_score and fitness> β_score and fitness< δ_score then
            δ_score = fitness;
            δ_Pos = P(k,:);
        end
    end
    for k = 1:POPLAT do
        for n = 1:Appliances do
            Randomly generate r_1 and r_2 by rand command;
            Compute fitness coefficients D and E using Equations (23) and (24);
            Update 3 vectors (α, β, δ) by Equations (29)–(31);
        end
    end
end
```

To evaluate the best hunting leader fitness function, compare the fitness of α, β, and δ. The positions are updated according to Equations (29)–(31).

5.3. HGWmEDE

In this section, the proposed hybrid algorithm is demonstrated in detail. Initial population in mEDE is generated by four phases, i.e., initialization phase, mutation phase, crossover phase, and selection phase and the population are updated by comparatively analyzing trial vector with the target vector. The procedure of trail vector selection is effective in choosing the best trial vector from available vectors. The GWO comprised of three steps, i.e., encircling prey, hunting, and wolves position update within the pack. All search agents' positions are updated according to the leader α within the pack. In GWO, unlike the mEDE agents, α with β, and δ are not compared. However, the β

and δ are closer to the prey as compared to α. To conduct a comparison of all search agents a crossover phase of mEDE is adopted. Thus, the best search agent is selected according to the crossover phase of mEDE and search agents' position is updated according to GWO. The HGWmEDE is proposed by combining the taking key characteristics of both mEDE and GWO.

In Algorithm 3, the detailed stepwise procedure of HGWmEDE is presented. The main stages of the proposed HGWmEDE algorithm are initialization stage, encircling prey stage, best search agent selection stage, and wolves position update stage. Initially, the population of wolves is generated randomly using Equation (20). The best search agent is selected by following the steps presented in Algorithm. The mutant vector m is generated using Equation (14). The fitness of m, α, β, and δ is computed by Equation (33). Crossover phase is conducted to select the best search agent using the following Equations:

$$\alpha_{new} = \begin{cases} m_n & if \text{ fitness of} m_n \leq \alpha \\ \alpha & Otherwise \end{cases} \tag{34}$$

$$\beta_{new} = \begin{cases} m_n & if \text{ fitness of} m_n \leq \beta \\ \beta & Otherwise \end{cases} \tag{35}$$

$$\delta_{new} = \begin{cases} m_n & if \text{ fitness of} m_n \leq \delta \\ \delta & Otherwise \end{cases} \tag{36}$$

Algorithm 3: HGWmEDE

```
Parameters initialization Mx.itr, POPLAT, A_n^t, α, β, δ;
Initial gray wolves population generation Pk(k = 1, 2, 3, ..., Pn);
P(k, n) = rand(POPLAT, A_n^t);
while itr < Mx.itr do
    for k = 1:POPLAT do
        Create a mutant vector by Equation (14) from mEDE;
        Compute mutant vector fitness as cost × v_n;
        Generate randomly α, β, and δ;
        Compute α, β, and δ fitness by Equation (33);
        if fitness of m(n) < α_score then
            α_position = m_n;
        end
        if fitness m_n > α_score and m_n < β_score then
            β_position = m(j);
        end
        if fitness v_n > α_score and fitness v_n > β_score and fitness v_n < δ_score then
            δ_position = m_n);
        end
    end
    for k = 1:POPLAT do
        for n = 1:Appliances do
            Randomly generate r1 and r2 by rand command;
            Compute coefficients D and E fitness by Equations (23) and (24);
            Update three vectors (α, β, δ) by Equations (29)–(31);
        end
    end
end
```

When the best search agents are selected then search agents' position is updated according to GWO. The position is updated using Equation (25).

The detail description of Algorithm 3 for each step is as follows. In 1st step, parameters are initialized. In 2nd step, randomly population is generated, and the counter is adjusted to maximum epochs. Crossover phase of mEDE is conducted to compare the fitness of the mutant vector with α, β, and δ. The search agent status is updated using GWO. The procedure is repeated for several epochs until the termination criteria are reached.

6. Simulation Results and Discussion

Simulation results and discussions of the proposed modular framework are demonstrated, in this section. The pricing signals used for load scheduling is forecasted using RBM. The aim is to evaluate the performance (peaks in demand and electricity cost reduction) of proposed and existing schemes under RBM-based forecasted RTPS and CPPS for different OTI. The HEMC is responsible for scheduling the appliances using the forecasted RTPS and CPPS pricing signals. The performance is evaluated for different OTI such as 15, 30, and 60 timeslots. The home appliances and their parameters such as OTI, LOT, starting time, ending time, and power rating are adopted from [39]. Simulation results and discussion of the proposed and existing algorithms are demonstrated in the succeeding sections in terms of peaks in the demand and electricity cost reduction with affordable appliances waiting time. The detail description is as follows.

6.1. Electricity Cost Evaluation under Price-Based DR

The power company provides various pricing schemes such as RTPS, CPPS, TOUPS, DAPS, and FPS for electricity cost calculation; however, we adopted RTPS and CPPS for the proposed framework. The RTPS and CPPS are the Midwest independent system operator (MISO) daily electricity pricing signals taken from the federal energy regulatory commission (FERC). The electricity cost using RTPS and CPPS is individually discussed in the succeeding sections.

6.2. Electricity Cost Evaluation Using RTPS

To evaluate the cost parameters of the proposed scheme simulations are conducted using different OTI, i.e., 15, 30, and 60 min. The proposed HGWmEDE algorithm reduced electricity cost as compared to GWO and mEDE by scheduling smart home appliances using forecasted RTPS. The scheduled appliances sustain coordination among pricing scheme and the consumption pattern in a particular timeslot of a day to alleviate the electricity cost. The proposed algorithm shifts smart home appliances from ON-peak timeslots to OFF-peak timeslots in an optimal manner to alleviate electricity cost and peaks in demand.

The energy consumption pattern in terms of electricity cost of the proposed and existing algorithms with 15 min OTI is illustrated in Figure 3a. In this Figure 3a, both scheduled and unscheduled scenarios are observed. The peaks in demand are high in case of unscheduled load, which reveals that prices are high in these particular hours. Thus, the use of appliances in these hours results in high electricity cost. However, scheduling smart home appliances using the proposed and existing algorithms eliminate these peaks in demand and reduce the electricity cost. Thus, the proposed HGWmEDE-based framework outperforms both GWO and mEDE in terms of peaks in demand and electricity cost reduction. The electricity cost pattern of the proposed and existing algorithms with 30 min OTI under RTPS is illustrated in Figure 3b. Both proposed and existing algorithms-based HEMC can schedule smart home appliances. However, the electricity cost of GWO is high at the starting timeslots, while HGWmEDE has minimum electricity cost throughout the 24 h. Likewise, in Figure 3c, the electricity cost pattern of the proposed and existing algorithms for 60 min OTI under RTPS is depicted. The proposed HGWmEDE algorithm has reduced the electricity cost by optimally scheduling smart home appliances of resident's, which is one of our main objectives.

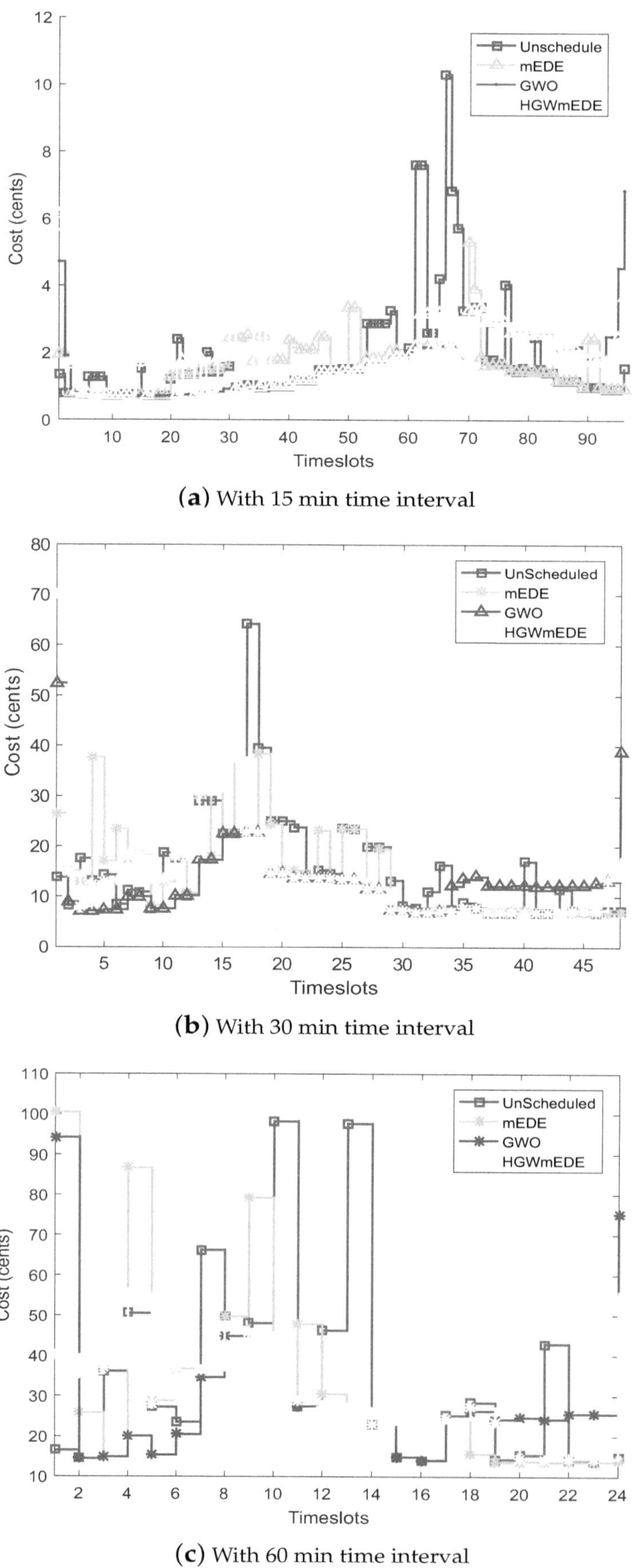

(**a**) With 15 min time interval

(**b**) With 30 min time interval

(**c**) With 60 min time interval

Figure 3. Electricity cost per timeslot evaluation for different OTI under RTPS.

The Figure 4 illustrate that the proposed framework optimally scheduled the smart home appliances as compared to mEDE and GWO under forecasted RTPS and CPPS and reduced the overall aggregated electricity cost of the residents.

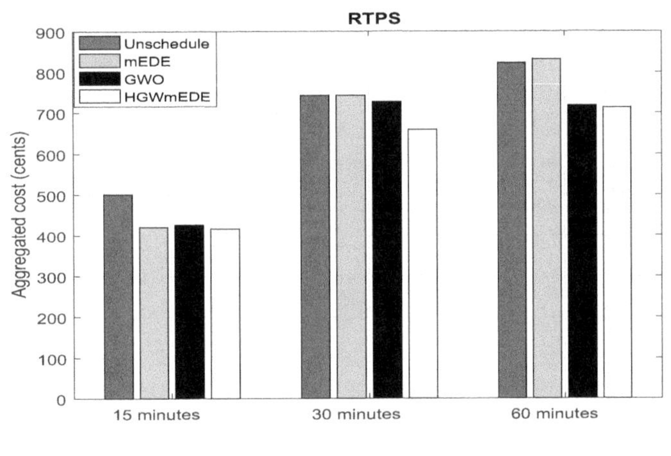

(**a**) With forecasted RTPS

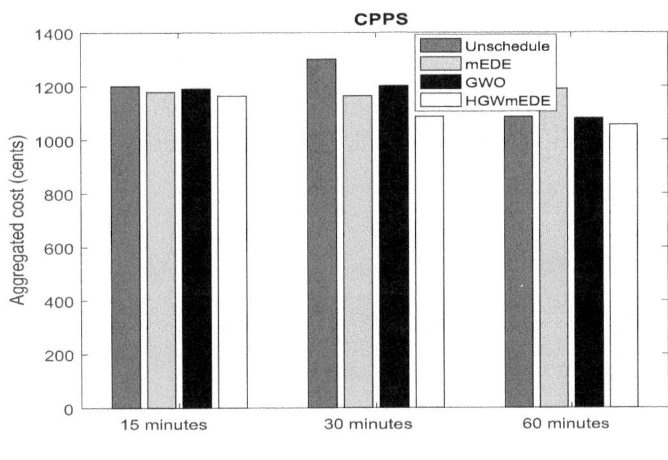

(**b**) With forecasted CPPS

Figure 4. Aggregated electricity cost evaluation under forecasted RTPS and CPPS.

Figure 4a presents the overall electricity bill of 15, 30, and 60 OTI under forecasted RTPS. The electricity cost of unscheduled load for 15 min OTI is measured as 500.4827 cents. However, with scheduling smart home appliances using mEDE and GWO reduced the overall electricity cost to 420.5381 cents and 426.0508 cents, respectively. The proposed HGWmEDE scheme reduced electricity cost up to 416.7468 cents, which is the maximum reduction as compared to mEDE and GWO. In a similar fashion, electricity cost reduction behavior of the proposed and existing schemes can be observed for both 30 and 60 min OTI.

Electricity Cost Evaluation under CPPS for Different OTI

The load scheduling is favorable for both residents and power company because it reduces electricity cost, which is favorable for residents; and peaks in the demand, which is favorable for the power company. Cost reduction facilitates residents to deposit less electricity bill and peaks reduction in the demand facilitate power company in the optimal management of supply with demand. In this subsection, electricity cost evaluation is conducted under the forecasted CPPS profile. The electricity bill reduction evaluation of the proposed and existing algorithms are performed using 15, 30, and 60 min OTI under CPPS and is illustrated in Figure 5. The electricity cost profile of 15 min OTI is depicted in

Figure 5a. Generally, the forecasted CPPS remains constant except during critical peak hours where the electricity price reaches to its maximum value [46]. The timeslots from 40 to 65 are critical periods, where the electricity price is high. In unscheduled load scenario, the maximum peak is at 181.55 cents. The scheduled load scenario, where smart home appliances are scheduled, and the peak is reduced to 83.07 cents. The electricity cost for 30 min OTI is illustrated in Figure 5b. The electricity cost is varying for 48 timeslots and the remaining electricity cost profile is as the same as observed in Figure 5a. In our proposed HGWmEDE algorithm-based scenario, no peaks in demand are emerged except at the starting time of day, which is 56 cents. The electricity cost for 60 min OTI is depicted in Figure 5c. The unscheduled appliance electricity cost reaches to 766.8 cents, which is reduced to 203.46 cents when these smart home appliances are scheduled using our proposed HGWmEDE algorithm. This means that the proposed HGWmEDE algorithm optimally scheduled the smart home appliances. The overall cost for the proposed and existing optimization schemes is illustrated in Figure 4b. The overall unscheduled cost is 1300.891 cents, which is reduced to 1085.91 cents when smart home appliances are scheduled using the proposed HGWmEDE algorithm. The proposed HGWmEDE algorithm outperforms both mEDE and GWO algorithms in terms of electricity cost reduction. The overall electricity cost reduction for 30 and 60 min OTI is depicted in Figure 4b. A brief comparison of electricity cost under forecasted RTPS and CPPS is listed in Table 3 for 15, 30, and 60 min OTI.

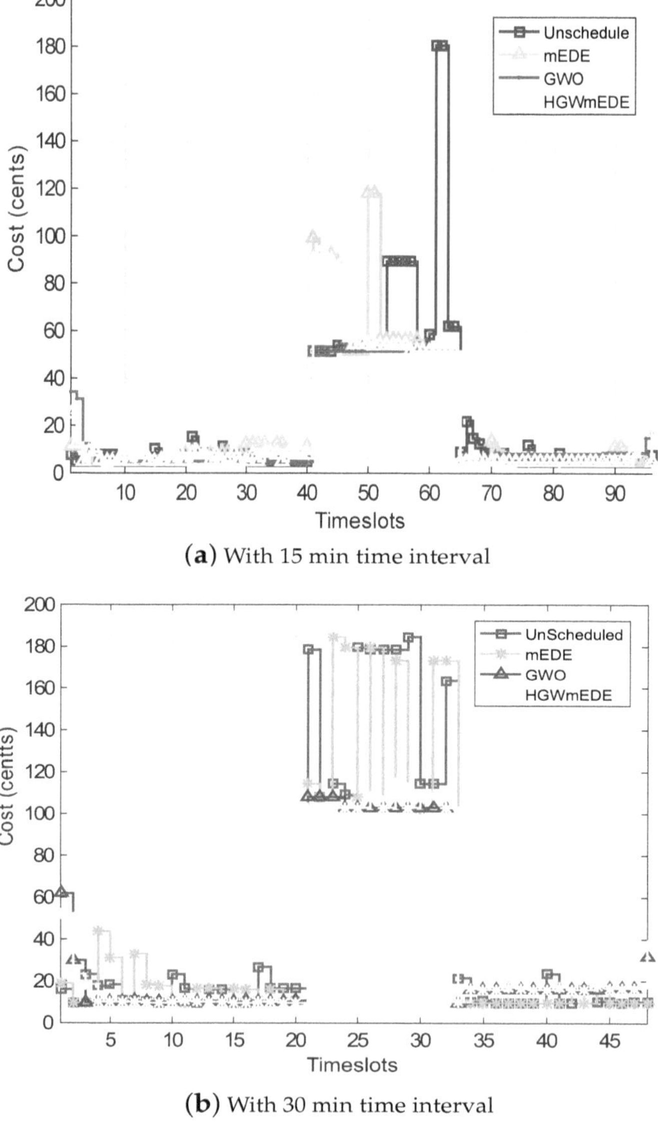

(a) With 15 min time interval

(b) With 30 min time interval

Figure 5. *Cont.*

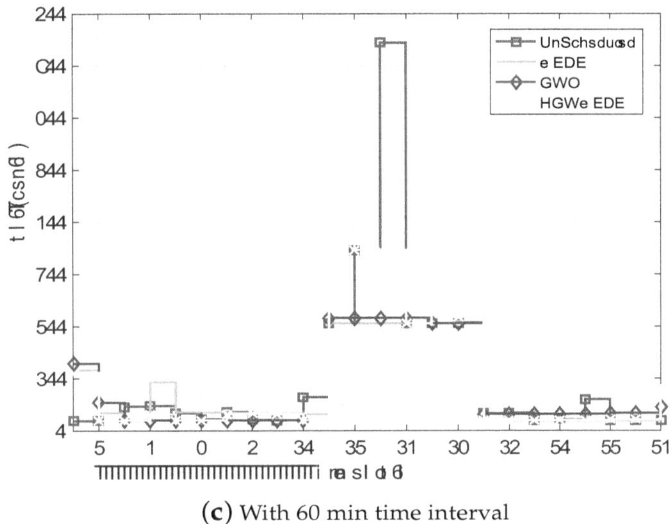

(**c**) With 60 min time interval

Figure 5. Electricity cost per timeslot evaluation under forecasted CPPS for different OTI.

Table 3. Overall electricity cost comparative evaluation for 24 h time horizon under forecasted RTPS and CPPS.

Scenarios	Electricity Cost (Cents) under RTPS			Electricity Cost (Cents) under CPPS		
	15 min	**30 min**	**60 min**	**15 min**	**30 min**	**60 min**
Without scheduling	500.4821	743.4871	822.1561	1200.1561	1300.8910	1085.6481
mEDE	420.5381	743.1951	831.2132	1178.0461	1164.4901	1190.6901
GWO	426.0507	727.1431	717.9402	1190.5122	1200.9612	1080.4091
HGWmEDE	416.7468	658.6502	712.7292	1164.4901	1085.9022	1056.7891

6.3. Smart Home Energy Consumption

The smart home appliances energy consumption for both RTPS and CPPS are discussed in detail in the following subsection:

6.3.1. Smart Home Energy Consumption Using RTPS

Energy consumed by smart home appliances under RTPS in each timeslot for 15, 30, and 60 min OTIs is illustrated in Figure 6. In Figure 6a, the smart home energy consumption profile for 15 min OTI is depicted. From the figure it is obvious that at the start and end timeslots of the day have low per unit electricity price; Thus, HEMC based on our proposed HGWmEWDE shifted the load to these low pricing timeslots. In this manner, the proposed scheme optimally curtailed peak load on the power company.

The smart home energy consumption profile for 30 min time interval is illustrated in Figure 6b. The HMEC based on our proposed HGWDE algorithm results in optimal energy consumption profile as compared to GWO and mEDE-based HEMC. The HGWmEDE eliminated load peaks, which curtailed burden on both power company in terms of peak power generation and on residents in terms reduce electricity bill deposit. The mEDE and GWO also reduced peaks in demand and reduced the burden on both power company and residents as compared to without scheduling scenario. The burden reduction of the proposed scheme is more as compared to the existing schemes (GWO and mEDE); thus, the proposed scheme outperforms the existing schemes.

In Figure 6c, energy consumption pattern for 60 min OTI is depicted. The electricity prices are low during starting timeslots and ending timeslots in forecasted RTPS profile. Thus, the proposed and existing schemes shifted most of the load to these low-price timeslots. The prices are maximum from 3 to 7 p.m., so, our proposed scheme not scheduled appliance in these timeslots because the operation of appliances during these timeslots results in high electricity cost. Load shifting from ON-peak timeslots to OFF-peak timeslots results in user discomfort as the residents must stay to switch on a particular smart home appliance because of the trade-off between electricity cost and user-comfort.

The smart home overall energy consumption for 15 min OTI is 56.3108 kWh, which remains the same before and after scheduling subjected to a fair comparison. The scheduled and unscheduled energy consumption for 30 min OTI is same and recorded as 57.7656 kWh. Likewise, for 60 min OTI, smart home overall energy consumption for both scheduled and unscheduled scenarios is the same and recorded as 64.5661 kWh.

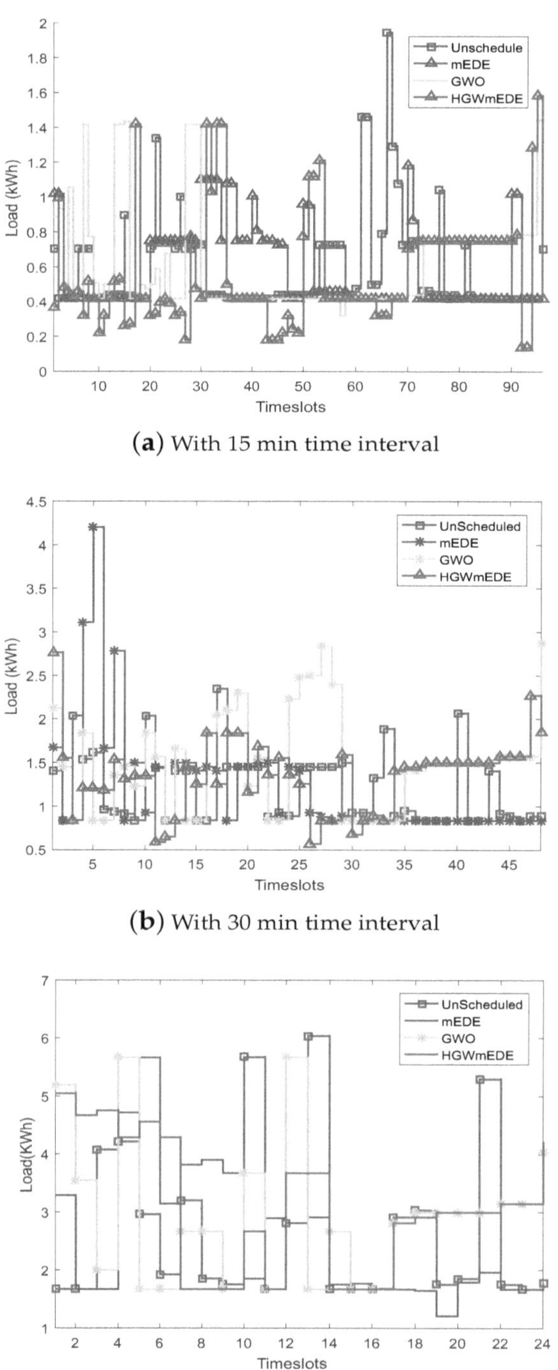

(**a**) With 15 min time interval

(**b**) With 30 min time interval

(**c**) With 60 min time interval

Figure 6. Electricity energy consumption evaluation per timeslot under forecasted RTPS.

6.3.2. Energy Consumption Using CPPS

In this subsection, the proposed HGWmEDE scheme is comparatively evaluated under forecasted CPPS for 15, 30, and 60 min OTI. The comparison is illustrated in Figure 7. The smart home energy consumption profile for 96 timeslots is depicted in Figure 7a. The proposed scheme shifted load to the timeslots where electricity price is low to reduce cost and peaks in demand. In CPPS, at starting and ending timeslots electricity price is constant, while during 40 to 65 timeslots price is maximum as illustrated in Figure 2b. The profile GWO and HGWmEDE is almost similar, except some peaks of GWO are much higher than HGWmEDE at starting timeslots. The proposed HGWmEDE eliminates the peaks in demand for 30 min OTI case; however, the existing schemes have high peaks as compared to the proposed HGWmEDE scheme.

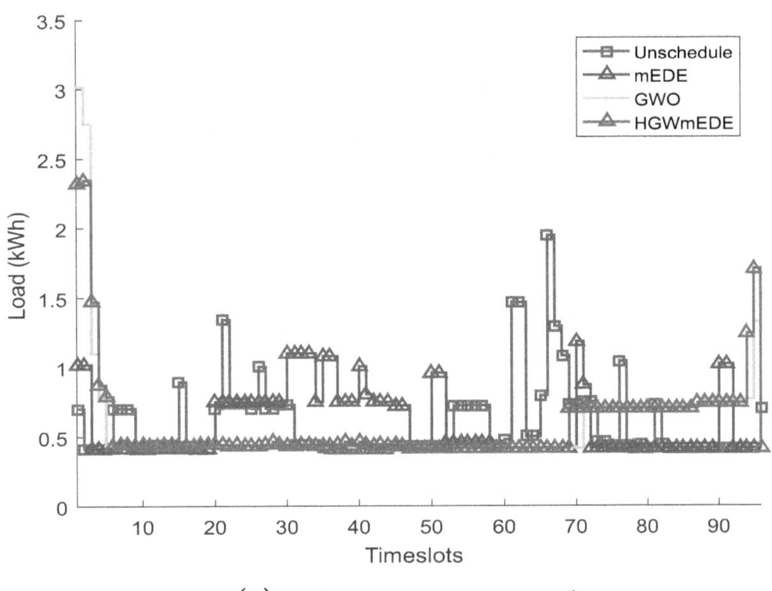

(**a**) With 15 min time interval

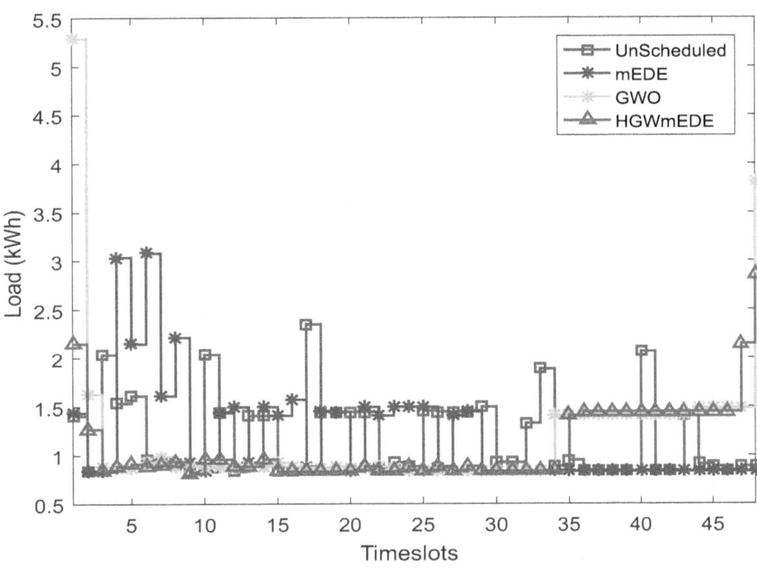

(**b**) With 30 min time interval

Figure 7. *Cont.*

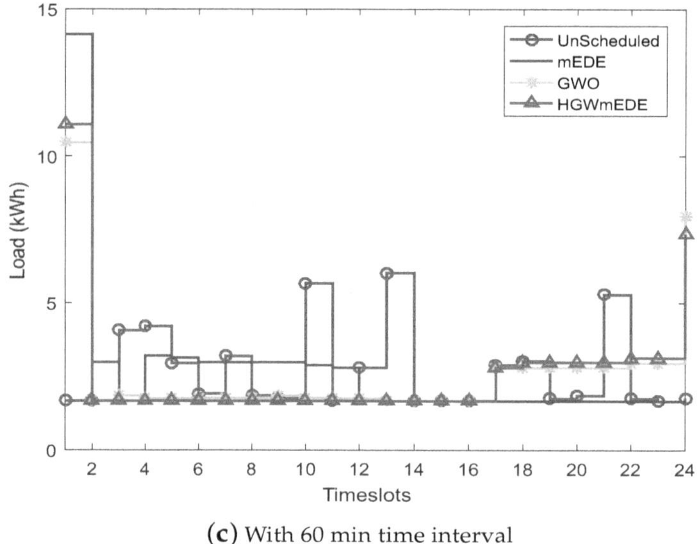

(c) With 60 min time interval

Figure 7. Smart home energy consumption per timeslot evaluation using CPPS.

The energy consumption for 60 min OTI is illustrated in Figure 7c. The energy consumption for all cases remain same but the status of appliances vary according to the schedule of HEMC based on our proposed HGWmEDE algorithm and existing algorithm.

The overall energy consumption with and without scheduling must remain same subjected to a fair comparison. The overall energy consumption for 15 min OTI is 56.3107 kWh for the proposed (HGWmEDE) and existing (GWO and mEDE) schemes. Similarly, the energy consumption for 30, and 60 min OTI are 57.7656 kWh and 64.5661 kWh, respectively. Thus, both electricity cost and peaks in demand are reduced by scheduling the smart home appliance while keeping the energy consumption constant.

6.4. Peaks in Demand

Peaks in demand are defined as the value of peak load emerged during a time horizon of 24 h or the maximum load switched on by user during 24 h time horizon. Our objective is to reduce the peaks in demand and to ensure smooth demand curve for 24 h time horizon. Various DSM programs can be applied to alleviate peaks in demand such as peak clipping, load shifting, and price-based DR to eliminate peaks in demand and smooth the demand curve. Peaks elimination in the demand reduces the electricity cost and burden on the power company. The peaks reduction in demand is evaluated for both RTPS and CPPS in the succeeding section.

6.4.1. Peaks Reduction in Demand Evaluation under RTPS

The peaks reduction in demand evaluation under RTPS for different OTI is depicted in Figure 8a. In case, when the load is not scheduled the peak emerged in demand is 10.9697. In case, when the load is scheduled based on mEDE and GWO, the peaks emerged in demand are 8.1722 and 5.6750, respectively. The peak emerged for proposed HGWmEDE scheme is 5.1530, which is low as compared to the both unscheduled and scheduled (GWO and mEDE) cases. The percentage reduction in peaks of the proposed HGWmEDE scheme is 53.02%, while the percentage reduction in peaks of the GWO and mEDE is 25.50%, 48.26%, respectively. Thus, the proposed HGWmEDE scheme outperforms the other schemes in terms of peaks reduction in demand.

In case of unscheduled load for 30 min OTI, the peak emerged in demand is recorded as 6.0257; however, the peak has reduced to 5.8424 and 5.9335 when the load is scheduled using mEDE and GWO algorithms. The proposed HGWmEDE scheme outperforms both mEDE and GWO schemes by reducing peak in demand to 3.6209. Similarly, for 60 min OTI, mEDE, GWO, and HGWmEDE reduced the peaks in demand by a value of 3.6559, 4.3510 and 2.5370 as compared to without scheduling case, which is 5.0245. Thus, it is obvious from the aforementioned statistical analysis that the proposed HGWmEDE scheme outperforms both mEDE and GWO schemes.

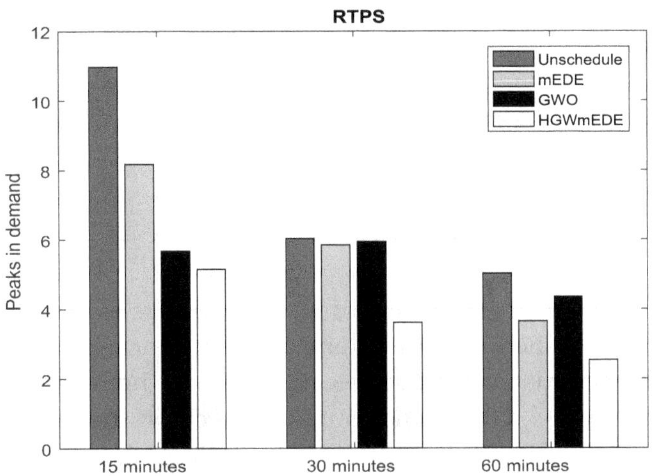

(**a**) Peaks in demand with RTPS

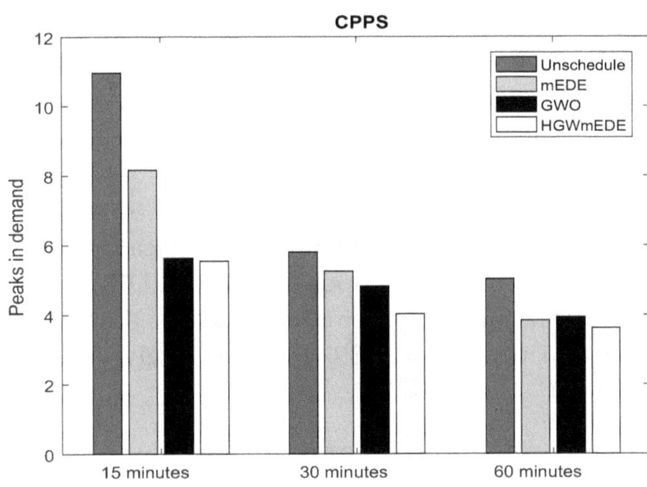

(**b**) Peaks in demand with CPPS

Figure 8. Peaks in demand evaluation under forecasted RTPS and CPPS for different OTI.

6.4.2. Peaks in Demand Evaluation under CPPS for Different OTI

Evaluation of peaks in demand under CPPS is illustrated in Figure 8b. In case of unscheduled load for 15 min OTI, peak in demand is recorded as 10.9697. However, after performing load scheduling, peak in demand is alleviated to 5.5416 with HGWmEDE, 5.627 with GWO and 8.1723 with mEDE. Peaks reduction in terms of percentage for the proposed HGWmEDE, mEDE, GWO, are 49.4836%, 48.7120%, and 25.5012%, respectively.

In case of unscheduled load for 30 min OTI, the peak emerged is 5.86. After load scheduling with mEDE, GWO, and HGWmEDE, peaks in demand are recorded as 5.2536, 4.8165, and 4.0215, respectively. Thus, the proposed HGWmEDE scheme outperforms the existing (mEDE and GWO) schemes in terms of peaks reduction in demand. The comparative evaluation of the proposed HGWmEDE and existing (mEDE and GWO) schemes in terms of peaks reduction in demand under RTPS and CPPS for different OTI is listed in Table 4.

Table 4. Peaks in demand evaluation of the proposed and existing schemes for 24 h.

Scenarios	Peaks in Demand under RTPS with Different OTI			Peaks in Demand under CPPS with Different OTI		
	15 min	**30 min**	**60 min**	**15 min**	**30 min**	**60 min**
Without scheduling	10.9698	6.0258	5.0258	10.9698	5.8035	5.0258
mEDE	8.1723	5.8425	3.6558	8.1723	5.2537	3.8425
GWO	5.676	5.9336	4.3509	5.6265	4.8166	3.9336
HGWmEDE	5.1531	3.6210	2.5369	5.5416	4.0264	3.6210

6.5. User-Comfort Evaluation in Terms of Waiting Time

In this subsection, the user-comfort in terms of waiting time is evaluated. In nature, always there is a trade-off between different conflicting parameters. In this paper, a trade-off between electricity cost and waiting time (user-comfort) exist. To reduce electricity cost, the residents must wait for timeslots where the electricity price is low, i.e., OFF-peak hours to switch on their load. Thus, user-comfort and electricity cost are directly related to [47]. In without scheduling scenario, waiting time is almost zero because the smart home appliances are operated according to the resident choice and priority. However, the case when the load is scheduled based on mEDE, GWO, and the proposed HGWmEDE, the user switched on their appliances according to the schedule provided by HEMC to reduce electricity cost. Thus, user-comfort is compromised while reducing the electricity cost due to the trade-off. Evaluation of appliances waiting time under RTPS and CPPS are as follows:

6.5.1. Smart Home Appliances Waiting Time Evaluation under RTPS

The appliance waiting time for 15, 30, and 60 min OTI is illustrated in Figure 9a. Waiting time of the proposed HGWmEDE, mEDE, and GWO for 15 min OTI are calculated as 10.4 h, 4.3 h, and 9.7 h, respectively. Waiting time for 30 min OTI of the proposed HGWmEDE, mEDE, and GWO are calculated as 12.7007 h, 4.5394 h, and 10.0262 h. Similarly, for 60 min OTI of the proposed HGWmEDE, mEDE, and GWO, waiting time is 3.8 h, 2.6560 h, and 2.2397 h, respectively.

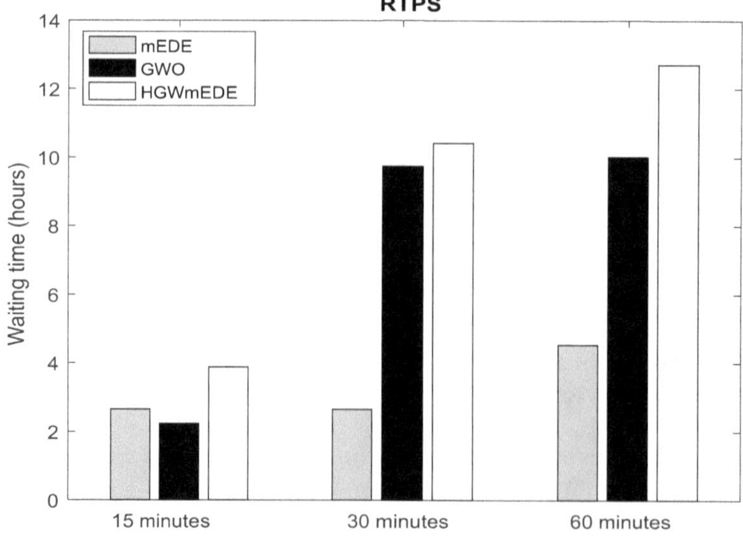

(**a**) User-comfort (waiting time) using RTPS

Figure 9. *Cont.*

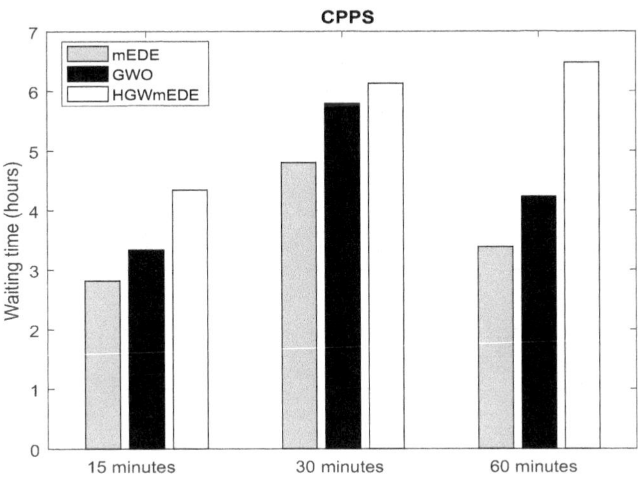

(**b**) User-comfort (waiting time) using CPPS

Figure 9. User-comfort (waiting time) of the scheduled load based on the proposed HGWmEDE, mEDE, and GWO using RTPS and CPPS.

It is concluded from the above results and discussion that user frustration in terms of waiting time always reduces by keeping the OTI smaller. For example, for 15 min OTI, if the operational time of a toaster is 12 min, then HEMC will allocate 15 min timeslot to toaster; in this way, only 3 min of a timeslot is wasted because this slot is not allocated to other appliances. In contrast, in case of 60 min OTI, the HEMC will allocate 60 min timeslot to an electric kettle, which has the operational time of only 5 min, then rest of 55 min timeslots will be wasted. Thus, in larger OTI, the user frustration increase in terms of waiting time.

6.5.2. Smart Home Appliances Waiting Time Evaluation under CPPS

Waiting time of the proposed HGWmEDE and existing (mEDE and GWO) under CPPS is illustrated in Figure 9b. The recorded value of waiting time for mEDE, GWO, and the proposed HGWmEDE is as 3.39 h, 4.23 h, and 6.49 h, respectively. It is obvious that the load schedule return by HMEC based on HGWmEDE algorithm has more waiting time, which indicates that user-comfort is compromised for the purpose to reduce electricity cost. The statistical analysis of the proposed and existing algorithms in terms of waiting time for different OTI under RTPS and CPPS is listed in Table 5.

Table 5. Comparative evaluation of the proposed HGWmEDE and existing (mEDE and GWO) algorithm in terms of waiting time under RTPS and CPPS for different OTI.

Scenarios	Evaluation of Waiting under RTPS for Different OTI			Evaluation of Waiting Time under CPPS for Different OTI		
	15 min	**30 min**	**60 min**	**15 min**	**30 min**	**60 min**
mEDE	4.3781 h	4.5394 h	2.6560 h	3.3826 h	4.8012 h	2.8158 h
GWO	9.7494 h	10.0262 h	2.2397 h	4.2293 h	5.7853 h	3.3346 h
HGWmEDE	10.4249 h	12.7007 h	3.8793 h	6.4814 h	6.1335 h	4.3408 h

6.6. *Convergence Evaluation of the Proposed HGWmEDE Algorithm-Based Fitness Function*

The convergence evaluation of the fitness function of the proposed HGWDE algorithm is depicted in Figure 10. X-label of the plot is the number of iterations and Y-label is the value of the fitness function. Figure 10 illustrates that the solution converges after 100 iterations, which indicates that the global maximum is achieved. The proposed HGWmEDE algorithm converging behavior for each iteration is plotted. The value of cost is constantly decreasing from 0 to 10 iterations and after these iterations, the graph slight varies. The behavior of convergence of the proposed HGWDE algorithm iterations is observed for 100 iterations. Finally, a straight line is achieved, which means that the solution is converged, and this is the most optimal point.

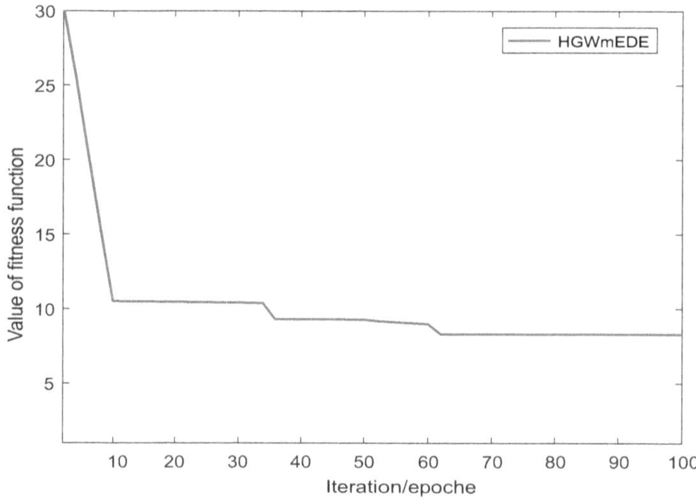

Figure 10. Convergence evaluation of the fitness function.

6.7. Performance Trade-Off

The performance trade-off exists in various conflicting parameters of the system. In the proposed modular framework, a trade-off between electricity cost and user-comfort exist. When the residents wants to reduce their electricity cost, he must face frustration in the form of user discomfort. Residents frustration increasing when the difference between the resident preferred time and the HEMC scheduled time is more. It is observed in Figure 4a,b that the electricity cost is reduced by applying HGWmEDE for scheduling at the expense of maximum average waiting time.

7. Conclusions and Future Research Directions

In this paper, first a modular framework is introduced, and then an algorithm HGWmEDE is proposed, which with the help of forecasted price-incentive DR scheme schedule the household load to maximize the aggregate utility of both resident and power company. The proposed approach is beneficial for residents because it reduces the electricity bill with an affordable waiting time of smart home appliances. In addition, it is beneficial for the power company because it reduces peaks in demand, which smooth out the demand curve and increase the stability of the power system. The integration of the forecaster module to the home energy management framework provides optimal load schedule, which not only facilitates residents but also power companies. Simulation analysis validated the proposed approach by comparing with two other approaches, a system without HEMC and a system with HEMC based on GWO and mEDE algorithms. Surely, the proposed framework can be applied for DSM and reliable operation of the SG. The idea of this paper in the future can be extended to various directions:

1. A system with renewable and non-renewable energy provider can be considered.
2. A system with a deep neural network can be used to optimize home energy management.
3. A system with multiple power companies and multiple homes and effect malicious residents can be considered.
4. To implement fog and cloud concept for household load scheduling instead of using a HEMC.
5. The same framework can be extended for other heuristics, deterministic, and stochastic techniques under RES.

Author Contributions: Conceptualization, G.H.; Formal analysis, S.A. and M.U.; Investigation, G.H.; Methodology, G.H. and K.S.A.; Software, G.H.; Supervision, K.S.A.; Validation, K.S.A., N.I. and A.A.; Writing—original draft, G.H.; Writing—review and editing, K.S.A., S.A., and M.U.

Nomenclature

Acronym	Meaning
RTPS	Real-Time Pricing Scheme
TOUPS	Time of Use Pricing Scheme
CPPS	Critical Peak Pricing Scheme
FRPS	Flat-Rate Pricing Scheme
DAPS	Day-Ahead Pricing Scheme
HEMC	HEM controller
DSM	Demand-Side Management
HGWmEDE	Hybrid Gray Wolf-Modified Differential Evolution
RBM	Restricted Boltzmann Machine
LOT	Length of Operational Time
OTI	Operation Time Interval
GA	Genetic Algorithm
BPSOA	Binary-Particle Swarm Optimization Algorithm
DEA	Differential Evolution Algorithm
IDSS	Intelligent Decision Support System
WDO	Wind-Driven Optimization
K-WDO	Algorithm with Knapsack
ACOA	Ant Colony Optimization Algorithm
MOEA	Multi-Objective Evolutionary Algorithm
PS	Pareto Sets
GMI	Generalized Mutual Information
TLBOA	Teaching and Learning-Based Optimization Algorithm
SFL	Shuffled-Frog Leaping
ANN-GA	Artificial Neural Network
LP	Linear Programming
NLP	Non-LP
BFA	Bacterial Foraging Algorithm
DR	Demand Response
AMI	Advanced Metering Infrastructure
ICT	Information and Communication Technology
RERs	Renewable Energy Resources
SG	Smart Grid

References

1. Mhanna, S.; Chapman, A.C.; Verbi, G. A fast distributed algorithm for large-scale demand response aggregation. *IEEE Trans. Smart Grid* **2016**, *7*, 2094–2107. [CrossRef]
2. Energy Reports. Available online: http://www.enerdata.net/enerdatauk/press-and-publication/energy-features/enerfuture-2007.php (accessed on 10 January 2019).
3. Logenthiran, T.; Srinivasan, D.; Shun, T.Z. Demand side management in smart grid using heuristic optimization. *IEEE Trans. Smart Grid* **2012**, *3*, 1244–1252 [CrossRef]
4. Shirazi, E.; Jadid, S. Optimal residential appliance scheduling under dynamic pricing scheme via HEMDAS. *Energy Build.* **2015**, *93*, 40–49. [CrossRef]
5. Bradac, Z.; Kaczmarczyk, V.; Fiedler, P. Optimal scheduling of domestic appliances via MILP. *Energies* **2014**, *8*, 217–232. [CrossRef]

6. Katz, J.; Andersen, F.M.; Morthorst, P.E. Load-shift incentives for household demand response: Evaluation of hourly dynamic pricing and rebate schemes in a wind-based electricity system. *Energy* **2016**, *115*, 1602–1616. [CrossRef]
7. Rabiya, K.; Javaid, N.; Rahim, M.H.; Aslam, S.; Sher, A. Fuzzy energy management controller and scheduler for smart homes. *Sustain. Compu. Inf. Syst.* **2019**, *21*, 103–118.
8. Adika, O.C.; Wang, L. Autonomous appliance scheduling for household energy management. *IEEE Trans. Smart Grid* **2014**, *5*, 673–682. [CrossRef]
9. Li, C.; Yu, X.; Yu, W.; Chen, G.; Wang, J. Efficient computation for sparse load shifting in demand side management. *IEEE Trans. Smart Grid* **2017**, *8*, 250–261. [CrossRef]
10. Javaid, N.; Naseem, M.; Rasheed, M.B.; Mahmood, D.; Khan, S.A.; Alrajeh, N.; Iqbal, Z. A new heuristically optimized Home Energy Management controller for smart grid. *Sustain. Cities Soc.* **2017**, *34*, 211–227. [CrossRef]
11. Bilal, H.; Javaid, N.; Hasan, Q.; Javaid, S.; Khan, A.; Malik, S. An inventive method for eco-efficient operation of home energy management systems. *Energies* **2018**, *11*, 3091.
12. Huang, Y.; Wang, L.; Guo, W.; Kang, Q.; Wu, Q. Chance constrained optimization in a home energy management system. *IEEE Trans. Smart Grid* **2018**, *9*, 252–260. [CrossRef]
13. Ogwumike, C.; Short, M.; Abugchem, F. Heuristic Optimization of Consumer Electricity Costs using a Generic Cost Model. *Energies* **2016**, *9*, 6. [CrossRef]
14. Rasheed, M.B.; Javaid, N.; Ahmad, A.; Khan, Z.A.; Qasim, U.; Alrajeh, N. An efficient power scheduling scheme for residential load management in smart homes. *Appl. Sci.* **2015**, *5*, 1134–1163. [CrossRef]
15. Tuaimah, F.M.; Abd, Y.N.; Hameed, F.A. Ant Colony Optimization based Optimal Power Flow Analysis for the Iraqi Super High Voltage Grid. *Int. J. Comput. Appl.* **2013**, *67*, 13–18.
16. Logenthiran, T.; Srinivasan, D.; Vanessa, K.W.M. Demand side management of smart grid: Load shifting and incentives. *J. Renew. Sustain. Energy* **2014**, *6*, 033136. [CrossRef]
17. Islam, S.M.; Das, S.; Ghosh, S.; Roy, S.; Suganthan, P.N. An adaptive differential evolution algorithm with novel mutation and crossover strategies for global numerical optimization. *IEEE Trans. Syst. Man Cybern. Part B* **2012**, *42*, 482–500. [CrossRef] [PubMed]
18. Li, H.; Zhang, Q. Multiobjective optimization problems with complicated Pareto sets, MOEA/D and NSGA-II. *IEEE Trans. Evol. Comput.* **2009**, *13*, 284–302. [CrossRef]
19. Setlhaolo, D.; Xia, X. Optimal scheduling of household appliances incorporating appliance coordination. *Energy Proc.* **2014**, *61*, 198–202. [CrossRef]
20. Osório, G.J.; Matias, J.C.O.; Catalão, J.P.S. Electricity prices forecasting by a hybrid evolutionary-adaptive methodology. *Energy Convers. Manag.* **2014**, *80*, 363–373. [CrossRef]
21. Shayeghi, H.; Ghasemi, A.; Moradzadeh, M.; Nooshyar, M. Simultaneous day-ahead forecasting of electricity price and load in smart grids. *Energy Convers. Manag.* **2015**, *95*, 371–384. [CrossRef]
22. Derakhshan, G.; Shayanfar, H.A.; Kazemi, A. The optimization of demand response programs in smart grids. *Energy Policy* **2016**, *94*, 295–306. [CrossRef]
23. Soares, J.; Ghazvini, M.A.F.; Vale, Z.; de Moura Oliveira, P.B. A multi-objective model for the day-ahead energy resource scheduling of a smart grid with high penetration of sensitive loads. *Appl. Energy* **2016**, *162*, 1074–1088. [CrossRef]
24. Ghasemi, A.; Shayeghi, H.; Moradzadeh, M.; Nooshyar, M. A Novel hybrid algorithm for electricity price and load forecasting in smart grids with demand-side management. *Appl. Energy* **2016**, *177*, 40–59. [CrossRef]
25. Jayabarathi, T.; Raghunathan, T.; Adarsh, B.R.; Suganthan, P.N. Economic dispatch using hybrid grey wolf optimizer. *Energy* **2016**, *111*, 630–641. [CrossRef]
26. Asif, K.; Javaid, N.; Khan, M.I. Time and device based priority induced comfort management in smart home within the consumer budget limitation. *Sustain. Cities Soc.* **2018**, *41*, 538-555.
27. Geem, Z.W.; Kim, J.H.; Loganathan, G.V. A new heuristic optimization algorithm: Harmony 578 search. *Simulation* **2001**, *76*, 60–68. [CrossRef]
28. Storn, R.; Price, K. *Differential Evolution—A Simple and Efficient Adaptive Scheme for Global 580 Optimization over Continuous Spaces*; International Computer Science Institute: Berkeley, CA, USA, 1995.
29. Vardakas, J.S.; Zorba, N.; Verikoukis, C.V. Performance evaluation of power demand scheduling scenarios in a smart grid environment. *Appl. Energy* **2015**, *142*, 164–178. [CrossRef]

30. Awais, M.; Javaid, N.; Ullah, I.; Abdul, W.; Almogren, A.; Alamri, A. An intelligent hybrid heuristic scheme for smart metering based demand side management in smart homes. *Energies* **2017**, *10*, 1258.
31. Yuce, B.; Rezgui, Y.; Mourshed, M. ANN–GA smart appliance scheduling for optimised energy management in the domestic sector. *Energy Build.* **2016**, *111*, 311–325. [CrossRef]
32. Reka, S.S.; Ramesh, V. A demand response modeling for residential consumers in smart grid environment using game theory based energy scheduling algorithm. *Ain Shams Eng. J.* **2016**, *7*, 835–845. [CrossRef]
33. Erdinc, O. Economic impacts of small-scale own generating and storage units, and electric vehicles under different demand response strategies for smart households. *Appl. Energy* **2014**, *126*, 142–150. [CrossRef]
34. Agnetis, A.; de Pascale, G.; Detti, P.; Vicino, A. Load scheduling for household energy consumption optimization. *IEEE Trans. Smart Grid* **2013**, *4*, 2364–2373. [CrossRef]
35. Setlhaolo, D.; Xia, X.; Zhang, J. Optimal scheduling of household appliances for demand response. *Electr. Power Syst. Res.* **2014**, *116*, 24–28. [CrossRef]
36. Pradhan, M.; Roy, P.K.; Pal, T. Grey wolf optimization applied to economic load dispatch problems. *Int. J. Electr. Power Energy Syst.* **2016**, *83*, 325–334. [CrossRef]
37. Vardakas, J.S.; Zorba, N.; Verikoukis, C.V. Power demand control scenarios for smart grid applications with finite number of appliances. *Appl. Energy* **2016**, *162*, 83–98. [CrossRef]
38. Ogunjuyigbe, A.S.O.; Ayodele, T.R.; Akinola, O.A. User satisfaction-induced demand side load management in residential buildings with user budget constraint. *Appl. Energy* **2017**, *187*, 352–366. [CrossRef]
39. Javaid, N.; Ullah, I.; Akbar, M.; Iqbal, Z.; Khan, F.A.; Alrajeh, N.; Alabed, M.S. An intelligent load management system with renewable energy integration for smart homes. *IEEE Access* **2017**, *5*, 13587–13600. [CrossRef]
40. Yao, E.; Samadi, P.; Wong, V.W.S.; Schober, R. Residential demand side management under high penetration of rooftop photovoltaic units. *IEEE Trans. Smart Grid* **2016**, *7*, 1597–1608. [CrossRef]
41. Nadeem, J.; Hafeez, G.; Iqbal, S.; Alrajeh, N.; Alabed, M.S.; Guizani, M. Energy efficient integration of renewable energy sources in the smart grid for demand side management. *IEEE Access* **2018**, *6*, 77077–77096.
42. Ghulam, H.; Javaid, N.; Iqbal, S.; Khan, F. Optimal residential load scheduling under utility and rooftop photovoltaic units. *Energies* **2018**, *11*, 611.
43. Ashfaq, A.; Javaid, N.; Guizani, M.; Alrajeh, N.; Khan, Z.A. An accurate and fast converging short-term load forecasting model for industrial applications in a smart grid. *IEEE Trans. Ind. Inform.* **2017**, *13*, 2587–2596.
44. Muqaddas, N.; Iqbal, Z.; Javaid, N.; Khan, Z.; Abdul, W.; Almogren, A.; Alamri, A. Efficient power scheduling in smart homes using hybrid grey wolf differential evolution optimization technique with real time and critical peak pricing schemes. *Energies* **2018**, *11*, 384.
45. Singh, N.; Singh, S.B. Hybrid Algorithm of Particle Swarm Optimization and Grey Wolf Optimizer for Improving Convergence Performance. *J. Appl. Math.* **2017**, *2017*, 2030489. [CrossRef]
46. Waterloo North Hydro. Available online: https://www.wnhydro.com/en/your-home/time-of-use-rates.asp (accessed on 9 January 2019).
47. Muralitharan, K.; Sakthivel, R.; Shi, Y. Multiobjective optimization technique for demand side management with load balancing approach in smart grid. *Neurocomputing* **2016**, *177*, 110–119. [CrossRef]

5

Wind Energy Generation Assessment at Specific Sites in a Peninsula in Malaysia Based on Reliability Indices

Athraa Ali Kadhem [1,*], Noor Izzri Abdul Wahab [2] and Ahmed N. Abdalla [3]

1 Center for Advanced Power and Energy Research, Faculty of Engineering, University Putra Malaysia, Selangor 43400, Malaysia
2 Advanced Lightning, Power and Energy Research, Faculty of Engineering, University Putra Malaysia, Selangor 43400, Malaysia
3 Faculty of Electronics Information Engineering, Huaiyin Institute of Technology, Huai'an 223003, China
* Correspondence: athraaonoz2007@yahoo.com

Abstract: This paper presents a statistical analysis of wind speed data that can be extremely useful for installing a wind generation as a stand-alone system. The main objective is to define the wind power capacity's contribution to the adequacy of generation systems for the purpose of selecting wind farm locations at specific sites in Malaysia. The combined Sequential Monte Carlo simulation (SMCS) technique and the Weibull distribution models are employed to demonstrate the impact of wind power in power system reliability. To study this, the Roy Billinton Test System (RBTS) is considered and tested using wind data from two sites in Peninsular Malaysia, Mersing and Kuala Terengganu, and one site, Kudat, in Sabah. The results showed that Mersing and Kudat were best suitable for wind sites. In addition, the reliability indices are compared prior to the addition of the two wind farms to the considered RBTS system. The results reveal that the reliability indices are slightly improved for the RBTS system with wind power generation from both the potential sites.

Keywords: reliability indices; wind farms; Sequential Monte Carlo Simulation; Malaysia

---

1. Introduction

Recent environmental impacts and the depletion of fossil fuel reserves are the main concerns that have stimulated the integration of renewable energy power plants using solar power, wind power, biomass, biogas, etc. as alternative sources of electrical generation. This has inspired global concerns in energy balance, sustainability, security, and environmental preservation [1].

Wind energy is non-depletable, free, environmentally friendly, and almost available globally [2]. It is intermittent, though very reliable from a long-term energy policy viewpoint [3]. In the measure of adequacy, wind energy is regarded as a better choice compared with other energies.

Electric power systems continue to witness the penetration of high-level wind power into the system as a global phenomenon [4], due to the problems associated with power system planning and operation. This makes the assessment of wind power generation system capacities, and their impacts on reliability in the system by appropriate planning, in line with their power utilization and environmental benefits. Thus, high penetration of intermittent wind energy resources into the electric power system requires the need to investigate the system reliability while adding a large amount of varying wind power generation to the system [5].

Owing to the industrial development and growth in the economy, an increase in the demand for electricity is one of the major challenges faced by both developed and developing countries like Malaysia. This has precipitated the Malaysian electrical utility to integrate wind generation based renewable energy into the grid. Many studies have been carried out by researchers to identify the potential location of wind energy systems in Malaysia. This process has been encouraged by both public and private institutions, with the aim of producing green energy [6,7]. In addition, the extraction of power from wind energy is optimized, even in location with average wind speed, by the proper design of wind turbine models that can effectively trap power due to the advancement in technologies [8].

In general, Malaysia experiences low wind speeds, but some particular regions experience strong winds in specific periods of the year [9]. Locations like Mersing experience higher wind speed variations throughout the year, with average wind speeds ranging from 2 m/s to 5 m/s [10]. According to the literature, the wind in Malaysia could be able to generate a great quantity of electric energy despite its lower average wind speeds, especially at the eastern coastal areas or its remote islands [11]. Researchers in [12] applied the Weibull function to investigate the characteristics of wind speed and subsequently evaluated the wind energy generation potential at Chuping and Kangar in Perlis, Malaysia. Furthermore, small capacity wind turbine plants (5–100 kW) have been installed by the Ministry of Rural and Regional development in Sabah and Sarawak [11]. Researchers in [13] stated that ten units of wind turbines with three different rated powers (6, 10, and 15 kW) were used in energy calculation for the area in the north part of Kudat. As the wind turbine units are principally dependent on wind velocity and location, wind speed forecast is essential for siting a new wind generating turbine in a prospective location [14], as the study in Kudat location reveals. Moreover, another study was performed by a research group of the University of Malaysia (UM) using the Weibull distribution function for the analyses of wind energy potential at the sites in Kudat and Labuan in the Sabah region in Malaysia [6]. The outcome of this research demonstrated that Kudat and Labuan are suitable for sitting small-scale wind generating units [15].

The question to ask here is whether it is probable to harness small-scale wind generating units at selected locations in Malaysia for the purpose of electricity generation. So far, studies on wind power characteristics in Malaysia are limited and wind speed depends on geographical and meteorological factors. This study discusses the effect of potential wind power from various locations in Malaysia for adequately reliable power systems. Analysis of the wind speed data characteristics and wind power potential assessment at three given locations in Malaysia was done. The main objective of the paper is to examine the capacity contribution of wind power in generating system adequacy and its impact on generation system reliability. The Sequential Monte Carlo simulation (SMCS) technique and Weibull models are employed to demonstrate the impact of wind power in power system reliability. Also, the results presented in the paper could serve as preliminary data for the establishment of a wind energy map for Malaysia.

This paper is structured in six sections. The introduction includes a brief introduction of the concept for the wind energy potential in Malaysia. The next section describes related work adapted to enable estimation of the wind power potential of the region. Section 3 shows the fundamental reliability indices evaluated in this work, which are used by assessment policy makers to exploit the wind power potential of the region. Section 4 describes the wind speed data analysis at specific sites in Malaysia. Section 5 shows the obtained results of the simulation in the case study, which are also discussed. Finally, Section 6 summarizes the main conclusions of the study.

2. Related Work

2.1. Weibull Distribution for the Estimation of Wind Power and Energy Density

The Weibull distribution is the most well recognized mathematical description of wind speed frequency distribution. The value of the scale parameter c of the Weibull distribution is close to the mean wind speed in actual wind speed data, and because of that, the Weibull distribution is a reasonable fit for the data. Consequently, using the two parameters (shape parameter k and scale parameter c), the Weibull distribution can be used with acceptable accuracy to present the wind speed frequency distribution and to predict wind power output from wind energy conversion system (WECS) [16].

Many numerical methods are employed to estimate the values of the shape parameter k and scale c. The Empirical Method (EM) is used in this paper for calculating the Weibull parameters. The EM can be calculated by employing mean wind speed and the standard deviation, where the Weibull parameters c and k are given by the following equations [17].

$$k = \left(\frac{\sigma}{\overline{\mathrm{v}}}\right)^{-1.089} \tag{1}$$

$$c = \frac{\overline{\mathrm{v}}}{\Gamma\left(1 + \frac{1}{k}\right)} \tag{2}$$

where σ is standard deviation, $\overline{\mathrm{v}}$ is the mean wind speed, Γ is the gamma function, and k can be determined easily from the values of σ and $\overline{\mathrm{v}}$, which are computed from the wind speed data set provided with the following formulation [18].

$$\overline{\mathrm{v}} = \frac{1}{n}\sum_{i=1}^{n} v_i \tag{3}$$

$$\sigma = \left[\frac{1}{n-1}\sum_{i=1}^{n}\left(v_i - \overline{\mathrm{v}}\right)^2\right]^{\frac{1}{2}} \tag{4}$$

where $\overline{\mathrm{v}}$ is the mean wind speed (m/s), n is the number of measured data, v_i is wind speed of the observed data in the form of time series of wind speed (m/s), and σ is standard deviation. Once k is obtained from the solution of the above numerical expression (1), the scale factor c can be calculated by the above Equation (2).

Wind power density is a beneficial way of evaluating wind source availability at a potential height. It indicates the quantity of energy that can be used for conversion by a wind turbine [19]. The power that is available in the wind flowing at mean speed, $\overline{\mathrm{v}}$, through a wind rotor blade with sweep area, An (m^2), at any particular site can be projected as

$$P(v) = \frac{1}{2}\rho A\left(\overline{\mathrm{v}}\right)^3.$$

The monthly or annual mean wind power density per unit area of any site on the basis of a Weibull probability density function can be displayed in [20] as follows:

$$P_D(w) = \frac{p(v)}{A} = \frac{1}{2}\rho c^3\left(1 + \frac{3}{k}\right) \tag{5}$$

where $p(v)$ is the wind power (Watts), $P_D(w)$ is the mean wind power density (Watts/m^2), ρ is the air density at the site (1.225 kg/m^3), A is the sweep area of the rotor blades (m^2), and Γ (x) is the gamma function.

The extractible mean energy density over a time period (T) is calculated as

$$E_D = \frac{1}{2}\rho c^3\Gamma\left(1 + \frac{3}{k}\right)T \tag{6}$$

where the time period (T) is expressed a daily, monthly or annual.

2.2. Estimation of Wind Turbine Output Power and Capacity Factor

The performance of how a wind machine located in a site performs can be assessed as mean power output $P_{e,ave}$ over a specific time frame and capacity factor, C_f, of the wind machine. $P_{e,ave}$ determines the total energy production and total income, whereas C_f is a ratio of the mean power output to the rated electrical power P_{rated} of the chosen wind turbine model [21]. Depending on the Weibull distribution parameters, the $P_{e,ave}$ and capacity factor C_f of a wind machine are computed according to the following equations.

$$P_{e,ave} = P_{rated}\left[\frac{e^{-\left(\frac{V_c}{c}\right)^k} - e^{-\left(\frac{V_r}{c}\right)^k}}{\left(\frac{V_r}{c}\right)^k - \left(\frac{V_c}{c}\right)^k}\right] - e^{-\left(\frac{V_c}{c}\right)^k} \quad (7)$$

$$C_f = \frac{P_{e,ave}}{P_{rated}} \quad (8)$$

where V_c is cut-in wind speed and V_r is the rated wind speed of the wind turbine generator (WTG). For an economical and viable investment in wind power, it is advisable that the capacity factor should exceed 25% and be maintained in the range of 25–45% [22].

2.3. Extrapolation of Wind Speed at Different Heights

Indirect wind speed estimation methods consist of measuring wind speed at a lower height and applying an extrapolation model to estimate the wind speed characterization at different elevations. The most commonly used model is the power law [23].

Wind speed increases significantly with the height above ground level, depending on the roughness of the terrain. Therefore, correct wind speed measurements must consider the hub height *(H)* for the WTG and the roughness of the terrain of the wind site. If measurements are difficult at high elevations, the standard wind speed height extrapolation formula, as in the power law Equation (9) [24], can be used to estimate wind speed at high elevations by using wind speed measured at a lower reference elevation, typically 10 m [25].

$$v = v_o\left[\frac{H}{H_O}\right]^n \quad (9)$$

where v is the wind speed estimated at desired height, H; v_o is the wind speed reference hub height H_o, and (n) is the ground surface friction coefficient. The exponent (n) is dependent on factors such as surface roughness and atmospheric stability. Numerically, it ranges from (0.05–0.5) [26]. The normal value of ground surface for every station is approximated 1/7 or 0.143, as suggested by [27] for neutral stability conditions.

3. Reliability Assessment for Generation Systems

3.1. Fundamental Reliability Indices

The Load and generation models are conjoined to produce the risk model of the system. Indices that evaluate system reliability and adequacy can be used to forecast the reliability of the power generating system.

The fundamental reliability indices evaluated in this work are adapted to enable the estimation of the reliability level of the power generating systems, comprised of Loss of Load Frequency (LOLF), Loss of Energy Expectation (LOEE), Loss of Load Duration (LOLD), and Loss of Load Expectation (LOLE).

At present, LOLE represents the reliability index of the electrical power systems used in many countries [3]. The standard level of LOLE is one-day-in ten years or less. This does not mean a full day of shortages once every ten years; rather, it refers to the total accumulated time of shortages, which should not exceed one day in ten years. Therefore, the level of LOLE in this study is used as a reliability index of the generation systems.

The combined Sequential Monte Carlo simulation (SMCS) method (or the Monte Carlo simulation method cooperate with Frequency and Duration method) in [28] enables accurate evaluation of reliability indices. To accurately evaluate the reliability assessment for the overall reliability of generating systems adequacy containing wind energy, an SMCS method was used alongside the Weibull distribution model to generate and repeat the wind speed. The Roy Billinton Test System (RBTS) is an essential reliability test system produced by the University of Saskatchewan (Canada) for educational and research purposes. The RBTS has 11 conventional generating units, each having a power capacity ranging of around 5–40 MW, with an installed capacity of 240 MW and a peak load of 185 MW. Figure 1 shows the single line diagram for the RBTS, and the detailed reliability data for the generating units in the test system are shown in Appendix A. The load model is generally represented as chronological Load Duration Curve (LDC), which is used along with different search techniques. The LDC will generate values for each hour, so there will be 8736 individual values recorded for each year. The chronological LDC hourly load model shown in Figure 2 was utilized, and the system peak load is 185 MW. Besides the traditional generators, the wind farm was comprised of 53 identical WTG units with a rated power of 35 kW, each of which was considered in the current study. A peak load of 1% penetrated wind energy in the RBTS system, which has a peak load of 185 MW.

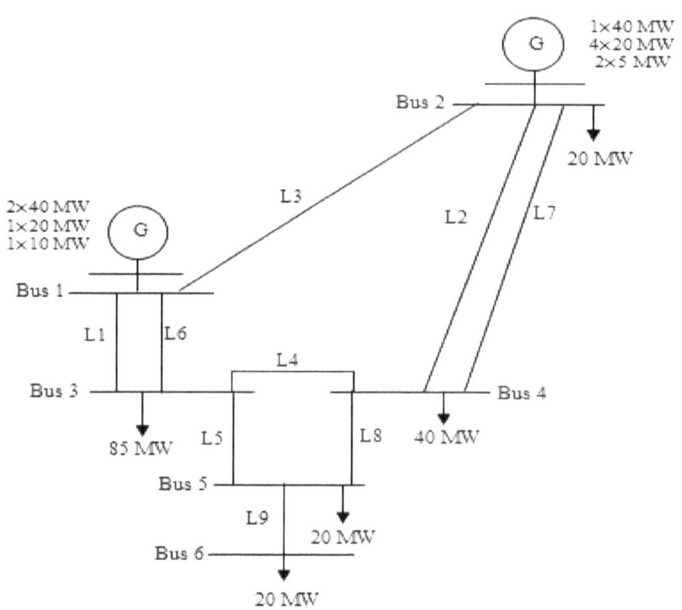

Figure 1. Single line diagrams of the Roy Billinton Test System (RBTS).

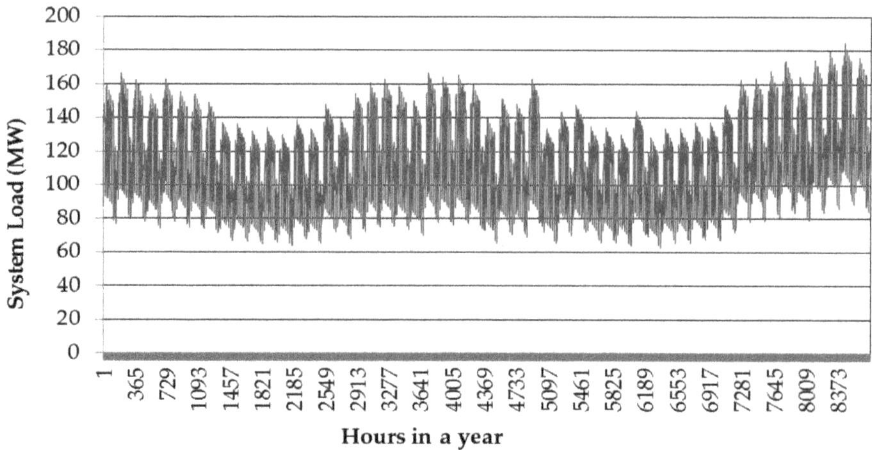

Figure 2. Chronological hourly load (Load Duration Curve (LDC)) model for the RBTS.

3.2. Proposed Methodology

The basic simulation procedures for applying the SMCS with the Weibull model in calculating reliability indices for the electrical power generating systems with wind energy penetration are based on the following steps.

Step 1: The generation of the yearly synthetic wind power time series employs a Weibull model, as follows:

- Set the Weibull distribution parameters k and c.
- Generate a uniformly distributed random number U between (0,1).
- Determined the artificial wind speed v with Equation (10).

$$v = c\left[-\ln(U)^{\frac{1}{k}}\right] \tag{10}$$

- Set the WTG's Vci, Vr, and Vco wind speeds.
- Determine the constants A, Bx, and Cx with the equations below.

$$A = \frac{1}{(V_{ci} - V_r)^2}\left\{V_{ci}\,(V_{ci} + V_r) - 4V_{ci}V_r\left[\frac{V_{ci} + V_r}{2V_r}\right]^3\right\}$$

$$Bx = \frac{1}{(V_{ci} - V_r)^2}\left\{4\,(V_{ci} + V_r)\left[\frac{V_{ci} + V_r}{2V_r}\right]^3 - (3V_{ci} + V_r)\right\}$$

$$Cx = \frac{1}{(V_{ci} - V_r)^2}\left\{2 - 4\left[\frac{V_{ci} + V_r}{2V_r}\right]^3\right\}.$$

- Calculate the WTG output power using Equation (11),

$$P_{WTG} = \begin{cases} 0 & ws < V_{ci} \\ (A + B_x + Cx^2) \times P_r & V_{ci} \le ws < V_r \\ P_r & V_r \le ws < V_{co} \\ 0 & ws > V_{co} \end{cases} \tag{11}$$

where ws = wind speed (m/s), V_{ci} = WTG cut-in speed (m/s), V_{co} = WTG cut-out speed (m/s), V_r = WTG rated speed (m/s), and P_r = WTG rated power output (MW). The constants A, Bx, and Cx have previously been calculated by [3].

Step 2: Create the total available capacity generation by a combination of the synthetic generated wind power time series with a conventional chronological generating system model by employing SMCS, as follows:

- Define the maximum number of years (N) to be simulated and set the simulation time (h), (usually one year) to run with SMCS.
- Generate uniform random numbers for the operation cycle (up-down-up) for each of the conventional units in the system by using the unit's annual MTTR (mean time to repair) and λ (failure rate) values.
- The component's sequential state transition processes within the time of all components are then added to create the sequential system state.
- Define the system capacity by aggregating the available capacities of all system components by combining the operating cycles of generating units and the operating cycles with the WTG available hourly wind at a given load level.

- Superimpose the available system capacity curve on the sequential hourly load curve to obtain the available system margin. A positive margin denotes sufficient system generation to meet the system load whereas a negative margin suggests system load shedding.
- The reliability indices for a number of sample years (N) can be obtained using Equations (12)–(18).

$$\Phi_{LOLE}(s) = \begin{cases} 0 & if \;\; sj \in \;\; s_{success} \\ 1 & if \;\; sj \in \;\; s_{failure} \end{cases} \tag{12}$$

$$\tilde{E}(\Phi_{LOLE}(s)) = \frac{\sum_{i=1}^{N}\left\{\sum_{j=1}^{nj(s)} \Phi_{LOLE}(sji)\right\}}{N} \tag{13}$$

where $i = 1, 2 \ldots N$, N = number of years simulated, $\phi(sji)$ = index function analogous to jth occurrence within the year i, $j = 1, 2 \ldots, nj(s)$, $nj(s)$ is the number of system state occurrences of (sj) in the year i, $sj = s_{success} \cup s_{failure}$ is the set of all possible states (sj) (i.e., the state-space), and the content of two subspaces s_{sucess} of the success state and $s_{failure}$ of the failure states.

$$\Phi_{LOEE}(s) = \begin{cases} 0 & if \;\; sj \in \;\; s_{success} \\ \Delta Pj \times T & if \;\; sj \in \;\; s_{failure} \end{cases} \tag{14}$$

$$\tilde{E}(\Phi_{LOEE}(s)) = \frac{\sum_{i=1}^{N}\left\{\sum_{j=1}^{nj(s)} \Phi_{LOEE}(sji)\right\}}{N} \tag{15}$$

where $\Delta Pj \times T$ is the amount of curtailing energy in the failed state (sji).

$$\Phi_{LOLF}(s) = \begin{cases} 0 & if \;\; sj \in \;\; s_{success} \\ \Delta\lambda j & if \;\; sj \in \;\; s_{failure} \end{cases} \tag{16}$$

$$\tilde{E}(\Phi_{LOLF}(s)) = \frac{\sum_{i=1}^{N}\left\{\sum_{j=1}^{nj(s)} \Phi_{LOLF}(sji)\right\}}{N}. \tag{17}$$

$\Delta\lambda j$, is the sum of the transition rates between sj and all the $s_{success}$ states attained from sj in one transition.

$$LOLD = \frac{LOLE}{LOLF}. \tag{18}$$

- If (N) is equal to the maximum number of years, stop the simulation; otherwise, set (N = N + 1), (h = 0), then return to move 2 and repeat the attempt.

Step 3: Evaluate and update the outcome of the test function for the reliability indices evaluation. The above procedure is detailed in the form of flowchart, as represented in Figure 3.

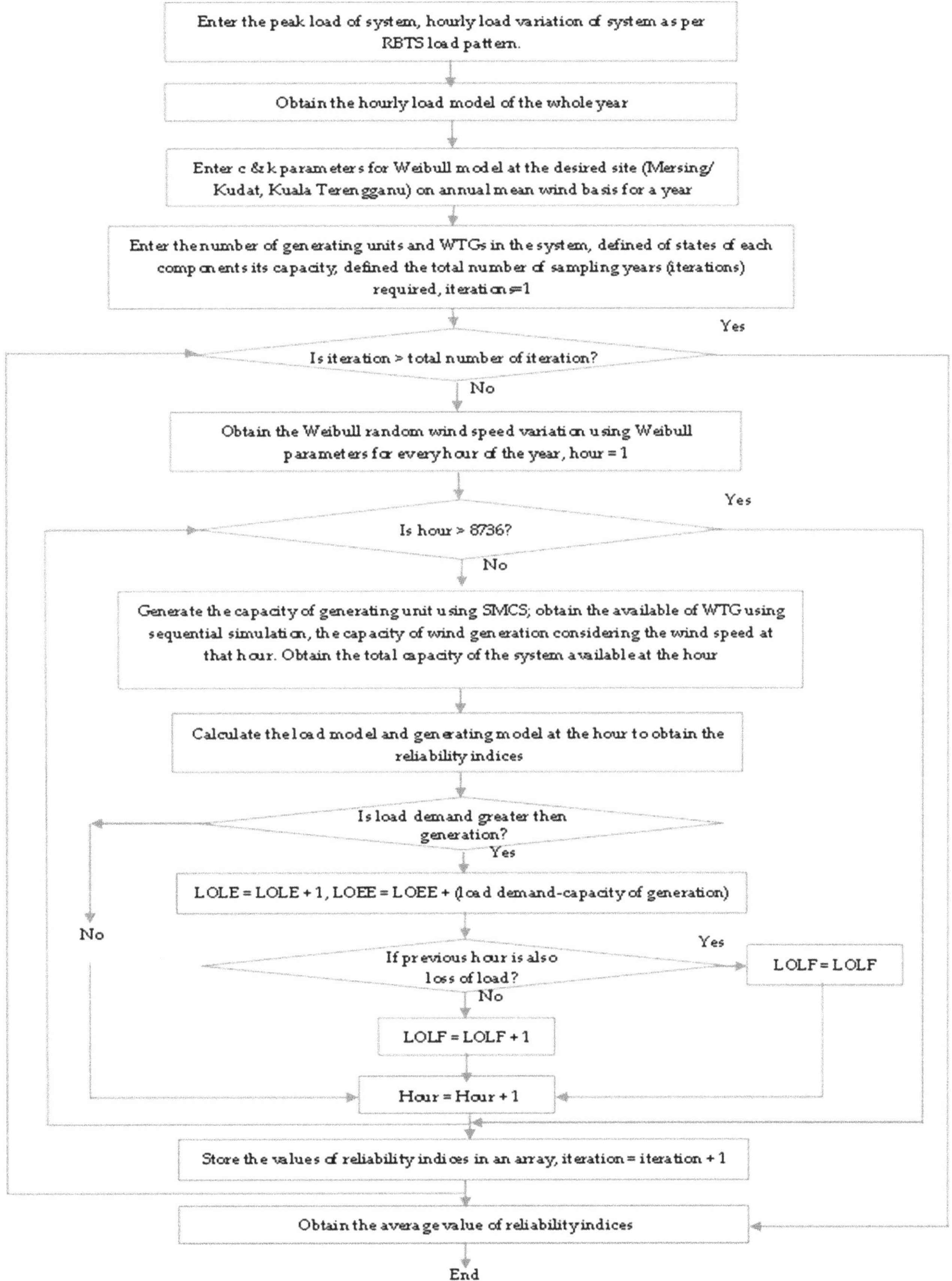

Figure 3. Flow chart showing reliability assessment for a generation system including wind generating sources.

4. Wind Speed Data Analysis at Specific Sites in Malaysia

To estimate the possible potential wind energy site in Malaysia, the analysis, correlation, and prediction of wind data from the location need to be done. It has been often recommended in the literature that making use of the wind data available from meteorological stations increases the vicinity of the proposed candidate site by preliminary estimates of the wind resource potential of the site. Meteorological data that are recorded for long periods need to be extrapolated to obtain an estimation of the wind profile of the site. In this study, wind speed data from Mersing, Kudat, and Kuala Terengganu have been statistically analyzed to propose the wind energy characteristics for these sites. The data for this study were gathered from the Malaysia Meteorological Department (MMD). The data recorded comprise three years of hourly mean surface wind speeds from 2013 to 2015 at three locations in Malaysia. The mean of the wind speed form the simulated process for each hour is calculated based on Weibull parameters. The hourly mean wind speed is then used in the sequential simulation process. Figure 4 shows the locations of MMD stations in Peninsular Malaysia. This map was drawn by using the Arc Graphical Information System (AGIS) software and depicts the strength of the wind speed distribution in Mersing, Kudat, and Kuala Terengganu. The area that showed the highest wind speed value is in red and orange, while other areas show moderate wind speeds. Table 1 presents a description of the selected regions in Malaysia, which consist of the latitude, longitude, and elevation of the anemometer.

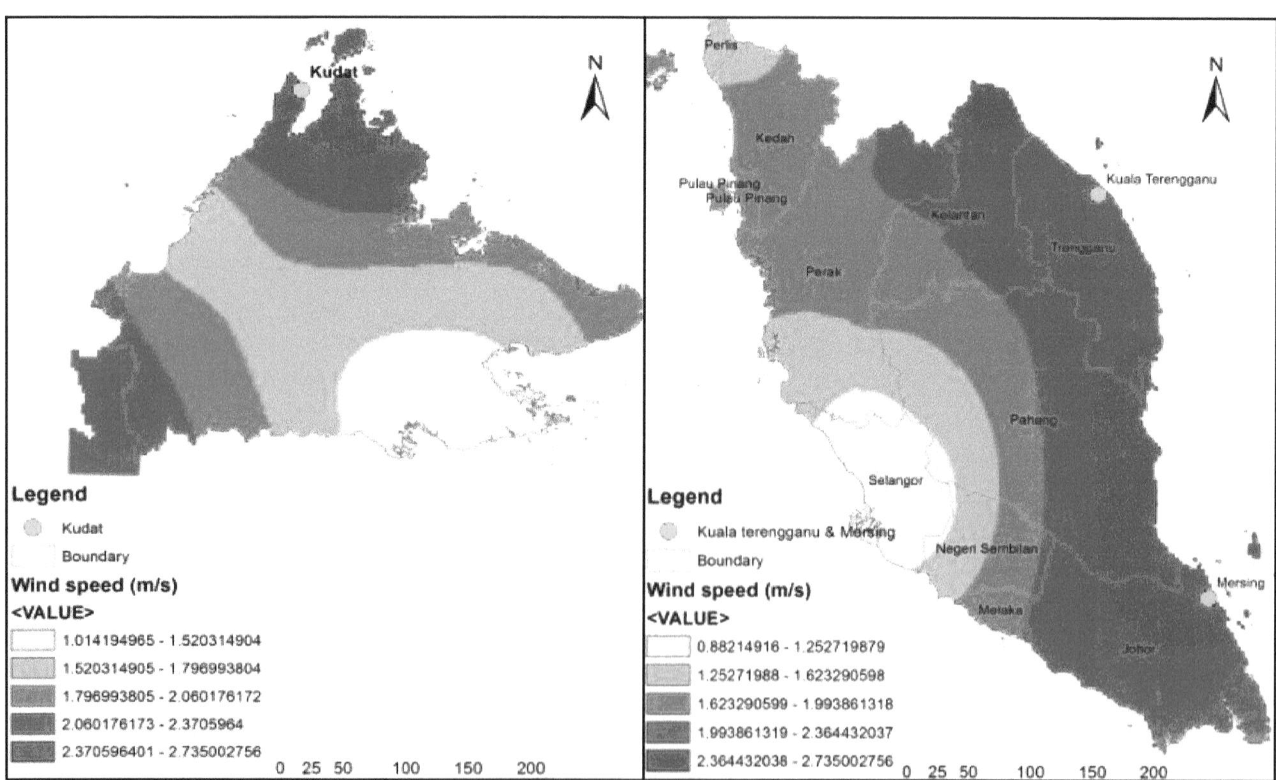

Figure 4. Wind speed distribution maps for station sites used in this study in Malaysia.

Table 1. Description of the wind speed stations at selected regions in Malaysia.

Station	Latitude	Longitude	Altitude (m)
Mersing	2°27′ N	103°50′ E	43.6
Kuala Terengganu	5°23′ N	103°06′ E	5.2
Kudat	6°55′ N	116°50′ E	3.5

4.1. Estimation of Average Wind Speed with Different Height

In this study, the scale parameter (c) of Weibull distribution was applied to test the generated wind speed data from the Weibull model at the Mersing, Kudat, and Kuala Terengganu sites. The monthly average wind speed data for the three sites within the three year period, with a scale parameter, are presented in Figure 5. It is clear from the figures that for the entire three years of the wind speed data and scale parameter c of the Weibull distribution they showed similar variations during one year.

Wind speed is typically measured at standard heights, such as 10 m, but there is the need to obtain wind speed values at a high level in cases where the electricity is generated by the mean power of a wind turbine [29]. The wind power law has been acknowledged to be a beneficial tool and is frequently employed in assessing the wind power where wind speed data at different elevations must be adjusted to a standard height before use. In this study, the exponent n of the power law is set to 0.143 for the Kudat and Kuala Terengganu sites, as suggested by [13]. Meanwhile, for the Mersing site, the exponent n of the power law is set to 0.5, according to the nature of the ground. Wind data taken from the MMD station were measured at a level height of 43.6 m for Mersing. However, wind data of the Kudat and Kuala Terengganu sites require the data to be converted at a height of 10 m above hub height, because the wind turbine always runs at elevations above 10 m height.

Figure 6 shows both monthly and annual mean wind speeds at the sites (Mersing, Kudat, and Kuala Terengganu) for the average years of 2013–2015; these results were extrapolated to a different height. In this study, the wind speed was extrapolated for various heights (mean wind observation station, 60 m, and 100 m) in the wind observation stations at Kudat, Kuala Terengganu, and Mersing (extrapolated). The obtained results are presented in Table 2.

All the tabulated values reveal that the wind speed increases with an increase in elevation. For Mersing, the annual average wind speed was 2.82 m/s, 3.31 m/s, and 4.27 m/s at wind observation station elevations of 60 m, and 100 m, respectively. Furthermore, the annual average wind speeds in Kudat and Kuala Terengganu were 2.45 m/s, 3.68 m/s, and 3.95 m/s; and 2.03 m/s, 2.89 m/s, and 3.10 m/s at wind observation station elevations of 60 m, and 100 m, respectively.

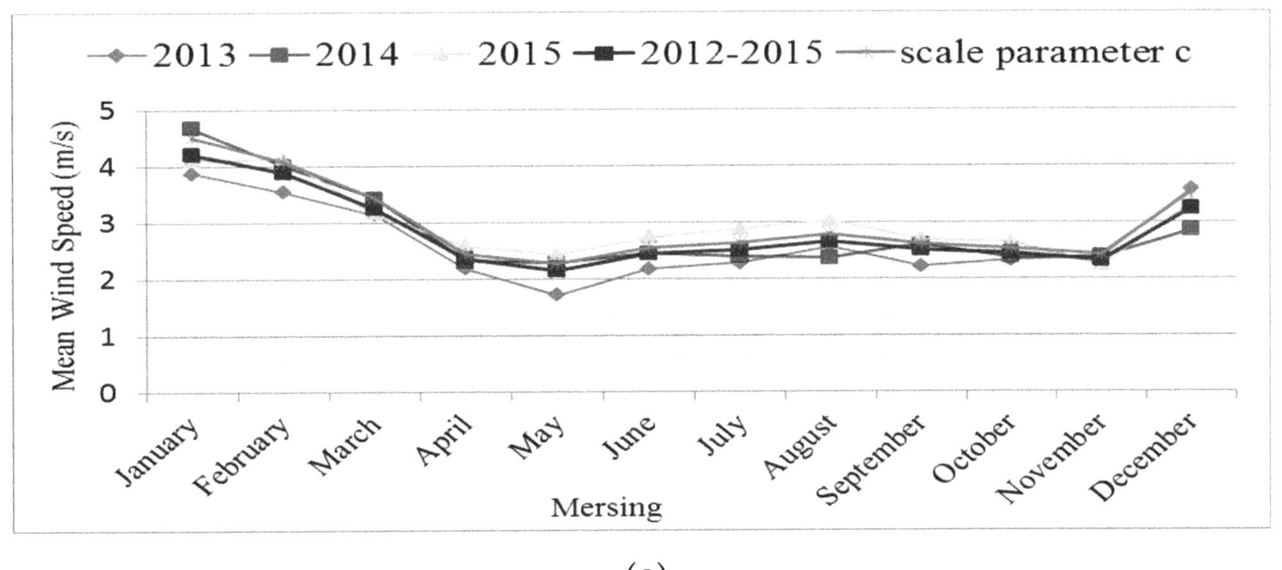

(a)

Figure 5. *Cont.*

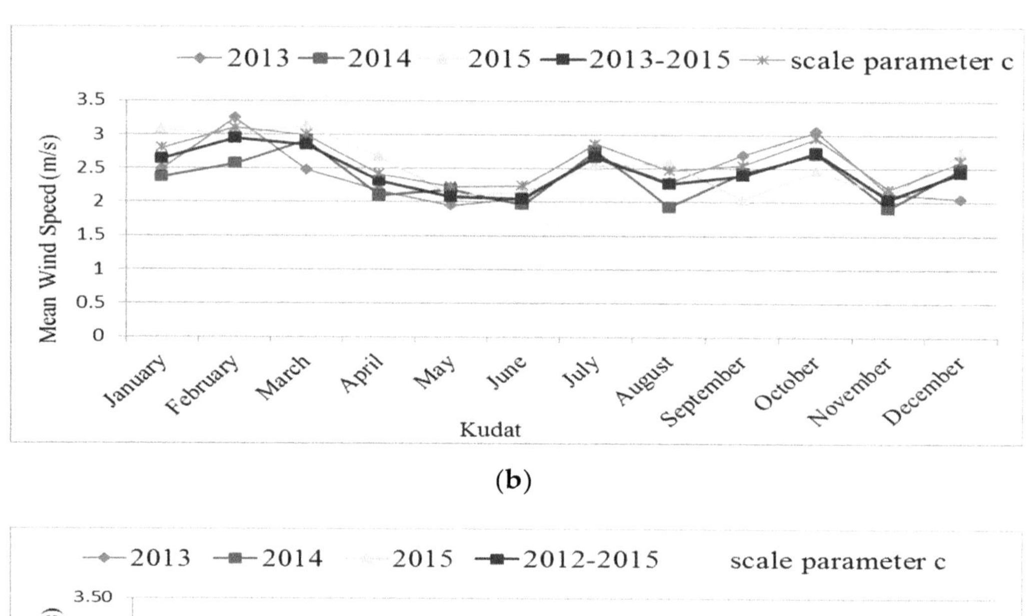

(b)

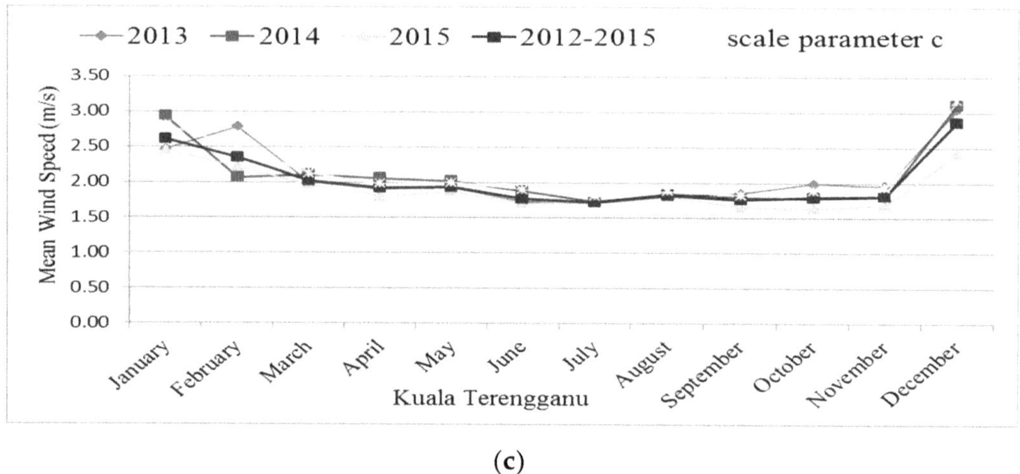

(c)

Figure 5. Comparison between monthly mean wind speed data and the scale parameter (**c**) over 3 years period at sites (**a**) Mersing, (**b**) Kudat, and (**c**) Kuala Terengganu.

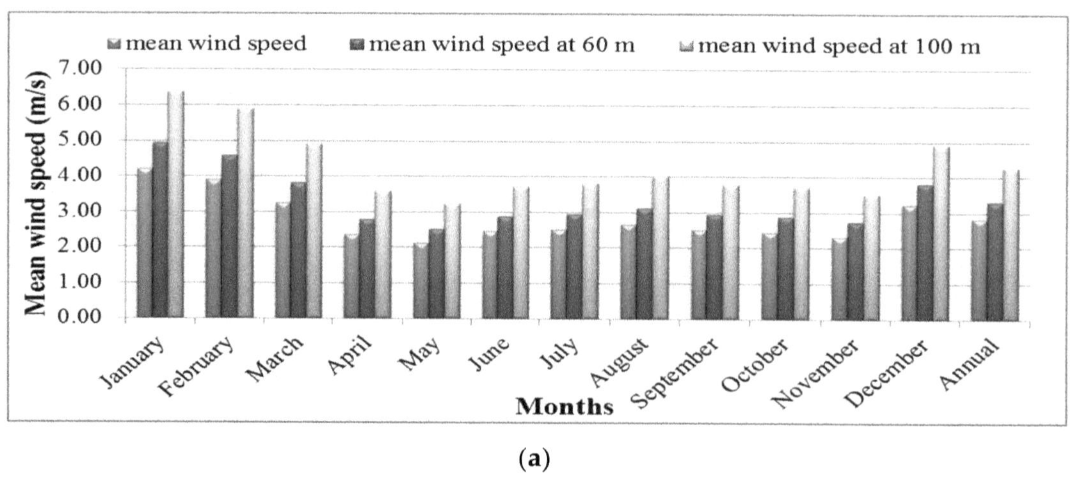

(a)

Figure 6. *Cont.*

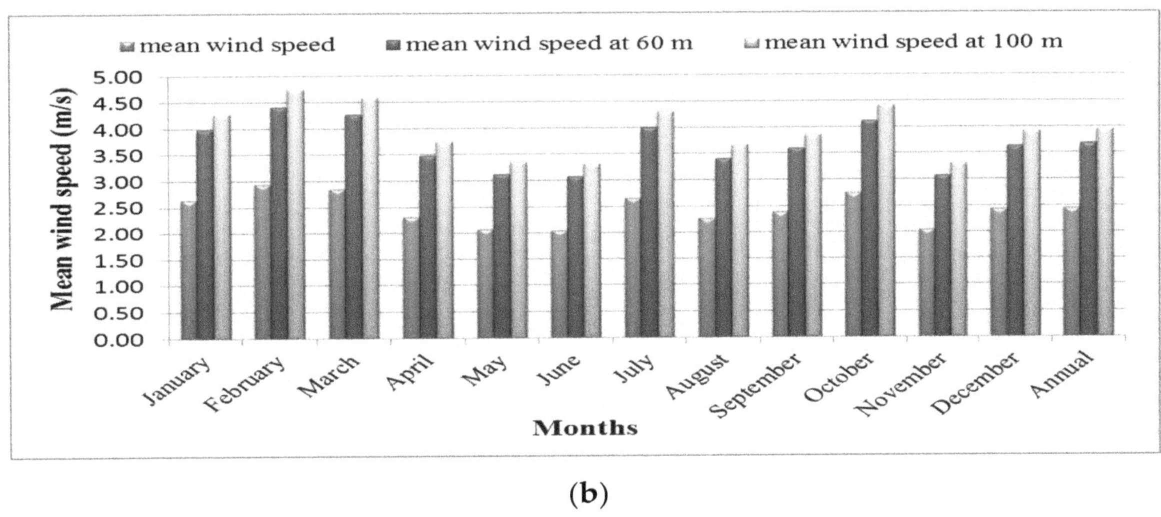

(b)

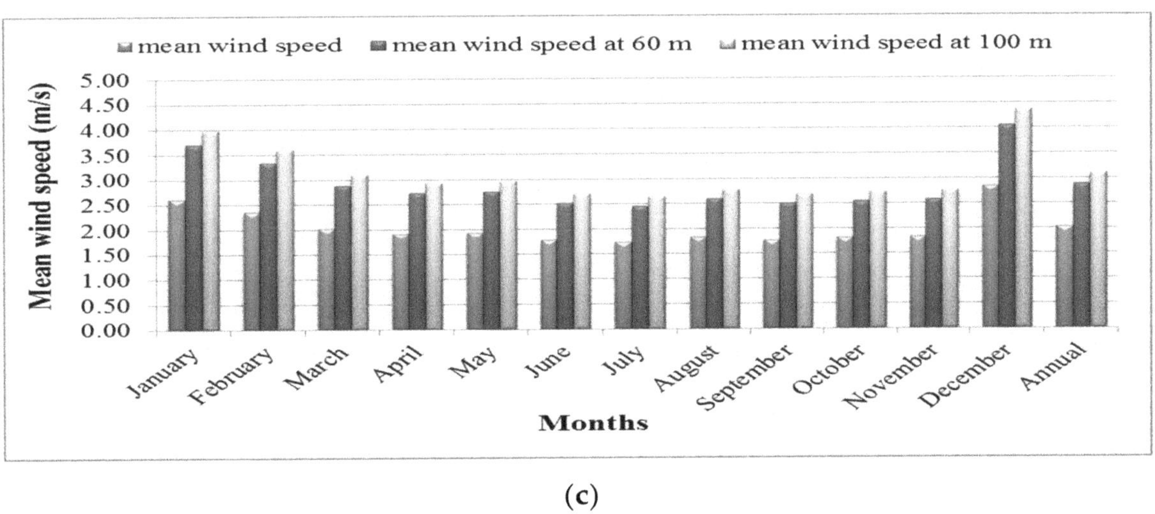

(c)

Figure 6. Monthly and annual mean wind speeds at sites (**a**) Mersing, (**b**) Kudat, and (**c**) Kuala Terengganu.

4.2. Estimation of Wind Power Density and Energy Density at Different Heights

The observed wind speed data at the station were converted to 100 m height wind speed data using Equation (9), and then the converted data were used to determine the wind potential. The scale c and shape k Weibull parameters were estimated using the EM. The wind power and energy density were measured, respectively, by Equations (5) and (6) at heights of (43.6 m) and (100 m) in Mersing, as shown in Table 3. The rest of the calculations for wind power and energy density at heights of 3.5 and 100 m and 5.2 and 100 m in Kudat and Kuala Terengganu, respectively, can be found in Appendix B.

From Table 3, it is observed that the maximum power density from the actual wind speed of Mersing, Kudat, and Kuala Terengganu was found to be 52 W/m^2, 19 W/m^2, and 20 W/m^2, respectively. However, the maximum power density of Mersing, Kudat, and Kuala Terengganu, when the actual wind speed data was converted to 100 m height, was calculated to be 180 W/m^2, 80 W/m^2, and 72 W/m^2, respectively. Here, it is evident that the Mersing site has a higher mean monthly power density compared to Kudat and Kala Terengganu under various heights.

Table 2. Monthly and annual mean wind speed (m/s) in Mersing, Kudat, and Kuala Terengganu at different heights above ground level.

Wind Observation Station	Months/Year												Annual Mean
	Jan.	Feb.	Mar.	Apr.	May	Jun.	Jul.	Aug.	Sep.	Oct.	Nov.	Dec.	
Mersing													
mean wind speed	4.22	3.89	3.24	2.36	2.13	2.45	2.51	2.65	2.50	2.44	2.32	3.24	2.82
60 m	4.94	4.57	3.81	2.77	2.50	2.87	2.95	3.11	2.94	2.87	2.73	3.80	3.31
100 m	6.38	5.89	4.91	3.58	3.23	3.71	3.80	4.01	3.79	3.70	3.52	4.91	4.27
Kudat													
mean wind speed	2.65	2.94	2.84	2.31	2.07	2.05	2.66	2.27	2.39	2.74	2.05	2.43	2.45
60 m	3.97	4.41	4.27	3.47	3.11	3.08	4.00	3.41	3.59	4.12	3.08	3.64	3.68
100 m	4.28	4.75	4.59	3.73	3.35	3.31	4.30	3.66	3.86	4.43	3.31	3.92	3.95
Kuala Terengganu													
mean wind speed	2.61	2.35	2.03	1.92	1.94	1.78	1.73	1.82	1.76	1.80	1.82	2.86	2.03
60 m	3.70	3.34	2.88	2.72	2.75	2.52	2.45	2.59	2.50	2.55	2.58	4.05	2.89
100 m	3.98	3.59	3.10	2.93	2.96	2.71	2.64	2.78	2.69	2.75	2.77	4.36	3.10

Table 3. Wind power and energy density characteristics at heights of 43.6 m and 100 m in Mersing.

	A Height of 43.6 m						A Height of 100 m					
Months/Year	$\bar{v}$	k	c	P_D (w/m²)	Hours	E_D (kWh/m²)	$\bar{v}$	k	c	P_D (w/m²)	Hours	E_D (kWh/m²)
January	4.22	5.37	4.572	52.070	747	38.896	6.38	5.37	6.922	180.702	747	134.9841
February	3.89	5.68	4.209	40.532	675	27.359	5.89	5.68	6.373	140.698	675	94.97137
March	3.24	3.84	3.588	26.214	747	19.582	4.91	3.84	5.432	90.960	747	67.94703
April	2.36	3.89	2.612	10.086	723	7.292	3.58	3.89	3.955	35.014	723	25.31527
May	2.13	3.26	2.381	8.010	747	5.984	3.23	3.26	3.605	27.803	747	20.76847
June	2.45	3.67	2.716	11.488	723	8.306	3.71	3.67	4.112	39.866	723	28.823
July	2.51	3.37	2.798	12.859	747	9.606	3.80	3.37	4.237	44.653	747	33.35577
August	2.65	3.00	2.968	16.014	747	11.962	4.01	3.00	4.494	55.591	747	41.52655
September	2.50	3.35	2.788	12.746	723	9.215	3.79	3.35	4.221	44.232	723	31.97939
October	2.44	3.67	2.710	11.412	747	8.525	3.70	3.67	4.103	39.605	747	29.58467
November	2.32	3.88	2.568	9.590	723	6.934	3.52	3.88	3.887	33.257	723	24.04445
December	3.24	3.45	3.604	27.287	747	20.384	4.91	3.45	5.456	94.674	747	70.72135
Annual	2.82	2.25	3.221	24.370	-	17.790	4.27	2.25	4.877	84.595	-	61.754

The annual mean power density of Mersing, Kudat, and Kuala Terengganu varies between 84.59 W/m^2, 79.28 W/m^2, and 33.36 W/m^2 at a height of 100 m. The annual power density is also less than 100 W/m^2 for all the locations, and, therefore, these locations can be categorized as a class 1 wind energy resource. This wind energy resource class, in general, is inappropriate for large-scale wind turbine applications. Nevertheless, the generation of small-scale wind energy at a turbine height of 100 m [6] is viable. However, for small-scale applications, and in the long-term with the development of wind turbine technology, the use of wind energy continues to hold great promise.

4.3. *Estimation of the Suitable Wind Turbine Units at Malaysia Sites*

The selection of the wind turbine should be made with a rated wind speed that corresponds to the maximum energy wind speed in order to maximize energy output. For the annual energy output, the selected wind turbine will have the maximum capacity factor, defined by the ratio of the actual power generated to the rated power output [30]. The average power output values, $P_{e,ave}$, and C_f, are crucial performance factors of the wind energy conversion system (WECS).

The technical data of six differently sized wind turbines are summarized in Table 4. The summarized information in Table 4 is obtained from [13,19]. The cut-in wind speed, or the speed at which the turbine commences power production, is 2.7 m/s for four of the six turbines, while for the other two turbines, the cut-in wind speed values are 2 and 3.5 m/s, respectively. The cut-out wind speed of 25 m/s applies to all the turbines. Table 4 represents the information pertaining to the rated speed, rated output power, hub height, and rotor diameter of the wind turbines analyzed.

Table 4. The technical data of wind turbines.

Characteristics	P10-20	591672E	P12-25	G-3120	P15-50	P25-100
Rated power (kw)	20	22	25	35	50	100
Hub height (m)	-	30	-	42.7	-	-
Rotor diameter (m)	10	15	12	19.2	15.2	25
Cut-in wind speed (m/s)	2.7	2	2.7	3.5	2.7	2.7
Rated wind speed (m/s)	10	10	10	8	12	10
Cut-off wind speed (m/s)	25	25	25	25	25	25

Depending on the turbine's characteristics in Table 4, and the Weibull parameters derived from applying EM using the Matlab toolbox, the electrical output of the wind turbines can be made available by using the formulation earlier defined in Equation (7).

Knowing the output power of the wind turbines, it is then possible to obtain a computation of the average output power value of each wind turbine. As the capacity factor of a wind turbine is the ratio of its average output power to its rated power, the energy output data are employed in calculating the capacity factor of the wind turbines, which are of sizes 20, 22, 25, 35, 50, and 100 kW. A comparison of the capacity factors computed for various wind turbines at different heights is presented in Figure 7.

From Figure 7, it can be seen that the capacity factor goes up as the hub height increases. Moreover, the capacity factor increases for wind turbines of a size of 35 kW. In Mersing, the maximum capacity factor is achieved as about 23.66% for the Endurance America model of the G-3120 kW wind turbine, whereas in Kuala Terengganu, the lowest capacity factor is achieved as approximately 7.82% for the Endurance America model of the G-3120 kW wind turbine. Kudat, with about 19.21%, ranks second in terms of capacity factors compared to the regions.

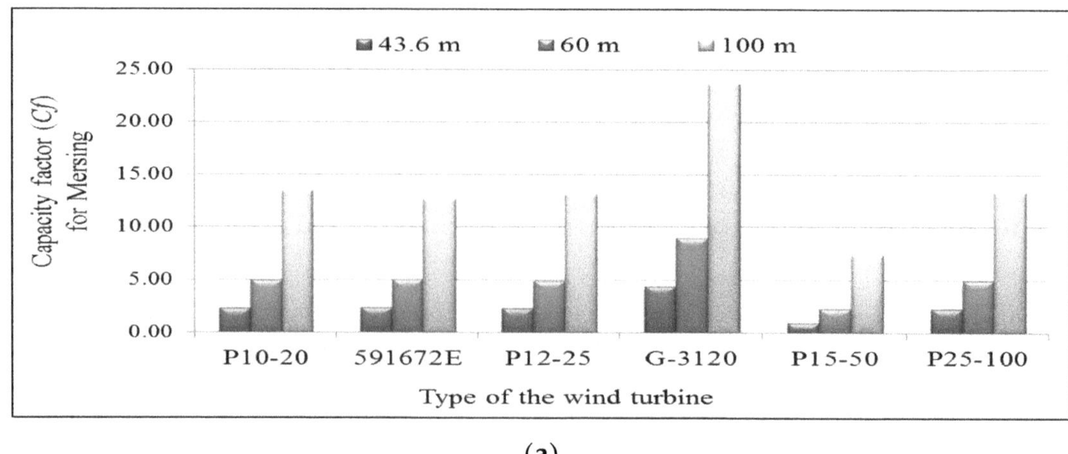

(a)

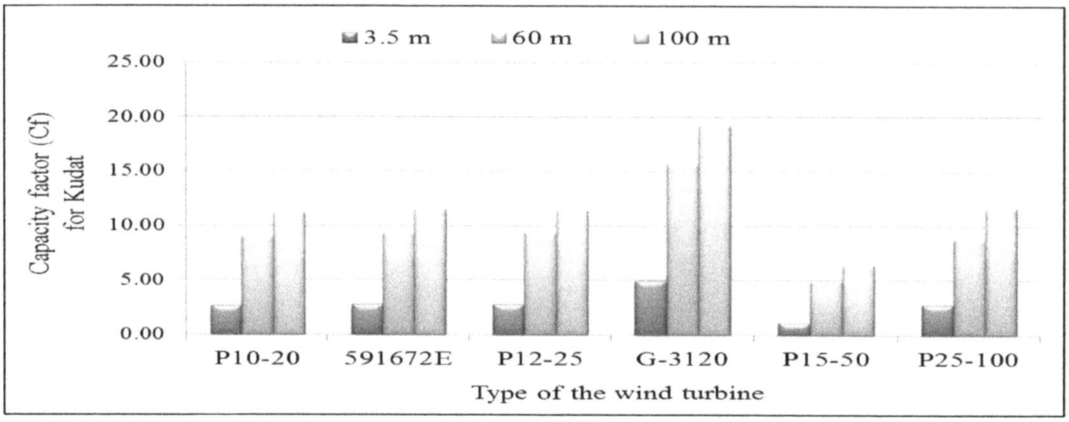

(b)

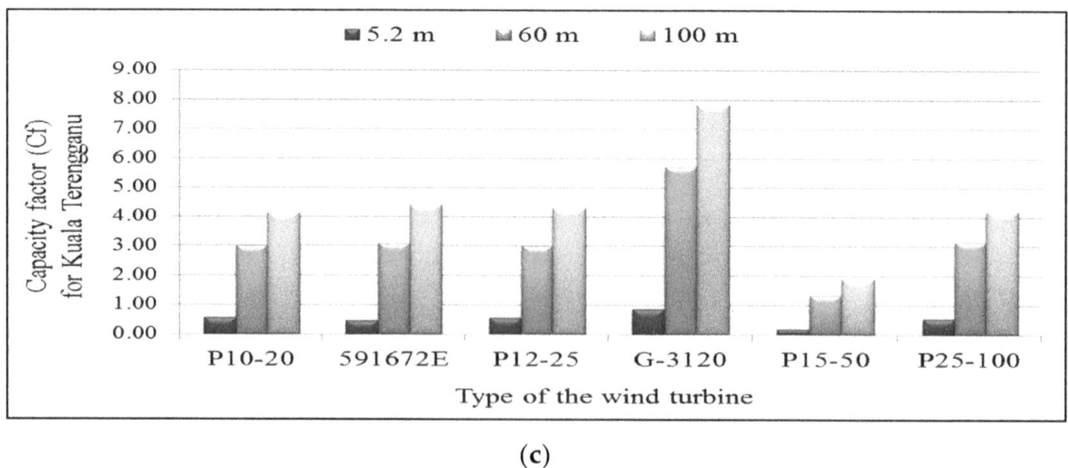

(c)

Figure 7. Comparison of the capacity factors obtained for different wind turbines at various heights for sites (**a**) Mersing, (**b**) Kudat, and (**c**) Kuala Terengganu.

The G-3120 (35 kW) wind turbine has the highest capacity factors of 23.66%, 19.21%, and 7.82%, at suggested heights of 100 m for Mersing, Kudat, and Kuala Terengganu, respectively, among the models considered. Therefore, the reliability analysis was carried out only for Mersing and Kudat, which have high capacity factors, whereas Kuala Terengganu was not considered as it has low capacity factors.

5. Results and Discussion

In this section, the reliability indices evaluation of generating systems for wind power generation using a sequential Monte Carlo simulation (SMCS) is presented. In addition, the strategies for wind farm operation at Malaysian sites (Mersing and Kudat) are presented and compared by assessing the reliability of wind energy generation when adding to the RBTS test system [31].

5.1. Case Studies

As reported in the literature, two wind generating stations suggested at the specific sites in Malaysia, Mersing and Kudatas, have low wind speed and thus require small-scale rated power wind turbines of around 35 kW for installation in two selected locations for reliability analysis.

Reliability analysis using the simulation technique suggested in this paper is applied to the RBTS, which contains the WECS. The hourly wind data obtained from the two locations—Mersing and Kudat—atre used for studying the hourly wind speed of the Weibull model considered for the simulation. Then, the Weibull parameters c and the k are obtained by the empirical method. The obtained values were used to generate hourly wind speed data for deducing the available wind power from the wind turbine generators (WTG) chosen for both of the sites for reliability assessment.

The values of c are around 4.88 and 4.46 m/s, and the values of k for wind speed distribution are 2.25 and 1.84 for Mersing and Kudat at the proposed height of 100 m, respectively; these values were obtained by simulation. The WTG unit that is selected for installation in the farm has the following specifications: Vci = 3.5 m/s, Vr = 8 m/s, and Vco = 25 m/s, and the rated power output of every WTG unit is Pr = 35 kW [19]. Figures 8 and 9 show the simulated output power with 35 kW for each WTG in the sampling year, and the simulation of the farm with output power is 1.85 MW for 53 WTG units in Mersing.

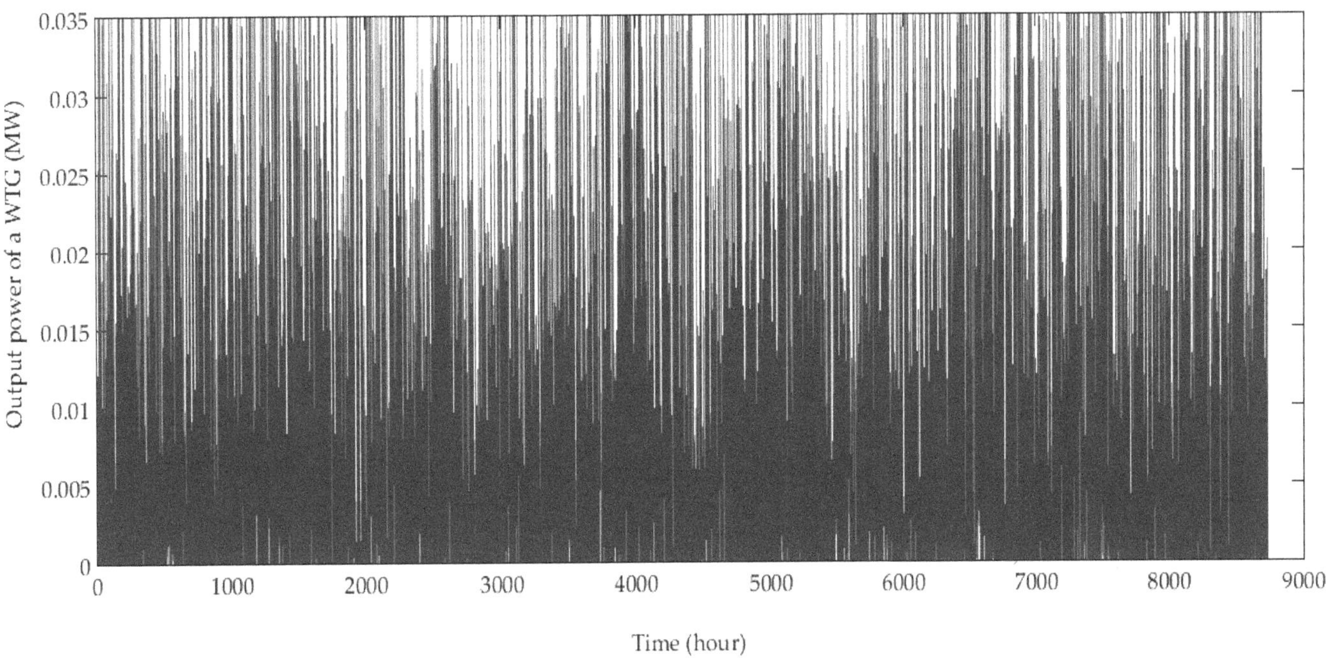

Figure 8. Simulation of the output power from WTG for the sampling year.

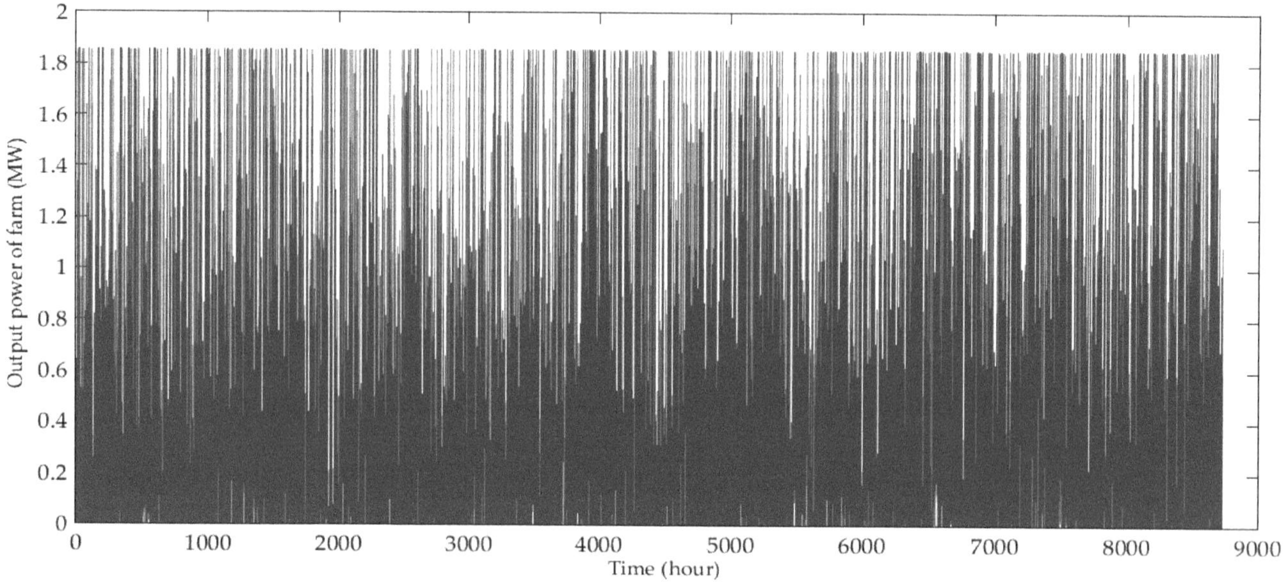

Figure 9. Simulation of the output power from wind farm for the sampling year.

The RBTS was simulated for 600 trials using the SMCS method. The simulation proceeded in chronological order from one hour to the next, repeatedly, using yearly samples until the specified convergence criteria were met. Figure 10 shows the available capacity for the power system containing wind power generation from the wind farm in Mersing during the simulated process for yearly samples and the superimposition of the available capacity with the chronological load model. It can be seen from this state of the system that the available capacity of the power generating system is not sufficient to meet the load demands. Thus, there are some intersections that are seen in the diagram. Figure 11 represents the reliability indices for simulation with (600) sampling years. The values of LOLE, the amount of the LOEE, and the frequency of losing a load during the simulation process are depicted in Figure 11 for wind power in the Mersing site.

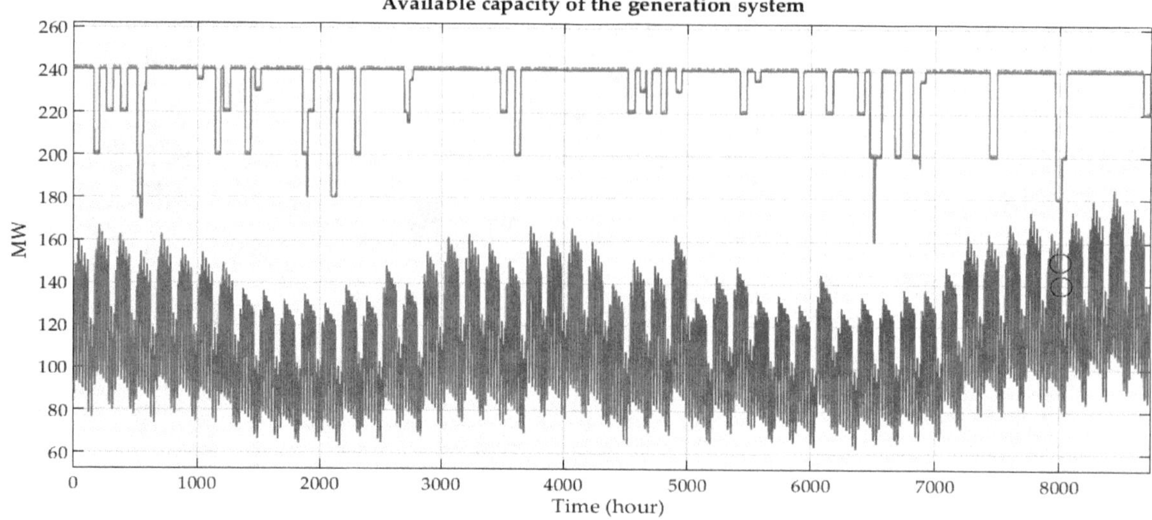

Figure 10. The available capacity of the generation system which is superimposed with the chronologically available load model.

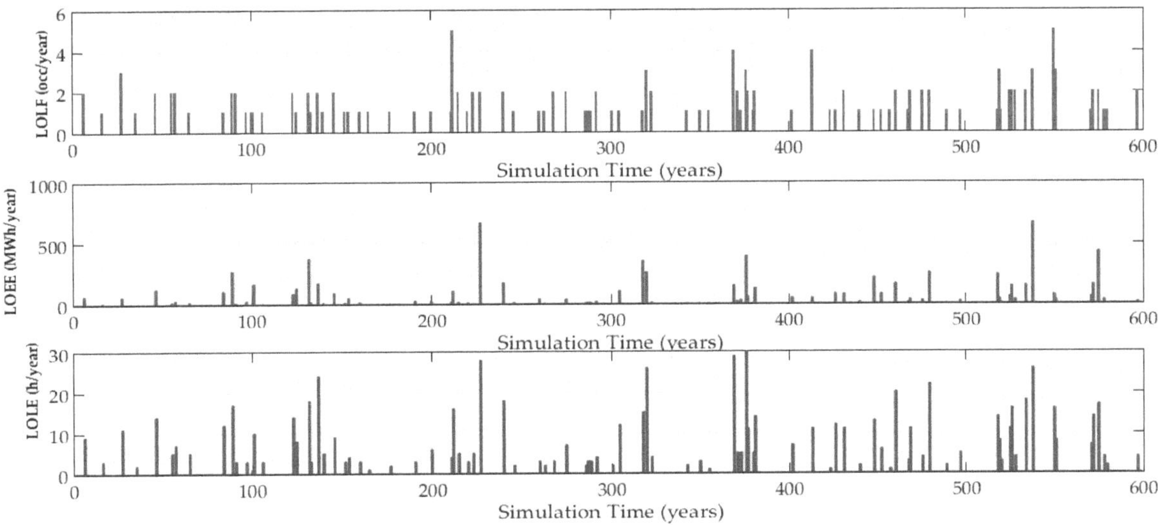

Figure 11. Simulation of reliability indices with (600) sampling years.

5.2. *Calculated Reliability Indices for RBTS Including Wind Power Generation*

To evaluate the contribution of wind energy to the overall reliability of the generating systems, Table 5 compares the reliability indices before and after adding the 53 WTGs and 106 WTGs to the conventional units of RBTS. The results obtained were compared with the results obtained from the SMCS method reported in [32]. The simulation process was terminated after a set number of samples (600 times) had been achieved. The results show that the reliability indices demonstrate a distinguished and slightly improved reliability of RBTS, including wind power from both locations (Mersing and Kudat) by the addition of 1.85 MW and 3.71 MW from the proposed wind farms. The LOLE and LOEE indices are typically employed to gauge the extent of benefits in assessing the wind energy of generating systems. Therefore, after adding the wind generating 1.85 MW to the system, the LOLE index was reduced to 1.115 and 1.131 h/year for Mersing and Kudat, respectively, when compared with the results from the base case, which shows the reliability assessment of the power generation system. Additionally, after adding the wind generating 3.71 MW to the system, the LOLE index wass reduced to 0.987 and 1.128 h/year for Mersing and Kudat, respectively, when compared with the results from the base case, which shows the reliability assessment of the power generation system.

Table 5. Reliability indices at various sites in Malaysia.

Name of Site	Reliability Indices			
	LOLE (hrs/year)	**LOEE (MWh/year)**	**LOLF (occ/year)**	**LOLD (hrs/occ)**
Basic RBTS system without wind generation (published)	1.152	11.78	0.229	4.856
Basic RBTS system without wind generation (computed)	1.161	10.191	0.230	5.05
Basic RBTS system and (53 × 0.035 = 1.85 MW) wind generators at Mersing site	1.115	9.744	0.225	4.944
Basic RBTS system and (106 × 0.035 = 3.71 MW) wind generators at Mersing site	0.987	7.357	0.220	4.486
Basic RBTS system and (53 × 0.035 MW) wind generators at Kudat site	1.131	10.948	0.225	5.012
Basic RBTS system and (106 × 0.035 = 3.71 MW) wind generators at Kudat site	1.128	10.018	0.236	4.779

Loss of Load Expectation (LOLE)/ LOLE (hour/year); Loss of Energy Expectation (LOEE)/ LOEE (MWh/year); Loss of Load Frequency (LOLF)/ LOLF (occurrence/year); Loss of Load Duration (LOLD)/ LOLD (hour/occurrence).

Further, it can be observed that this study was done with a small percentage of peak load reduction at around 1% and a small number of wind turbines, to demonstrate the primary effect of wind energy penetration from selected locations in Malaysia in the reliability of the generation system for the RBTS.

6. Conclusions

In this paper, analyses of the wind speed data characteristics and wind power potential assessment at three given locations in Malaysia were done. In addition, this study tests the effects of the potential wind power from different locations. An SMCS technique is used to show the effects of wind energy for the RBTS test system by a set of reliability indices. The results reveal that the wind power connected to the RBTS test system is only from two locations in Malaysia. Further, the reliability indices are compared prior to and after the addition of the two farms to the considered system. The results show that the reliability indices are slightly improved for RBTS, including wind power from both locations, as suggested. Moreover, the wind resources at specific sites in Malaysia are more suitable for small-scale standalone energy conversion systems and could also be hybrid energy systems.

Recommendations for future studies include extending the statistical analysis model used for different sites in Malaysia to include more relevant factors for wind farms and evaluating their impact on wind power potential for these sites, such as wind speeds at the installation site, types of wind turbine offshore and onshore, and numbers of wind turbines installed, according to the size of the farm.

Author Contributions: This work was part of the Ph.D. research carried out by A.A.K. The research is supervised by N.I.A.W. and A.N.A. The project administration and funding acquisition by University Putra Malaysia.

Appendix A

Table A1. The RBTS generating unit ratings and reliability data.

Units No.	Unit Size (MW)	FOR	MTTF (Hours)	MTTR (Hours)
1	5	0.01	4380	45
2	5	0.01	4380	45
3	10	0.02	2190	45
4	20	0.02	3650	55
5	20	0.02	3650	55
6	20	0.02	3650	55
7	20	0.02	3650	55
8	20	0.03	1752	45
9	40	0.02	2920	60
10	40	0.03	1460	45
11	40	0.03	1460	45

Forced outage rate (FOR); Mean time to failure (MTTF); Mean time to repair (MTTR).

Appendix B

Table A2. Wind power and energy density characteristics at heights of 3.5 m and 100 m in Kudat.

	A Height of 3.5 m						A Height of 100 m					
Months/Year	$\bar{v}$	k	c	P_D (w/m^2)	Hours	E_D (kWh/m^2)	$\bar{v}$	k	c	P_D (w/m^2)	Hours	E_D (kWh/m^2)
January	2.65	3.77	2.931	14.347	747	10.717	4.28	3.77	4.734	60.450	747	45.156
February	2.94	4.00	3.244	19.217	675	12.972	4.75	4.00	5.239	80.946	675	54.639
March	2.84	3.55	3.158	18.213	747	13.605	4.59	3.55	5.100	76.709	747	57.301
April	2.31	2.87	2.594	10.905	723	7.884	3.73	2.87	4.190	45.957	723	33.227
May	2.07	2.48	2.339	8.682	747	6.485	3.35	2.48	3.777	36.555	747	27.307
June	2.05	1.99	2.314	10.142	723	7.333	3.31	1.99	3.738	42.753	723	30.911
July	2.66	2.43	3.004	18.648	747	13.930	4.30	2.43	4.851	78.528	747	58.660
August	2.27	2.12	2.562	12.922	747	9.653	3.66	2.12	4.137	54.406	747	40.641
September	2.39	2.16	2.698	14.835	723	10.725	3.86	2.16	4.356	62.433	723	45.139
October	2.74	2.86	3.077	17.015	747	12.710	4.43	2.86	4.969	76.777	747	57.353
November	2.05	2.80	2.303	7.723	723	5.584	3.31	2.80	3.719	32.524	723	23.515
December	2.43	3.27	2.708	11.772	747	8.794	3.92	3.27	4.373	49.574	747	37.032
Annual	2.45	1.84	2.759	18.819	-	13.738	3.95	1.84	4.456	79.284	-	57.877

Table A3. Wind power and energy density characteristics at height of 5.2 m and 100 m in Kuala Terengganu.

	A Height of 5.2 m						A Height of 100 m					
Months/Year	$\bar{v}$	k	c	P_D (w/m^2)	Hours	E_D (kWh/m^2)	$\bar{v}$	k	c	P_D (w/m^2)	Hours	E_D (kWh/m^2)
January	2.6097	3.24	2.912	14.685	747	10.969	3.98	3.24	4.443	52.157	747	38.961
February	2.3543	3.10	2.632	11.020	675	7.439	3.59	3.10	4.017	39.177	675	26.445
March	2.0300	2.54	2.287	7.991	747	5.969	3.10	2.54	3.490	28.396	747	21.212
April	1.9203	2.63	2.161	6.601	723	4.772	2.93	2.63	3.298	23.463	723	16.964
May	1.9381	3.16	2.165	6.089	747	4.548	2.96	3.16	3.303	21.622	747	16.152
June	1.7760	2.68	1.997	5.154	723	3.726	2.71	2.68	3.048	18.324	723	13.248
July	1.7277	3.14	1.930	4.324	747	3.230	2.64	3.14	2.946	15.378	747	11.487
August	1.8242	3.61	2.024	4.774	747	3.566	2.78	3.61	3.088	16.954	747	12.664
September	1.7642	3.16	1.971	4.594	723	3.322	2.69	3.16	3.007	16.314	723	11.795
October	1.7993	2.81	2.020	5.202	747	3.886	2.75	2.81	3.083	18.496	747	13.816
November	1.8165	2.87	2.038	5.288	723	3.823	2.77	2.87	3.110	18.793	723	13.587
December	2.8554	2.88	3.203	20.496	747	15.310	4.36	2.88	4.887	72.799	747	54.381
Annual	2.0338	2.09	2.293	9.390	-	6.854	3.10	2.09	3.499	33.366	-	24.356

References

1. Ouammi, A.; Dagdougui, H.; Sacile, R.; Mimet, A. Monthly and seasonal of wind energy characteristics at four monitored locationals in Liguria region (Italy). *Renew. Sustain. Energy Rev.* **2010**, *14*, 1959–1968. [CrossRef]
2. Chaiamarit, K.; Nuchprayoon, S. Modeling of renewable energy resources for generation reliability evaluation. *Renew. Sustain. Energy Rev.* **2013**, *26*, 34–41. [CrossRef]
3. Shi, S.; Lo, K.L. An Overview of Wind Energy Development and Associated Power System Reliability Evaluation Methods. In Proceedings of the 2013 48th International Universities Power Engineering Conference (UPEC), Dublin, Ireland, 2–5 September 2013; pp. 1–6.
4. Benidris, M.; Mitra, J. Composite Power System Reliability Assessment Using Maximum Capacity Flow and Directed Binary Particle Swarm Optimization. In Proceedings of the IEEE Conference, Manhattan, KS, USA, 22–24 September 2013; pp. 1–6.
5. Padma, L.M.; Harshavardham, R.P.; Janardhana, N.P. Generation Reliability Evaluation of Wind Energy Penetrated Power System. In Proceedings of the International Conference on High Performance Computing and Applications (ICHPCA), Odisha, India, 22–24 December 2014; pp. 1–4.
6. Islam, M.R.; Saidur, R.; Rahim, N.A. Assessment of wind energy potentiality at Kudat and Labuan, Malaysia using Weibull distribution function. *Energy* **2011**, *36*, 985–992. [CrossRef]
7. Taylor, P.; Khatib, T.; Sopian, K.; Ibrahim, M.Z. Assessment of electricity generation by wind power in nine costal sites in Malaysia. *Int. J. Ambient Energy* **2013**, 37–41. [CrossRef]
8. Gebrelibanos, K.G. *Feasibility Study of Small Scale Standalone Wind Turbine for Urban Area: Case Study: KTH Main Campus*; KTH School of Industrial Engineering and Management, Energy Technology EGI: Stockholm, Sweden, 2013.
9. Siti, M.R.S.; Norizah, M.; Syafrudin, M. The Evaluation of Wind Energy Potential in Peninsular Malaysia. *Int. J. Chem. Environ. Eng.* **2011**, *2*, 284–291.
10. Kadhem, A.A.; Abdul, W.N.I.; Aris, I.; Jasni, J.; Abdalla, A.N. Advanced Wind Speed Prediction Model Based on Combination of Weibull Distribution and Artificial Neural Network. *Energies* **2017**, *10*, 1744. [CrossRef]
11. Borhanazad, H.; Mekhilef, S.; Saidur, R.; Boroumandjazi, G. Potential application of renewable energy for rural electrification in Malaysia. *Renew. Energy* **2013**, *59*, 210–219. [CrossRef]
12. Irwanto, M.; Gomesh, N.; Mamat, M.R.; Yusoff, Y.M. Assessment of wind power generation potential in Perlis, Malaysia. *Renew. Sustain. Energy Rev.* **2014**, *38*, 296–308. [CrossRef]
13. Albani, A.; Ibrahim, M.Z.; Yong, K.H. Wind Energy Investigation in Northern Part of Kudat, Malaysia. *Int. J. Eng. Appl. Sci.* **2013**, *2*, 14–22.
14. Goh, H.H.; Lee, S.W.; Chua, Q.S.; Teo, K.T.K. Wind energy assessment considering wind speed correlation in Malaysia. *Renew. Sustain. Energy Rev.* **2016**, *54*, 1389–1400. [CrossRef]
15. Hwang, G.H.; Lin, N.S.; Ching, K.B.; Wei, L.S. Wind Farm Allocation In Malaysia Based On Multi-Criteria Decision Making Method. In Proceedings of the National Postgraduate Conference (NPC), Seri Iskandar, Malaysia, 19–20 September 2011; pp. 1–6.
16. Chang, T.P. Performance comparison of six numerical methods in estimating Weibull parameters for wind energy application. *Appl. Energy* **2011**, *88*, 272–282. [CrossRef]
17. Kaoga, D.K.; Serge, D.Y.; Raidandi, D.; Djongyang, N. Performance Assessment of Two-parameter Weibull Distribution Methods for Wind Energy Applications in the District of Maroua in Cameroon. *Int. J. Sci. Basic Appl. Res.* **2014**, *17*, 39–59.
18. Kidmo, D.K.; Danwe, R.; Doka, S.Y.; Djongyang, N. Statistical analysis of wind speed distribution based on six Weibull Methods for wind power evaluation in Garoua, Cameroon. *Rev. Des. Energ. Renouv.* **2015**, *18*, 105–125.
19. Adaramola, M.S.; Oyewola, O.M.; Ohunakin, O.S.; Akinnawonu, O.O. Performance evaluation of wind turbines for energy generation in Niger. Sustain. *Energy Technol. Assess.* **2014**, *6*, 75–85. [CrossRef]
20. Ohunakin, O.S.; Adaramola, M.S.; Oyewola, O.M. Wind energy evaluation for electricity generation using WECS in seven selected locations in Nigeria. *Appl. Energy* **2011**, *88*, 3197–3206. [CrossRef]
21. Oyedepo, S.O.; Adaramola, M.S.; Paul, S.S. Analysis of wind speed data and wind energy potential in three selected locations in south-east Nigeria. *Int. J. Energy Environ. Eng.* **2012**, *3*, 1–11. [CrossRef]

22. Anurag, C.; Saini, R.P. Statistical Analysis of Wind Speed Data Using Weibull Distribution Parameters. In Proceedings of the 1st International Conference on Non-Conventional Energy (ICONCE 2014), Kalyani, India, 16–17 January 2014; pp. 160–163.
23. Molina-García, A.; Fernández-Guillamón, A.; Gómez-Lázaro, E.; Honrubia-Escribano, A.; Bueso, M.C. *Vertical Wind Profile Characterization and Identification of Patterns Based on a Shape Clustering Algorithm*; IEEE: New York, NY, USA, 2019; Volume 7, pp. 30890–30904.
24. Ahmed, A.S. Wind energy as a potential generation source at Ras Benas, Egypt. *Renew. Sustain. Energy Rev.* **2010**, *14*, 2167–2173. [CrossRef]
25. Hussain, M.Z.; Roy, D.K.; Khan, S.; Sharma, P.K.; Talukdar, R. Wind Energy Potential at Different Cities of Assam Using Statistical Models. *Int. J. Adv. Res. Innov.* **2018**, *6*, 38–43.
26. Ayodele, T.R.; Ogunjuyigbe, A.S.O.; Amusan, T.O. Wind power utilization assessment and economic analysis of wind turbines across fifteen locations in the six geographical zones of Nigeria. *J. Clean. Prod.* **2016**, *129*, 341–349. [CrossRef]
27. Chandel, S.S.; Ramasamy, P.; Murthy, K.S.R. Wind power potential assessment of 12 locations in western Himalayan region of India. *Renew. Sustain. Energy Rev.* **2014**, *39*, 530–545. [CrossRef]
28. Shi, S. Operation and Assessment of Wind Energy on Power System Reliability Evaluation. Ph.D. Thesis, Department of Electronic and Electrical Engineering, University of Strathclyde, Glasgow, Scotland, 2014.
29. Dursun, B.; Alboyaci, B. An Evaluation of Wind Energy Characteristics for Four Different Locations in Balikesir. *Energy Sour. Part A: Recovery Util. Environ. Eff.* **2011**, *33*, 1086–1103. [CrossRef]
30. Ayodele, T.R.; Jimoh, A.A.; Munda, J.L.; AgeeWind, J.T. distribution and capacity factor estimation for wind turbines in the coastal region of South Africa. *Energy Convers Manag.* **2012**, *64*, 614–625. [CrossRef]
31. Heshmati, A.; Najafi, H.R.; Aghaebrahimi, M.R.; Mehdizadeh, M. Wind Farm Modeling For Reliability Assessment from the Viewpoint of Interconnected Systems. *Electr. Power Compon. Syst.* **2012**, *40*, 257–272. [CrossRef]
32. Kadhem, A.A.; Abdul Wahab, N.I.; Aris, I.; Jasni, J.; Abdalla, A.N. Computational techniques for assessing the reliability and sustainability of electrical power systems: A review. *Renew. Sustain. Energy Rev.* **2017**, *80*, 1175–1186. [CrossRef]

Reliability Evaluation Method Considering Demand Response (DR) of Household Electrical Equipment in Distribution Networks

Hongzhong Chen [1], Jun Tang [1], Lei Sun [2], Jiawei Zhou [3], Xiaolei Wang [1,*] and Yeying Mao [1]

[1] State Grid Suzhou Power Supply Company, Suzhou 215004, China; chenhongzhong@js.sgcc.com.cn (H.C.); tj_sz@js.sgcc.com.cn (J.T.); myy_sz@js.sgcc.com.cn (Y.M.)
[2] School of Electrical and Automation Engineering, Hefei University of Technology, Hefei 230009, China; leisun@hfut.edu.cn
[3] State Grid Suzhou Power Supply Company Suzhou Electric Power Design Institute Co., Ltd., Suzhou 215004, China; jwzhou1990@163.com
[*] Correspondence: boboball.wang@hotmail.com

Abstract: The load characteristic of typical household electrical equipment is elaborately analyzed. Considering the electric vehicles' (EVs') charging behavior and air conditioning's thermodynamic property, an electricity price-based demand response (DR) model and an incentive-based DR model for two kinds of typical high-power electrical equipment are proposed to obtain the load curve considering two different kinds of DR mechanisms. Afterwards, a load shedding strategy is introduced to improve the traditional reliability evaluation method for distribution networks, with the capacity constraints of tie lines taken into account. Subsequently, a reliability calculation method of distribution networks considering the shortage of power supply capacity and outages is presented. Finally, the Monte Carlo method is employed to calculate the reliability index of distribution networks with different load levels, and the impacts of different DR strategies on the reliability of distribution networks are analyzed. The results show that both DR strategies can improve the distribution system reliability.

Keywords: demand response; household electrical equipment; real-time electricity price; incentive mechanism; capacity constraint; reliability evaluation

1. Introduction

China's electricity generation and demands have been growing rapidly in recent years, but the annual utilization hours of power generation equipment are decreasing year-by-year, and the peak-valley difference of power system loads is gradually expanding. Statistically, China's total electricity consumption in 2018 was 6840 billion kW h, up 8.5% year-on-year, which is the highest growth rate since 2012. The maximum cooling loads in summer have reached 260 million kW, with a year-on-year growth of 10.5%, accounting for 27.8% of the maximum loads. However, the average utilization hours of power generation equipment in power plants of 6000 kW or above in China in 2019 were only 3862 h. In general, increasing generation capacity to meet peak load electricity demands will lead to an increase in investment cost and a decrease in equipment utilization hours, which cannot meet the requirements of economic operation of power grid and optimal allocation of resources.

As a prospective solution to the above issues, demand response (DR) has been widely proposed in many researches [1,2], where DR is regarded as an important measure to facilitate the penetration of renewable energy [3], realize friendly source-load interaction, balance the fluctuations of load profile, and consequently improve the system reliability and operation economy [4,5]. With the rapid development and application practice of ubiquitous power Internet of Things (IOT) technology, bi-directional communication and intelligent control between power grid and user side can be supported [6]. DR, normally including electricity price-based DR and incentive-based DR [7,8], could regulate consumers' electricity behaviors to reduce peak loads and improve load curve [9–12]. Real-time electricity price can reflect the relationship between power supply and demand, and thus a reasonable electricity price mechanism could efficiently guide users to participate in peak shaving and valley filling [13,14]. For incentive-based DR, users can be encouraged to adjust load demands by signing an agreement on condition of certain economic compensation [15,16].

Residential loads can be regarded as a resource that can be flexibly dispatched in the distribution network, which can be divided into temperature-controlled equipment, non-temperature-controlled equipment, and uncontrollable equipment according to the operation characteristics of household power loads [17]. Temperature-controlled equipment, such as air conditioning, can participate in DR by adjusting comfort interval [18]. Non-temperature-controlled equipment, such as electric vehicles (EVs), can change electricity consumption behavior by responding to real-time electricity price [19]. In [20], an optimal scheduling model, which considers the reliability of microgrid and the customers' electricity cost, is established to regulate the electricity consumption of home appliances and EVs. In [21], a home energy management system is proposed and an energy consumption optimization model is presented to minimize the electricity cost, and dynamic programming is employed to solve the real-time rescheduling model for determining the on/off status of home appliances. In [22], a customer utility function is defined and a novel economic model is presented to determine the customers' reaction to electricity price change. Reference [23] points out that the implementation of DR is limited by the accurate forecast of demand and price elasticity, and therefore presents a novel DR model based on consumers' information while avoiding predicting these two items. Both air conditioning loads and EV loads account for a large proportion in the daily load, and therefore enjoy a large potential for DR.

The reliability of distribution systems changes nonlinearly with the increase or decrease of the loads. By far, this issue has been studied by many researchers. A time-of-use electricity price-based DR model is established in [24,25] to identify the influence of electricity price-based DR on distribution network reliability. In [26], the incentive-based DR mechanism is systematically investigated to design the DR contract based on load transfer and load reduction, respectively, and then a bidding decision optimization model is presented to maximize the benefits of load aggregators. The influence of incentive-based DR on distribution network reliability is analyzed. However, the aforementioned works evaluate the reliability of distribution networks under a single DR mechanism. Actually, it is necessary to consider the comprehensive effects of two kinds of DR mechanism. In order to evaluate the performance of the implementation of DR, the customer baseline load is introduced in [27] with two different calculating methods, the day matching method and regression analysis. In [28], the concept of a virtual power plant is introduced, and the cost models considering the incentive- and electricity-based DR are respectively proposed to quantitatively analyze the influence of the uncertainty of DR on the expected losses of energy in distribution networks. It should be mentioned that the tie line capacity is generally assumed to be infinity when distribution network reliability is analyzed, which means the loads in non-fault areas could be entirely transferred after fault isolation, leading to overoptimistic reliability results. The line capacity constraint is considered in [29] when the outage is caused by a failure, assuming the power supply is sufficient. Nevertheless, the quantity of power users is rapidly increasing and the network structure is gradually becoming complex. The equipment in distribution networks that has not been upgraded, in time, may lead to electricity supply shortage in distribution systems. Therefore, these models may not perform well for calculation accuracy.

In this paper, two typical items of household equipment, EV and air conditioning, are taken into account. The DR models are presented based on different DR mechanisms, while considering load characteristic, charging behavior, and thermodynamic property. Also, it is necessary to present an improved distribution network reliability evaluation method considering tie line capacity constraint and power supply capacity shortage constraint, and therefore the accuracy of distribution network reliability index can be improved. Given the aforementioned reviewed literature, the contributions can be summarized as:

- The load characteristic of two typical items of household electrical equipment is elaborately analyzed.
- An electricity price-based DR model and an incentive-based DR model are proposed for two typical items of high-power electrical equipment, considering charging behavior and thermodynamic property.
- A load shedding strategy is introduced to improve the traditional reliability evaluation method for distribution networks, while taking into account the capacity constraints.
- A reliability calculation method of distribution networks with shortage of power supply capacity and faults taken into consideration is presented.

The remainder of this paper is given as follows. Section 2 present an electricity price-based DR model and an incentive-based DR model for two kinds of typical high-power electrical equipment. A load shedding strategy is proposed in Section 3, and the reliability index is detailed in Section 4. Section 5 proposes an improved reliability evaluation method of distribution networks, with shortage of power supply capacity and faults taken into account. The numerical results of the developed model are discussed in Section 6. The final section of the paper outlines conclusions based on this study.

2. DR Modeling

DR can be categorized into two types: Electricity price-based DR and incentive-based DR. The former guides users' electricity behavior by varying electricity price, while the latter adjusts electricity loads by contracts and compensation terms signed by customers. In this paper, two kinds of household electrical equipment, i.e., EVs and air conditioners, are chosen to establish DR models.

2.1. DR Modeling of EVs Based on Electricity Price

The charging loads of EVs depend on the user's habits, driving distance, status of the battery charge, the time when EVs plug in and out of distribution systems. Generally, EV owners use their cars during the day and charge them at night for the next day's trip. Real-time electricity price can reflect the relationship between power supply and demand, and therefore, real-time electricity price could be employed to guide users to adjust charging behavior of EVs. Given the development of communication technology, intelligent switch technology, and Internet of Things industry, the DR strategy for EV proposed in this paper can be realized in the near future.

2.1.1. Optimization Objective

When EVs plug into the grid, the owners are required to set the time of departure and the expected state of charge (SOC). The DR strategy is presented to minimize the charging cost of the owners by optimizing the charging power of EVs at each time interval according to the real-time electricity price. It should be mentioned that the owners are willing to participate in the DR because the cost of EV charging can be reduced while the owners' demands for the next day's travel are met. For a single EV, the objective function is defined as follows:

$$Min \sum_{t=1}^{T} P_t^{\mathrm{EV}} C_t \Delta t \tag{1}$$

The objective Function (1) is devoted to minimize the charging cost of a single EV. P_t^{EV} and C_t represent the charging power of EV and time-of-use electricity price at time t, Δt is the length of one time interval, and T is the number of time intervals within the scheduling time.

2.1.2. Constraints

(1) SOC constraint

Considering the charging efficiency of the battery in an EV, the recursive formula of the SOC can be described as:

$$Soc_{t+1} = Soc_t + \frac{\varepsilon_{ch} P_t^{\mathrm{EV}} \Delta t}{E_{\mathrm{B}}} \tag{2}$$

$$Soc_{\min} \le Soc_t \le Soc_{\max} \tag{3}$$

where Soc_{t+1} and Soc_t denote the SOC of the battery at time $t+1$ and t, respectively; ε_{ch} represents the charging efficiency of the battery; E_{B} is the energy capacity of battery, and the unit is kW·h; $Soc_{\max}$ and $Soc_{\min}$ represent the upper and lower limit of SOC, respectively.

(2) Travel demand constraint

The SOC of the battery in an EV when it plugs out of the grid should be restricted to meet the owner's demand for travel.

$$Soc_{t_d} \ge Soc_{\exp} \tag{4}$$

where Soc_{td} and $Soc_{\exp}$ respectively represent the actual and expected SOC of the battery at t_d; t_d denotes the time when EV plugs out, which is with strong uncertainty and can be approximately estimated according to the historical data of EVs.

(3) Battery safety constraint

SOC of the battery and the charging power in the charging process are described as:

$$0 \le P_t^{\mathrm{EV}} \le P_{\max}^{\mathrm{EV}} \tag{5}$$

where ${P_{\max}}^{\mathrm{EV}}$ represents the maximum of charging power of EV.

(4) Undispatched time constraint

$$P_t^{\mathrm{EV}} = 0 \; t \notin [t_s, t_d] \tag{6}$$

where t_{s} represents the time when an EV plugs into the grid.

2.2. DR Modeling of Air Conditioners Based on Incentive

Air conditioning is chosen as the temperature-controlled load to participate in DR due to the widespread utilization of air conditioning and the large proportion of air conditioning loads in summer peak loads. Air conditioning could satisfy users' demands for temperature through cooling or heating equipment. The acceptable temperature range for a human body is relatively wide. Therefore, air conditioning can be regarded as a DR resource with great potential and flexibility.

2.2.1. Optimization Objective

The air conditioning load could be adjusted within the time period declared by consumers to smooth the load curve and realize peak load shifting. The air conditioning loads are controlled by adjusting the setting of the temperature. The scheduling model of air conditioning loads can be formulated as follows.

$$\min \frac{1}{T-1} \sum_{t=1}^{T} \left(P_t^{\mathrm{D}} - \sum_{j=1}^{N} \Delta P_{k,t}^{air} x_{k,t} - \overline{P}^{\mathrm{D}}\right)^{2.} \tag{7}$$

$$\overline{P}^{\mathrm{D}} = \frac{1}{T} \sum_{t=1}^{T} \left(P_t^{\mathrm{D}} - \sum_{j=1}^{N} \Delta P_{k,t}^{air} x_{k,t}\right) \tag{8}$$

The objective Function (7) aims to minimize the load variance, where P_t^{D} represents the base load of time t; $\Delta P_{k,t}^{air}$ denotes the load reduction of the k_{th} air conditioner in time t; $\overline{P}^{\mathrm{D}}$ is the mean value of the total load in the dispatching period; N represents the total number of air conditioners; T is the number of time intervals within the scheduling time; $x_{k,t}$ is the binary decision variable indicating whether the k_{th} air conditioner participates in load reduction at time t; $x_{k,t}$ is equal to 1 if the air conditioner k participates in scheduling at time t, and 0 otherwise.

2.2.2. Constraints

(1) Electrical constraints of air conditioners

The energy efficiency ratio of air conditioners can be defined as the ratio of power and capacity of refrigerating. It should be noted that the energy efficiency ratio of inverter air conditioning cannot be simply modeled as a constant. The mathematical relationship among power, refrigerating capacity, and frequency in actual operation can be formulated as follows:

$$P_A = af_A + b \tag{9}$$

$$Q_A = m{f_A}^2 + nf_A + q \tag{10}$$

Equation (9) describes the relationship between the power and frequency of air conditioners, where P_A and f_A represent the operating power and frequency of air conditioners, respectively. a and b are constants. Equation (10) denotes the relationship between refrigerating capacity and frequency of air conditioners, where Q_A represents the refrigerating capacity of air conditioners. m, n, and q are coefficients.

(2) Thermodynamic model of air conditioners

The indoor temperature is generally chosen as the control variable, which is affected by many factors, such as the indoor and outdoor heat exchange, outdoor temperature, the heat transfer between the air conditioner and the indoor air, the air heat capacity, and the air conditioning refrigerating capacity. It should be mentioned that there are many other affecting factors; therefore, precisely modeling air conditioning thermodynamic characteristics can be quite complex. The thermodynamic model of air conditioning can be approximately described as follows:

$$\theta_{in,t+1} = \theta_{out,t+1} - Q_t R(1 - \mathrm{e}^{-\Delta t/RC}) - (\theta_{out,t} - \theta_{in,t})\mathrm{e}^{-\Delta t/RC} \tag{11}$$

where $\theta_{in,t+1}$ and $\theta_{in,t}$ represent the indoor temperature at time t + 1 and t, respectively; $\theta_{out,t+1}$ denotes the outdoor temperature at time t + 1; Q_t is the air conditioning refrigerating capacity at time t; R and C are the equivalent heat resistance and heat capacity of air conditioning, with the units of °C/kW kW h/°C, respectively.

(3) Inverter air conditioning operating frequency constraint

The frequency of the inverter air conditioning can be adjusted by the compressor, and the operating frequency should meet the following constraint:

$$f_{\min,k} \leq f_k \leq f_{\max,k} \tag{12}$$

where f_k represents the operating frequency of the k_{th} air conditioner; $f_{max,k}$ and $f_{min,k}$ denote the maximum and minimum operating frequency of the k_{th} air conditioner, respectively.

(4) Maximum dispatchable time constraints

The maximum dispatchable number of air conditioners in the scheduling period should be restricted in order to meet the users' comfort requirements.

$$\sum_{t=1}^{T} x_{k,t} \leq t_{c,k} \tag{13}$$

where $t_{c,k}$ is the maximum dispatchable number of the k_{th} air conditioner.

2.2.3. Control Method

The control methods of inverter air conditioning [30] can be divided into temperature control and frequency control. The changing curve of indoor temperature and frequency under the two control methods is shown in Figure 1. In the temperature control mode, it can be found that the frequency rapidly decreases to the lowest value if the set temperature of the inverter air conditioner raises. As the room temperature gradually increases, the frequency rises with fluctuations and tends to be stable. For the frequency control mode, the frequency is also rapidly reduced to the lowest value when the set temperature raises. The frequency continues to run at the lowest value until the temperature rises to the set temperature, and meanwhile, the frequency rises to a stable value. It can be found from Figure 1 that the temperature control method is simple but it is difficult to obtain an analytical solution of frequency deviation and control period, and therefore, the frequency control method is employed to control air conditioning loads; namely, when the user sets the comfort temperature range, i.e., from θ_1 to θ_2, the air conditioning would operate with the lowest frequency at the first time, and then the air conditioning frequency would be adjusted to stable value for θ_2.

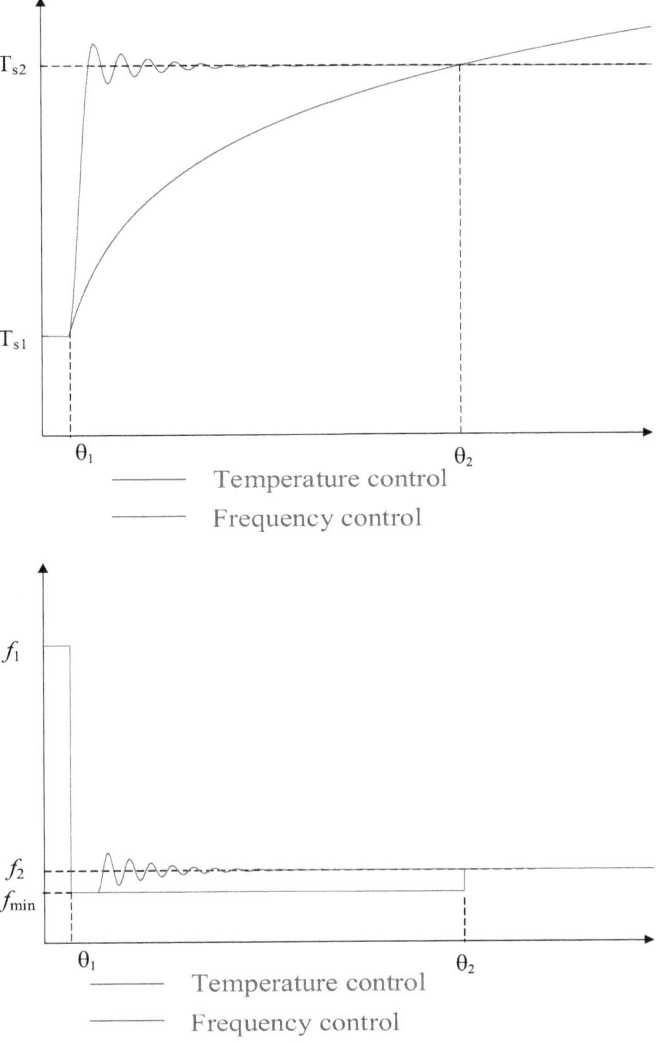

Figure 1. Schematic diagram of different control methods.

It is assumed that the indoor temperature at time $t+1$ is equal to the indoor temperature at time t in stable operation. When the air conditioning runs stably, its operation power can be calculated by Equations (9)–(11).

$$P_t^{air} = \frac{-n + \sqrt{n^2 - 4m[q - (\theta_{out,t} - \theta_1)/R]}}{2m} a + b \tag{14}$$

Therefore, load reduction of air conditioning can be described as:

$$\Delta P_{k,t}^{air} = P_{k,t}^{air} - a f_{\min,k} - b \tag{15}$$

where $P_{k,t}^{air}$ denotes the power demand of the k_{th} air conditioner in stable operation at time t.

3. Reliability Evaluation of Distribution Networks Considering DR

3.1. Model of Load Transfer Capacity

When a failure occurs, non-fault areas can be supplied through tie lines after fault areas are isolated. Considering the maximum transmission capacity limit of the tie line, the transfer capacity of the tie line l can be calculated as follows [31]:

$$P_{lz} = \frac{P_{l\max} - P_l(1+\beta_l)}{1+\beta_l} \tag{16}$$

where P_{lz} represents the power supplied by the tie line; P_l is the load of a line itself; $P_{l\max}$ is the maximum transmission power of the line l; β_l is the line loss ratio of the line l.

If the tie line has already supplied heavy loads, the transfer power will not be enough to meet the power supply of all the loads in the downstream area of the fault. In this situation, the load shedding strategy described in Section 3.2 should be applied in order to meet the power supply requirements of important loads in the downstream area of the fault.

3.2. Load Shedding Strategy

The over-load operation of the feeder may occur during in the peak load period if the maximum transmission capacity limit of the feeder in the distribution network is taken into consideration. It is necessary to reduce the load amount of the feeder by cutting parts of the loads to ensure the secure and stable operation of the distribution networks. Therefore, the load shedding model is proposed and shown in Equations (17) and (18):

$$\max \sum_{i \in S} \omega_i P_i(t) \tag{17}$$

$$\text{s.t.} \quad \begin{cases} (1+\beta) \sum\limits_{i \in S} P_i(t) \le P_{S\max} \\ U_{i,\min} \le U_i \le U_{i,\max} \end{cases} \tag{18}$$

The objective Function (17) should be maximized so as to supply as much load as possible, while taking the weight of the load points into consideration. ω_i is the importance coefficient of load point i for reliability requirement; S represents the collection set of all load points on the feeder; $P_i(t)$ is the power demand of load point i at time t. Equation (18) describes the power supply capacity constraint and node voltage constraint, where $P^{S\max}$ represents the upper limit of the power transmitted by the feeder and U_i, $U_{i,\max}$, and $U_{i,\min}$ are the voltage of node i and its maximum and minimum values, respectively.

4. Analysis of the Influence of DR on Distribution Network Reliability

The key of reliability evaluation is the selection and calculation of reliability index. In order to systematically study the influence of load changes caused by DR on the reliability of the distribution network, the reliability index influenced by load changes need to be firstly analyzed.

4.1. Load Point Reliability Index

(1) Load point average failure frequency index

The load point average failure frequency index is defined as the number of outages at load points in the statistical period. Long time overload operation of transformers will accelerate its aging, and then affect the average failure rate of the load point. DR can improve the load curve and reduce peak loads, thus reducing the failure frequency of load points.

(2) Load point average interruption frequency index

When the distribution transformer fails or the power supply is insufficient, load interruption occurs. The average interruption frequency index of load point i can be defined as:

$$P_{outage,i} = 1 - \frac{1}{T}\sum_{t=1}^{T} X(i,t) \tag{19}$$

where $P_{outage,i}$ denotes the outage probability of load point i; $X(I, t)$ is a binary decision variable for operation state of load point i at time t, which is equal to 1 in normally operating state, and 0 otherwise.

4.2. System Reliability Index

Three system reliability indexes are introduced in this paper as follows:

(1) Frequency index

Frequency index mainly refers to system average interruption frequency index (SAIFI).

(2) Time index

Time indexes mainly include customer average interruption duration index (CAIDI) and system average interruption duration index (SAIDI).

(3) Energy index

Energy not supplied (ENS) index mainly depends on the annual power outage time and load power of the system.

5. Improved Reliability Evaluation Method Based on Load Clustering

5.1. Improved Reliability Evaluation Method

According to the location of faults and their influence on other loads, the loads can be categorized into four types: (1) Type A: The loads in the fault area and their outage time depend on the time of fault isolation and repair; (2) Type B: The loads in the downstream of the fault area and their outage time depend on the load transfer time; (3) Type C: The loads in the upstream of the fault area, which can be supplied by the main transformer after fault isolation, and their outage time depend on the fault isolation time. It should be mentioned that if the supply capacity of the tie line is insufficient, the reliability indexes of Type B loads can be modified by applying the transferring capacity model and load shedding model described in Sections 3.1 and 3.2, and then the reliability index with and without DR implementation can be calculated, respectively.

The sequential Monte Carlo simulation method is applied to evaluate the reliability of distribution networks, and there are two situations that should be considered:

(1) When power supply capacity is insufficient in normal operation state, the load shedding strategy in Section 3.2 should be applied to supply power as much as possible with the feeder maximum capacity constraint respected. Calculate the system reliability indexes with and without DR, respectively.

(2) When a failure occurs in the distribution network, parts of the loads of Type B cannot get power supply due to restricted transfer capacity if the maximum capacity limit of feeders are considered.

5.2. *Reliability Calculation Method of Distribution Networks Considering Load Clustering*

The annual load peak is generally utilized to calculate the horizontal annual system reliability index in the traditional evaluation method. This method reflects the system reliability under the most severe situations, but ignores the impact of the load change on the system reliability. Calculating the reliability index only by the load peak value will greatly reduce the accuracy of evaluation results, since the annual load curve is composed of 8760 load points. However, the reliability calculation will be very time-consuming if each load point is substituted into the reliability evaluation. Therefore, it is necessary to cluster the annual load, then the reliability index can be calculated by employing the reliability calculation method in Section 5.1 based on the clustering results. The final system reliability index can be obtained according to the calculation results and weighted values of each cluster. In this way, the computational burden can be reduced under the premise of satisfying the accuracy.

6. Case Study

6.1. *Case 1*

A smart residential community in Suzhou is employed to illustrate the effectiveness of the proposed models. Residents' electricity consumption data at different time periods are collected by smart meters, and the residents' electricity consumption behavior is analyzed by the non-intrusive load monitoring, which could obtain the operation condition and loads of different types of electrical equipment by feature extraction and power decomposition technology.

6.1.1. Residential Electricity Load Analysis

The electricity consumption of a user in the smart community from 5–11 August in 2019 is shown in Figure 2, and the electricity consumption of each household equipment is depicted in Figure 3. It can be seen from Figure 2 that the electricity consumption of this user remained basically stable in a week, and the electricity consumption at the weekend increased compared to the working day. The maximum daily electricity consumption reached 23.44 kW h, since the residential community is mainly composed of villas. It can be concluded from Figure 3 that the electricity consumption of air conditioning accounts for up to 47% of the total electricity consumption, indicating that the air conditioning load has great potential for DR. Besides, there is little EV in this community, so only air conditioning is considered in this part.

6.1.2. Analysis on DR of Residential Load

In order to further analyze the DR potential of air conditioning load, the air conditioning load of 40 users in this smart community is selected for analysis. The air conditioning load curve can be roughly categorized into three classes, as depicted in Figure 4, and its daily load rate is shown in Table 1. Combined with Figure 4 and Table 1, it can be found that the air conditioning load characteristics of the three types of users are different from each other. The load curve of class A users have two peaks and two valleys, the peak of electricity consumption occurs at 14:00 and 22:00, and the daily load rate is relatively low, and therefore they have a great potential for DR. The power consumption of air conditioner of class B load in one day is almost zero, which has no potential of load control. Class C users are similar to class A, both of which show two peaks and two valleys, but the load rate of former one is relatively high, and they have a certain staggered peak response capability.

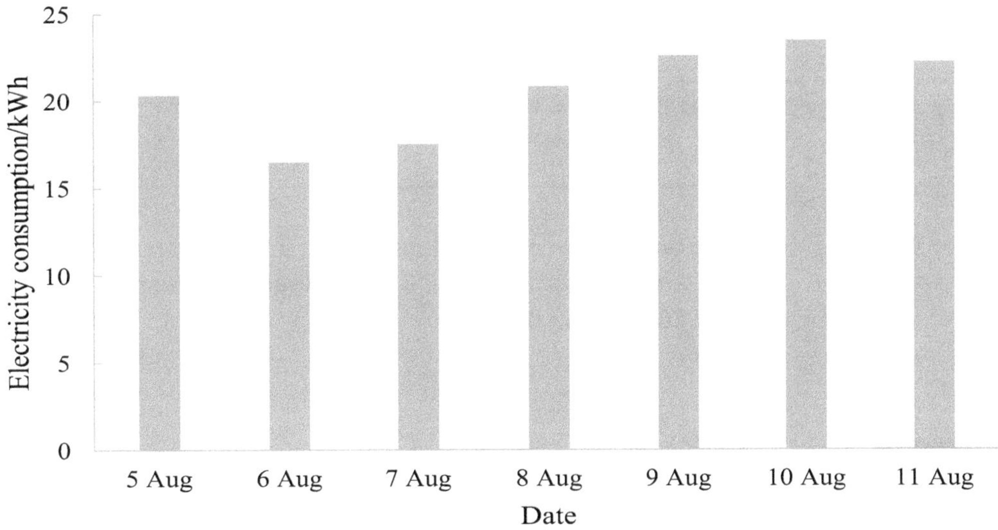

Figure 2. Residential electricity consumption in a week.

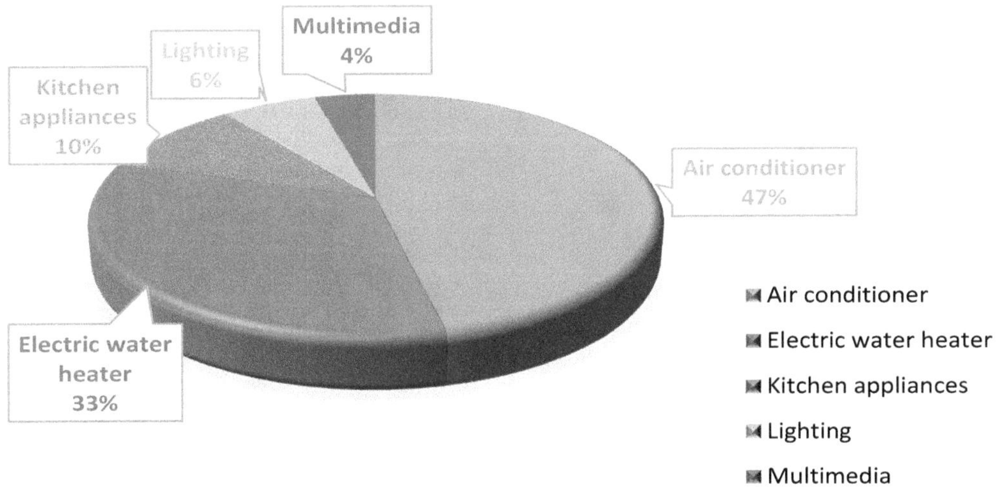

Figure 3. Household electricity composition analysis.

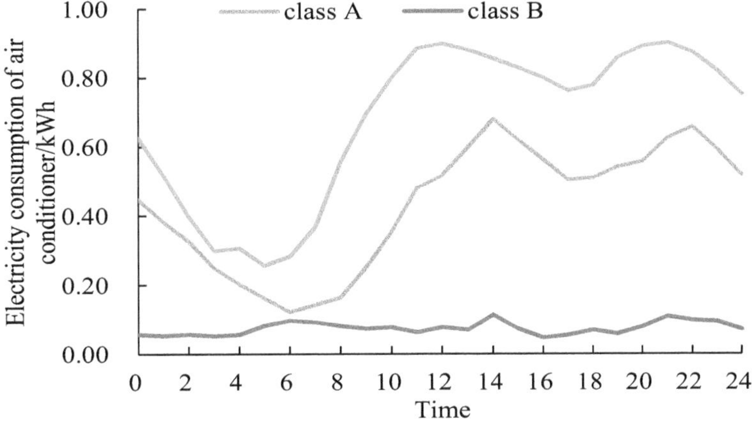

Figure 4. Air conditioning load curve of different classes.

Table 1. Daily load rate.

Type	Daily Load Rate	Percent
A	0.633253	56%
B	0.647609	11%
C	0.74984	33%

Among the 40 users in the smart community, the daily load curves of class A and class C users before and after participating in DR are shown in Figure 5. Both load peaks are reduced to a certain extent, the peak load decreases from 76.38 to 69.01 kW.

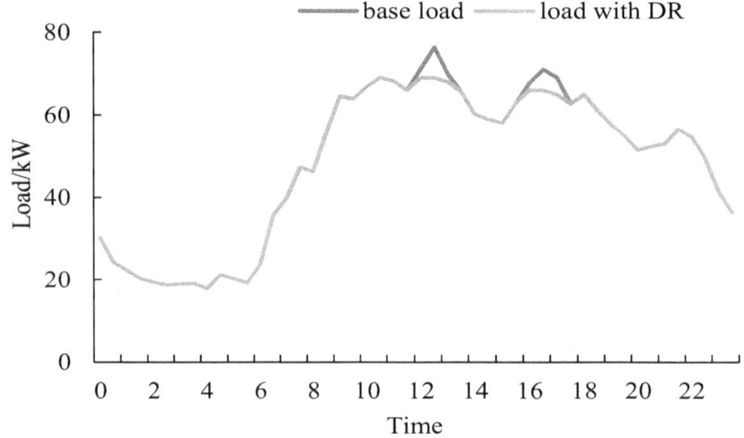

Figure 5. Comparison of load profile between base loads and loads with demand response (DR) taken into account.

6.2. Case 2

6.2.1. Simulation Settings

A modified RBTS-BUS 6 system is employed as the test system, in which the maximum power provided by the feeder F4 is 12 MW, and the transfer capacity of the tie line is 6 MW. As shown in Figure 6, the network frame includes 30 lines, 23 load points, 4 circuit breakers, and 21 load switches. The feeder length and load point data can be found in [32]. The circuit breakers, load switches, and fuses are assumed to be 100% reliable. The fault isolation time is 0.5 h, and the sum of fault isolation and load transfer time is 1 h. According to the current design requirements of distribution network capacity in residential areas, the capacity of each household is 8 kW, and the simultaneous factor is set at 0.5. The maximum power provided by the feeder F4 is 12 MW; therefore, there are about 3000 households in this area. The average EV inventory per household is set to 0.1, and all participate in DR. The air conditioning inventory per household is set to 2 and the proportion of air conditioning participating in DR is 0.2 considering the user's demand for comfort. The charging power and battery capacity of EVs is 5 kW and 40 kW·h, respectively. The energy consumption of EVs for traveling 100 km is 18 kW·h, and the battery charging efficiency is set to 0.9. It is assumed that the access time, departure time, and SOC of EVs are normally distributed, and could be obtained by employing Monte Carlo sampling. The highest and lowest running frequency of inverter air conditioning is 100 Hz and 20 Hz, respectively. R and C are specified as 5.56 °C/kW and 0.18 (kW h)/°C, respectively. The air conditioning power coefficient and the refrigerating capacity coefficient can be found in [33]. It is assumed that air conditioning cools in summer and heats in winter, and the comfort air conditioning interval is set to [22 °C, 28 °C]. The daily maximum dispatchable number is limited to 2 during the peak load period. The load data is shown in Appendix A, Table A1, and it is assumed that the load at each load point varies proportionally with the implementation of DR. The real-time electricity price curve is displayed in Appendix A, Figure A1. AMPL/CPLEX, an efficient commercial solver, is adopted to solve the proposed DR model, then the daily load curve considering DR of household equipment is obtained. Subsequently, the annual load curve considering DR could be determined according to the annual–weekly load curve and the weekly–daily load curve. Afterwards, the annual load curve is clustered, and the Monte Carlo simulation method is employed to calculate the reliability index of the distribution network [34].

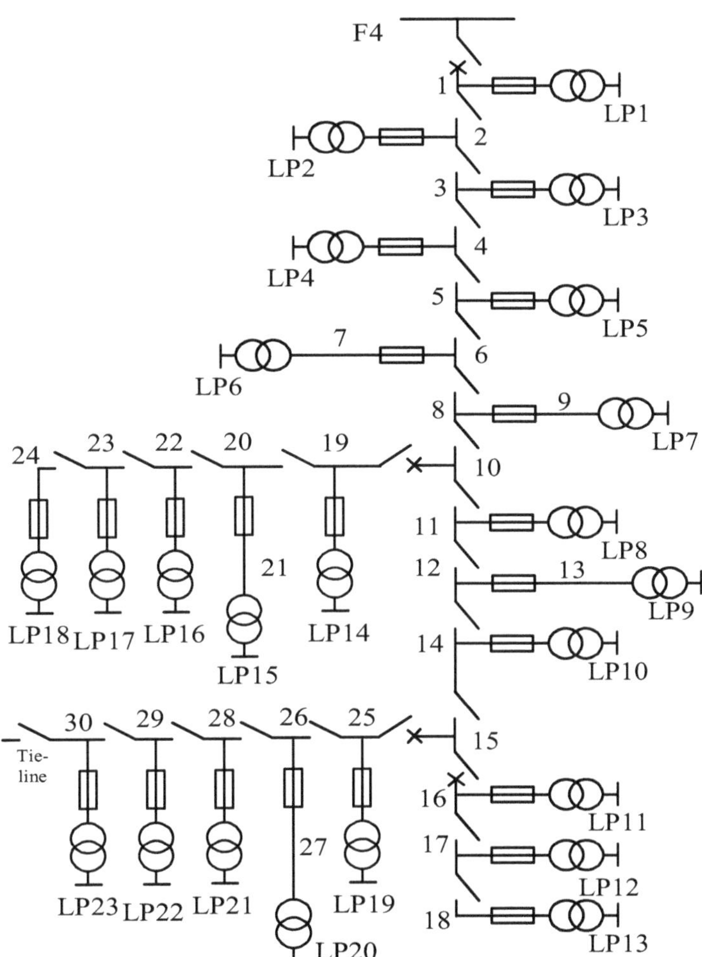

Figure 6. Topology of the test network.

6.2.2. Load Profile Considering DR

The daily load curve with and without EVs participating in DR is shown in Figure 7. It can be found that EV owners transfer the charging loads from the period of high electricity price to the period of low electricity price under the guidance of real-time electricity price, achieving peak load shaving and valley filling. The daily load curve with and without the air conditioning participating in DR is depicted in Figure 8. It can be found that the peak load is reduced and the load curve can be smoothed through the frequency reduction of air conditioning in the peak load period.

The daily load curve considering DR of EVs based on electricity price and air conditioning based on incentives is shown in Figure 9. It can be found that the load considering DR decreases during the peak period, and increases during the valley period compared to the base load, thus achieving peak load shaving and valley filling. Besides, the daily load peak without DR appears at 22:15, and the peak load is 12.27 MW. Considering the limitation of feeder capacity, part of the loads should be cut off. The daily load peak with DR appears at 22:30, and the maximum load is 11.57 MW, which satisfies the supply capacity constraints of feeders.

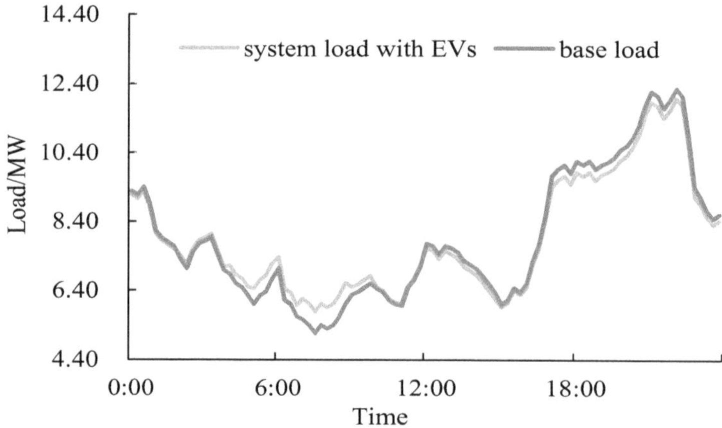

Figure 7. Comparison of load profile between base loads and loads with electric vehicles (EVs) taken into account

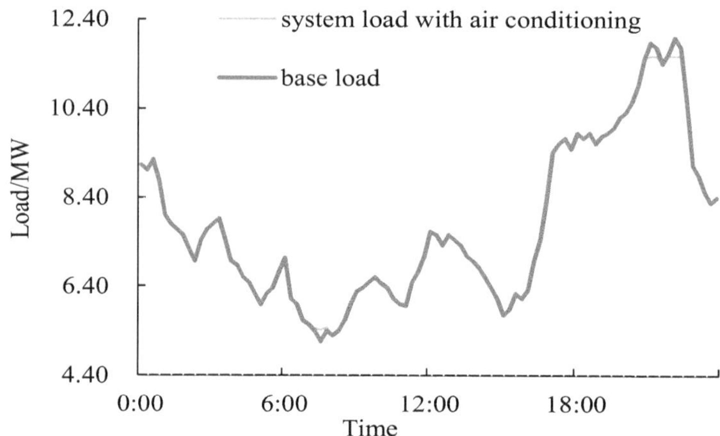

Figure 8. Comparison of load profile between base loads and loads with air conditioning taken into account.

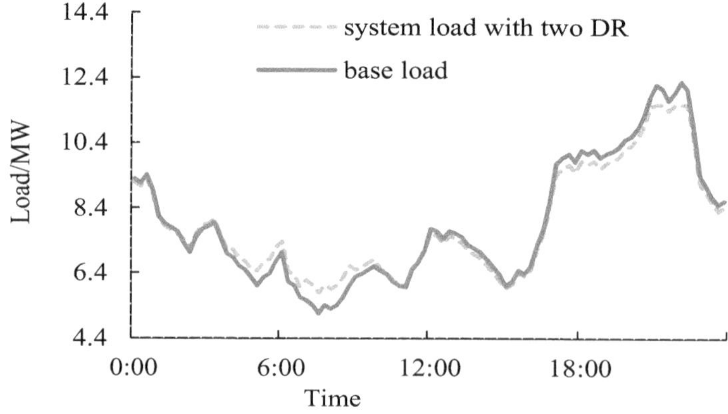

Figure 9. Comparison of load profile between base loads and loads with DR taken into account.

6.2.3. Influence of Real-Time Electricity Price on DR

The real-time electricity price, which has a certain randomness, is formulated according to the day-ahead load prediction, and plays the role of guiding electricity consumption. Considering the load uncertainty, a certain deviation between the actual electricity load and the predicted load may occur. Given this situation, applying the real-time electricity price shown in Appendix A, Figure A2, the daily load curve with and without EVs participating in the DR is shown in Figure 10. It can be found that the daily load peak appears at 21:00, because the real-time electricity in this period is relatively lower than others. Also, the daily load curve during the period of 17:00 to 18:45 is reduced, which is not the actual peak period needing to be clipped. Therefore, DR based on electricity price has a certain randomness and is greatly affected by real-time electricity price. Unreasonable real-time electricity price may lead to "peak-on-peak" or unnecessary load reduction.

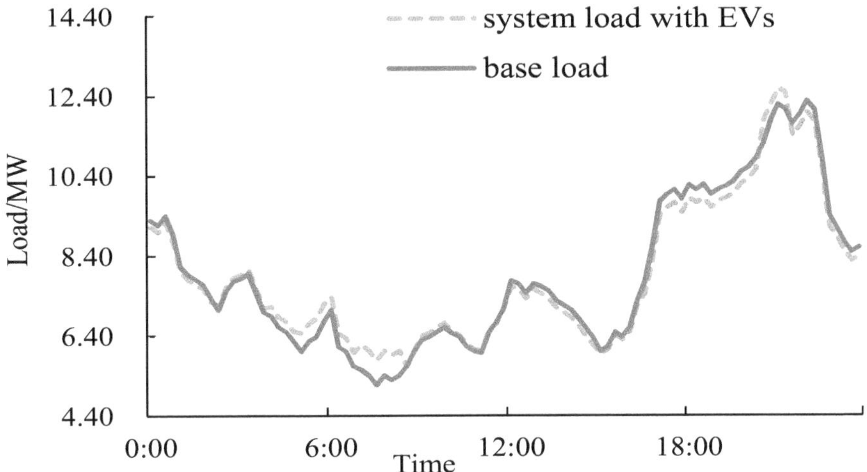

Figure 10. Comparison of load profile between base loads and loads with EVs taken into account.

6.2.4. Reliability Evaluation Considering DR

In this paper, four cases are conducted to evaluate the reliability of the distribution network considering DR. The annual load curve is composed of 8760 load points, which can be divided into 712 clusters, for there are 712 different points. The final system reliability evaluation indexes can be determined with the calculation results and weighted values of each cluster.

Case 1: DR is not taken into consideration.
Case 2: DR based on electricity price and incentive is considered.
Case 3: Case 1 with sufficient spare capacity of tie lines.
Case 4: Case 2 with sufficient spare capacity of tie lines.
The reliability index of the four cases is shown in Table 2.

Table 2. System reliability index (SAIFI, system average interruption frequency index; SAIDI, system average interruption duration index; CAIDI, customer average interruption duration index; ENS, energy not supplied).

Case	SAIFI/(times/a)	SAIDI/(h/a)	CAIDI/(h/times)	ENS/(MW h/a)
1	2.253	5.67	2.517	67.93
2	2.2052	5.2483	2.3798	59.31
3	2.2052	3.519	1.596	43.22
4	2.2052	3.519	1.596	40.76

By comparing the reliability indexes of cases 1 and 2, it can be concluded that DR can effectively improve the system reliability indexes. With DR, the peak load decreases 5.7%, the indexes of SAIDI, CAIDI, and ENS decrease 7.4%, 5.5%, and 12.7%, respectively. The main reason is that DR based on electricity price can transfer the loads from peak period to valley period to reduce the peak load of the system, and DR based on incentive can directly reduce part of the loads at the peak period, thus reducing the probability of power supply shortage of the distribution system. On the other hand, if a fault occurs during the peak period, the reduced loads considering DR can increase the transfer probability of the loads after fault isolation, thus reducing the average interruption frequency and interruption duration at the load point. Therefore the SAIFI, SAIDI, and ENS indexes are improved. By comparing cases 3 and 4, it can be found that when the capacity of the tie line is infinite, DR has no impact on the SAIFI and SAIDI indexes, and only the ENS index is improved due to the reduced peak load.

7. Conclusions

In this paper, two kinds of typical household electrical equipment are considered to participate in DR. DR models based on electricity price and incentive are proposed to obtain load curves with different DR mechanisms taken into account. Considering the transmission capacity limit of tie lines, a reliability evaluation method is proposed considering power supply capacity shortage. The following conclusions can be drawn from the simulation results:

(1) DR can improve the load curve and reduce the peak loads. DR based on electricity price has a certain randomness and is greatly affected by real-time electricity price. Unreasonable real-time electricity price may lead to "peak-on-peak" or unnecessary load reduction.

(2) Both DR based on electricity price and DR based on incentive can improve the reliability index of distribution networks. Compared with the reliability results attained by employing a single DR strategy, comprehensive DRs can improve reliability

(3) DR has no influence on the distribution network reliability index if transmission capacity of tie lines is assumed to be infinite. Under the premise of considering the tie line capacity limit, DR reduces the peak loads of the systems and decreases the probability of insufficient power supply capacity in normal operation, and meanwhile increases the possibility of the load point being transferred when a failure occurs.

Author Contributions: All the authors contributed to this work. H.C., J.T. and Y.M. conceived and structured the study, X.W. and J.Z. prepared the preliminary manuscript, L.S. reviewed and finalized the manuscript.

Acknowledgments: This work was supported by the Comprehensive Demonstration Project of Distribution Internet of Things in Suzhou Historic District.

Appendix A

Table A1. Load data.

Load Number (Class)	Average Load (kW)	Load Number (Class)	Average Load (kW)	Load Number (Class)	Average Load (kW)	Load Number (Class)	Average Load (kW)
LP1(II)	165.9	LP7(I)	210.1	LP13(III)	250.1	LP19(III)	155.4
LP2(III)	180.8	LP8(III)	155.4	LP14(III)	155.4	LP20(II)	186.1
LP3(III)	250.1	LP9(I)	283.1	LP15(II)	186.1	LP21(I)	283.1
LP4(III)	243.1	LP10(II)	158.5	LP16(II)	158.5	LP22(II)	158.5
LP5(I)	207.0	LP11(III)	155.4	LP17(III)	250.1	LP23(I)	210.1
LP6(II)	165.9	LP12(II)	158.5	LP18(III)	243.1		

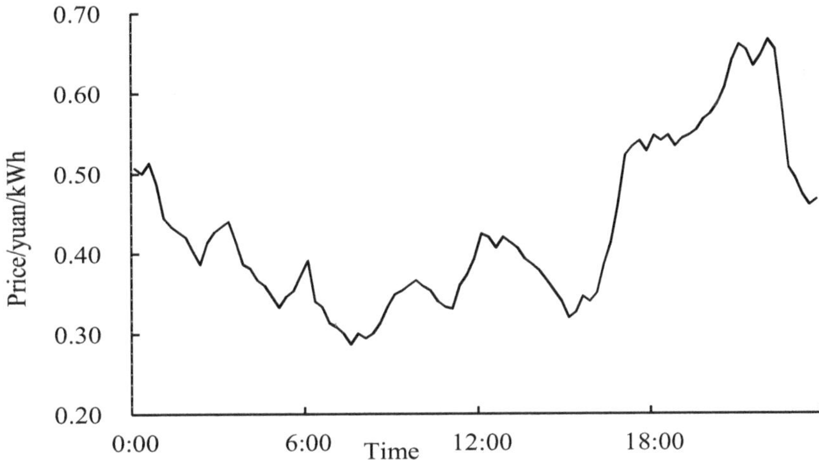

Figure A1. Real-time electricity price curve.

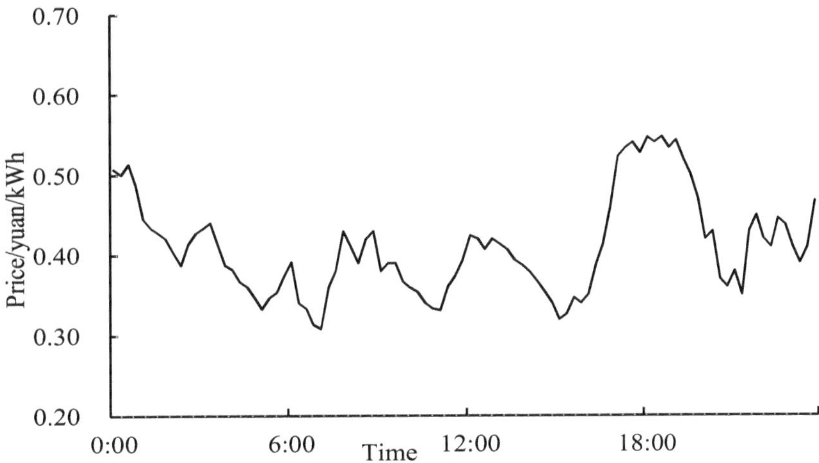

Figure A2. Real-time electricity price curve.

References

1. Eissa, M.M. Developing incentive demand response with commercial energy management system (CEMS) based on diffusion model, smart meters and new communication protocol. *Appl. Energy* **2019**, *236*, 273–292. [CrossRef]
2. Vahedipour-Dahraie, M.; Najafi, H.; Anvari-Moghaddam, A.; Guerrero, J. Study of the effect of time-based rate demand response programs on stochastic day-ahead energy and reserve scheduling in islanded residential microgrids. *Appl. Sci.* **2017**, *7*, 378. [CrossRef]
3. Pinson, P.; Madsen, H. Benefits and challenges of electrical demand response: A critical review. *Renew. Sustain. Energy Rev.* **2014**, *39*, 686–699.
4. Wang, F.; Zhou, L.; Ren, H.; Liu, X.; Talari, S.; Shafie-khah, M.; Catalão, J.P. Multi-objective optimization model of source-load-storage synergetic dispatch for building energy system based on TOU price demand response. *IEEE Trans. Ind. Appl.* **2018**, *54*, 1017–1028. [CrossRef]
5. Sharifi, R.; Anvari-Moghaddam, A.; Fathi, S.H.; Guerrero, J.M.; Vahidinasab, V. Economic demand response model in liberalised electricity markets with respect to flexibility of consumers. *IET Gener. Transm. Distrib.* **2017**, *11*, 4291–4298. [CrossRef]
6. Deng, R.; Yang, Z.; Chow, M.Y.; Chen, J.M. A survey on demand response in smart grids: Mathematical models and approaches. *IEEE Trans. Ind. Inform.* **2015**, *11*, 570–582. [CrossRef]
7. Yu, M.; Hong, S.H.; Ding, Y.; Ye, H. An incentive-based demand response (DR) model considering composited DR resources. *IEEE Trans. Ind. Electro.* **2018**, *66*, 1488–1498. [CrossRef]

8. Zhou, B.R.; Huang, Y.C.; Zhang, Y.J. Reliability analysis on microgrid considering incentive demand response. *Autom. Electr. Power Syst.* **2017**, *41*, 70–78.
9. Brahman, F.; Honarmand, M.; Jadid, S. Optimal electrical and thermal energy management of a residential energy hub, integrating demand response and energy storage system. *Energy Build.* **2015**, *90*, 65–75. [CrossRef]
10. Sharifi, R.; Moghaddam, A.A.; Fathi, S.H.; Vahidinasab, V. A flexible responsive load economic model for industrial demands. *Processes* **2019**, *7*, 147. [CrossRef]
11. Sharifi, R.; Fathi, S.H.; Vahidinasab, V. A review on demand-side tools in electricity market. *Renew. Sustain. Energy Rev.* **2017**, *72*, 565–572. [CrossRef]
12. Aalami, H.A.; Moghaddam, M.P.; Yousefi, G.R. Modelling and prioritizing demand response programs in power markets. *Electr. Power Syst. Res.* **2010**, *80*, 426–435. [CrossRef]
13. Smith, R.; Meng, K.; Dong, Z.; Simpson, R. Demand response: A strategy to address residential air-conditioning peak load in Australia. *J. Mod. Power Syst. Clean Energy* **2013**, *1*, 223–230. [CrossRef]
14. Reynolds, S.P.; Creighton, T.E. Time-of-use rates for very large customers on the pacific gas and electric company system. *IEEE Trans. Power Appar. Syst.* **1980**, *99*, 147–151. [CrossRef]
15. Kai, M.; Hu, G.; Spanos, C.J. A cooperative demand response scheme using punishment mechanism and application to industrial refrigerated warehouses. *IEEE Trans. Ind. Inform.* **2015**, *11*, 1520–1531.
16. Chen, C.; Kishore, S.; Wang, Z.; Alizadeh, M.; Scaglione, A. How will Demand Response aggregators affect electricity markets ?—A Cournot game analysis. In Proceedings of the 5th International Symposium on Communications Control and Signal Processing (ISCCSP), Roma, Italy, 2–4 May 2012; pp. 1–6.
17. Zeng, B.; Jiang, W.Q.; Yang, Z.; Li, G. Optimal load dispatching method based on chance-constrained programming for household electrical equipment. *Autom. Electr. Power Syst.* **2018**, *38*, 27–33.
18. Chen, Z.; Li, Y.Q.; Leng, Z.Y.; Lu, G.X. Refined modeling and energy management strategy of typical household high-power loads. *Autom. Electr. Power Syst.* **2018**, *42*, 135–143.
19. Sharifi, R.; Anvari-Moghaddam, A.; Fathi, S.H.; Guerrero, J.M.; Vahidinasab, V. An optimal market-oriented demand response model for price-responsive residential consumers. *Energy Eff.* **2019**, *12*, 803–815. [CrossRef]
20. Tushar, M.H.K.; Assi, C.; Maier, M.; Uddin, M.F. Smart microgrids: Optimal joint scheduling for electric vehicles and home appliances. *IEEE Trans. Smart Grid* **2014**, *5*, 239–250. [CrossRef]
21. Khalid, A.; Javaid, N.; Guizani, M.; Alhussein, M.; Aurangzeb, K.; Ilahi, M. Towards dynamic coordination among home appliances using multi-objective energy optimization for demand side management in smart buildings. *IEEE Access* **2018**, *6*, 19509–19529. [CrossRef]
22. Mohajeryami, S.; Moghaddam, I.N.; Doostan, M.; Vatani, B.; Schwarz, P. A novel economic model for price-based demand response. *Electr. Power Syst. Res.* **2016**, *135*, 1–9. [CrossRef]
23. Mnatsakanyan, A.; Kennedy, S.W. A novel demand response model with an application for a virtual power plant. *IEEE Trans. Smart Grid* **2015**, *6*, 230–237. [CrossRef]
24. Zhao, H.S.; Wang, Y.Y.; Chen, S. Impact of DR on distribution system reliability. *Autom. Electr. Power Syst.* **2015**, *39*, 49–55.
25. Zhang, Y.B.; Ren, S.J.; Yang, X.D.; Bao, K.K.; Xie, L.Y.; Qi, J. Optimal configuration considering price-based demand response for stand-alone microgrid. *Electr. Power Autom. Equip.* **2017**, *37*, 55–62.
26. Qi, X.J.; Cheng, Q.; Wu, H.B.; Yang, S.H.; Li, Z.X. Impact of incentive-based DR on operational reliability of distribution network. *Trans. China Electrotech. Soc.* **2018**, *33*, 5319–5326.
27. Sharifi, R.; Fathi, S.H.; Vahidinasab, V. Customer baseline load models for residential sector in a smart-grid environment. *Energy Rep.* **2016**, *2*, 74–81. [CrossRef]
28. Niu, W.J.; Li, Y.; Wang, B.B. Demand Response Based Virtual Power Plant Modeling Considering Uncertainty. *Proc. CSEE* **2014**, *34*, 3630–3637.
29. Lei, M.; Wei, W.; Zeng, J.H.; Mo, S.Y. Effect of load control on power supply reliability considering demand response. *Autom. Electr. Power Syst.* **2018**, *42*, 53–59.
30. Yang, J.R.; Shi, K.; Cui, X.Q.; Gao, C.W.; Cui, G.Y.; Yang, J.L. Peak Load Reduction Method of Inverter Air-conditioning Group Under DR. *Autom. Electr. Power Syst.* **2018**, *42*, 44–56.
31. Kwac, J.; Flora, J.; Rajagopal, R. Household energy consumption segmentation using hourly data. *IEEE Trans. Smart Grid* **2014**, *5*, 420–430. [CrossRef]
32. Billinton, R.; Jonnavithula, S. A test system for teaching overall power system reliability assessment. *IEEE Trans. Power Syst.* **1996**, *11*, 1670–1676. [CrossRef]

7

The Influence and Optimization of Geometrical Parameters on Coast-Down Characteristics of Nuclear Reactor Coolant Pumps

Yuanyuan Zhao, Xiangyu Si, Xiuli Wang *, Rongsheng Zhu, Qiang Fu and Huazhou Zhong

National Research Center of Pumps, Jiangsu University, Zhenjiang 212013, China; zyy-michelle@163.com (Y.Z.); sixiangyu01@163.com (X.S.); ujs_zrs@163.com (R.Z.); ujsfq@sina.com (Q.F.); huazhou.zhong@turbotides.com.cn (H.Z.)

* Correspondence: ujswxl@ujs.edu.cn

Abstract: Coast-down characteristics are the crucial safety evaluation factors of nuclear reactor coolant pumps. The energy stored at the highest moment of inertia of the reactor coolant pump unit is utilized to maintain a normal coolant supply to the core of the cooling loop system for a short period of time during the coast-down transition. As a result of the high inertia moment of the rotor system, the unit requires a high reliability of the nuclear reactor coolant pump and consumes considerable energy in the start-up and normal operation. This paper considers the operational characteristics of the coast-down transition process based on the existing hydraulic model of the nuclear reactor coolant pump. With the implementation of an orthogonal test, the hydraulic performance of the nuclear reactor coolant pump was optimized, and the optimal combination of impeller geometrical parameters was selected using multivariate linear regression to prolong the coast-down time of the reactor coolant pump and to avoid serious nuclear accidents.

Keywords: reactor coolant pump; coast-down characteristics; geometrical parameters; multiple linear regression; transition process

1. Introduction

The nuclear reactor coolant pump is the only rotating piece of equipment in the primary loop cooling system, and thus can be called the 'heart' of the nuclear power plant. In the unfortunate event of a power failure at the nuclear power plant, the reactor coolant pump loses its power source and it will enter a coast-down state. For a short period of time, the inertia of the inert wheel provides power for the reactor coolant pump and the coolant continues to cool the reactor core. The coast-down process of the nuclear reactor coolant pump can be seen as a typical transient process. The nuclear reactor has a short coast-down transient process and cooling circuit working time. As the heat of the reactor cannot be discharged in a short time the temperature tends to rise sharply, leading to the potential decomposition of the coolant and generating a large amount of hydrogen, which is not at all conducive to the safety of the system [1]. Hence, the study of the effect of dynamic characteristics on the nuclear reactor coolant pump during the coast-down transient process becomes imperative.

Very limited literature is available related to the transient process of domestic and foreign nuclear reactor coolant pumps. Nevertheless, the transient processes of the centrifugal pumps and the mixed flow pumps have been widely studied. The transient characteristics of the centrifugal pump in its acceleration and deceleration process were determined by Tsukamoto et al. [2] using a theoretical analysis. Wu and Li et al. [3,4] examined the starting and stopping transients of the centrifugal pumps and mixed flow pumps by combining experimental and numerical calculations. A systematic study on the start-up process of centrifugal pumps with different valve opening degrees, different impeller outside diameters, different blade widths, and different starting descending speeds was conducted by Elaoud et al. [5,6]. By establishing a mathematical start-up model for the primary circuit cooling system, Farhadi et al. [7–9] studied the effects of the ratio between the inertial energy of the unit and the fluid mass inertia energy of the pipe coolant during the start-up process of the nuclear reactor coolant pump. Whether the pumping capability meets the coast-down half time requirement as prescribed by safety analyses during the coast-down period was studied by Alatrash et al. [10] using experiments. Yonggang [11–13] studied the third and fourth generation nuclear main pumps, including gas–liquid two-phase flow and structural optimization. There have been numerous valuable studies on the coast-down characteristics of nuclear reactor coolant pumps, but comparatively little research details are available on the factors that affect the coast-down characteristics.

Geometric parameters of the impeller are one of the main factors affecting pump performance. Based on a numerical analysis of a 3D viscous flow, Hyuk et al. [14] designed a high-efficiency mixed-flow pump and the results suggested that the hydraulic efficiency of a mixed-flow pump at the design level can be improved by modifying its geometry. The impact of the geometry parameters of the impeller on the hydraulic performance of mixed flow pumps was studied by Varchola et al. [15], and several different designs were also compared. Long et al. [16] studied how the blade numbers of the impeller and the diffuser influence the reactor coolant pump performances using the numerical simulation method. The effect of the blade stacking lean angle on the hydraulic performance of a 1400 MW nuclear reactor coolant pump was studied by Zhou et al. [17], and it was determined that the geometric parameters such as the blade stacking lean angle highly influences the hydraulic efficiency of different flow intervals. Evidently, although the influence of the geometrical parameters on the pump has been studied, the research on the coupling effect of the nuclear reactor coolant pump has been very limited.

With regards to the factors influencing the coast-down characteristics of reactor coolant pumps, the indirect coupling effects between the different geometric parameters and combinations have been examined in depth by this paper. Changing the pump performance by changing the size of a certain geometric parameter virtually changes the direct impact of this parameter on performance, with an indirect impact on the performance of the other parameters simultaneously. This paper employs the multiple linear regression to analyze the optimal geometrical parameters of the impeller, based on the relationship between geometric parameters of the impeller and its efficiency, and the head.

2. Research Method

In the event of an unfortunate loss of power source to the nuclear reactor coolant pump, the unit utilizes its own moment of inertia to store energy to ultimately maintain the operation of the reactor coolant pump for a longer stretch of period, this phenomenon is known as the coast-down characteristic of the nuclear reactor coolant pumps [18]. In the coast-down transition process, the energy stored by the flywheel ensures that the time of the nuclear reactor coolant pump flow decreases by half within the specified safe time margin. The flywheel of an AP1000 nuclear reactor coolant pump is usually split into the upper flywheel and the lower flywheel while being fixed on the main shaft. To maximize improvement in the moments of inertia at a limited volume to maintain the coast-down characteristics of the reactor coolant pump, the flywheels were usually encompassed with high-density heavy metal tungsten alloy blocks and high-quality stainless-steel wheels. With respect to the issue of energy consumption, the main function of the flywheel was to provide energy to keep the nuclear reactor coolant pump running during the coast-down transition, and the flywheel should consume enough amounts of energy to maintain the start-up process and normal operation. It was suggested in the combination of Equations (1) and (2) that the time of coast-down was not only affected by the moment of inertia, but the efficiency of the rated operating point and the energy loss of the coast-down transition were also significant factors. As the rotational inertia of the impeller was obviously smaller than that of the flywheel, the moment of inertia of the rotor basically remains unchanged when the geometric parameters of the impeller were changed [18–20]. Accordingly, this study endeavors to increase the efficiency of the hydraulic model rated point by optimizing the impeller geometric parameters to reduce any extra energy loss in the coast-down transition, to extend the time of coast-down transition, to increase the system reliability, and to reduce the cost thereon.

$$E_T = 2\pi^2 \int (\oint \rho n^2(t) A dz) dt + E_f \tag{1}$$

$$t = \frac{P_0}{4\pi^2 I \eta_0 n_0^2} [(\frac{n_0}{n(t)} - 1)] \tag{2}$$

In Equations (1) and (2), $E_T = \frac{1}{2} I p \omega_0^2 + \frac{1}{2} \oint \rho \omega_0^2 A dz$ denotes the total energy stored by the moment of inertia of the unit and the inertia of the conveying liquid, E_f refers to the energy of various losses in the coast-down transition, ρ represents the density of the conveying liquid with units in kg/m^3, A denotes the average sectional area of the loop pipe with the unit of m^2, z refers to the effective pipeline length for the whole circuit with the unit of m, t represents the time since the outage began with the unit of s, and I refers to the total moment of inertia of the unit. With the unit of kg m^2, P_0, η_0, and n_0 are the effective power, efficiency, and rated speed of the nuclear reactor coolant pump under rated conditions, respectively, *n(t)* refers to the rotational speed at different points in the coast-down transition with the unit of r/min.

Figure 1 shows the three-dimensional fluid calculation domain of the reactor coolant pump. In Figure 2, γ denotes the outlet inclination of the impeller, β_2 denotes the outlet angle of the impeller, φ denotes the wrap angle of the impeller, Z denotes the blade numbers of the impeller, D_2 denotes the outlet diameter of the impeller in mm, b_2 denotes the outlet width of the impeller in mm, and D_j denotes the inlet diameter in mm. Area ratio Y represents the ratio of the impeller outlet area to the volute throat area.

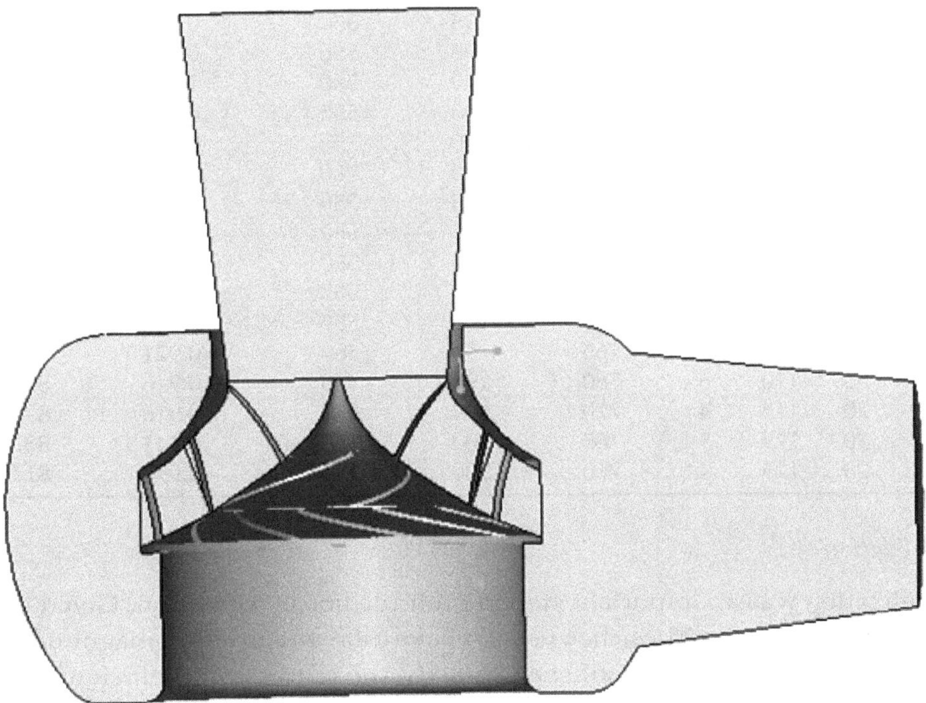

Figure 1. Three-dimensional fluid calculation domain of reactor coolant pump.

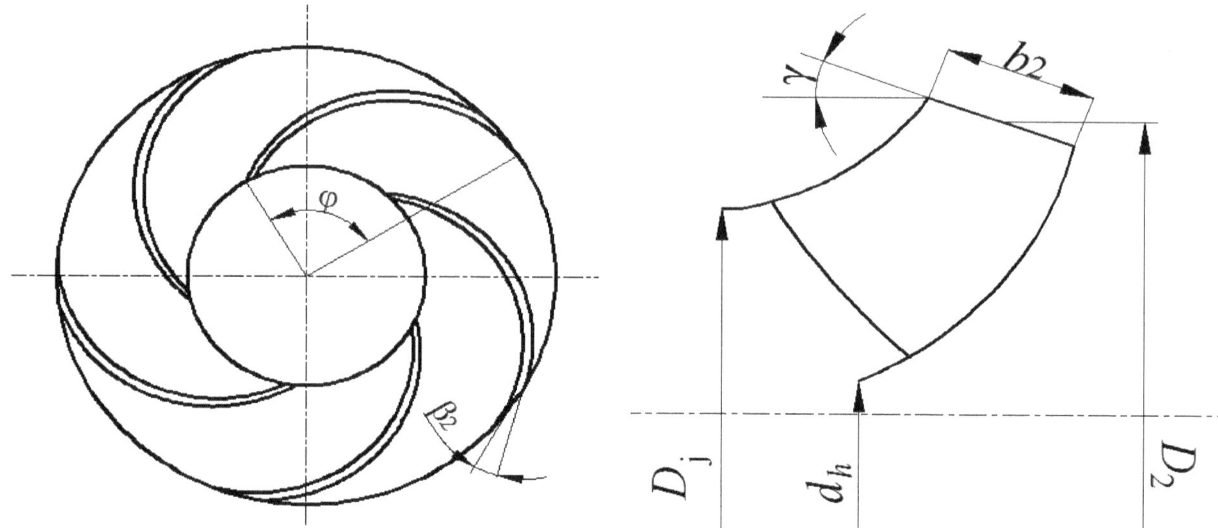

Figure 2. Schematic diagram of the main structural parameters of the impeller.

Table 1 presents that the efficiencies and heads obtained by the different combinations of eight different impeller geometric parameters, which suggest the influence of the different geometric parameters, parameter combination on efficiency, and the difference in the head [21]. Multiple linear regression was selected by this paper to analyze the relationship among each parameter, efficiency, and the head.

Table 1. Test scheme and performance calculation results.

	$\gamma/°$	$\beta_2/°$	$\varphi/°$	Z	D_2/mm	b_2/mm	D_j/mm	Area Ratio Y	Index	
									η_1	H
1	20	20	115	4	760	190	555	0.927	79.45	94.89
2	20	25	120	5	765	190	560	0.916	81.09	109.48
3	20	30	125	6	770	195	550	0.916	82.76	134.33
4	23	30	125	5	765	195	550	0.932	84.38	111.69
5	23	25	120	6	760	200	560	1.002	84.70	138.02
6	23	20	115	4	760	200	555	0.952	84.04	100.39
7	26	20	115	4	770	190	555	0.914	82.24	101.80
8	26	25	120	5	765	190	560	0.925	82.16	112.89
9	26	30	120	6	765	195	550	0.935	82.32	136.20
10	20	20	125	6	770	195	550	0.916	80.67	101.59
11	23	30	125	4	770	200	560	0.931	83.10	95.47
12	26	25	115	5	760	200	555	0.956	82.90	119.49
13	20	30	115	6	770	190	555	0.905	84.22	137.83
14	23	30	120	5	765	190	560	0.921	83.69	124.90
15	26	25	120	6	760	195	550	0.946	82.27	120.81
16	20	20	115	4	770	195	550	0.916	83.98	101.89
17	23	20	125	5	765	200	560	0.941	83.11	100.60
18	26	25	125	4	760	200	555	0.956	81.31	103.49

2.1. Data Normalization

Data normalization was an important step in multiple linear regression. Given that each variable was different in the physical properties, they usually have different orders of magnitude and dimensions. When the regression equation was established directly, the regression coefficient was very often not directly comparable [21]. When variables vary largely in level, the role of the value of the higher numerical variable in the comprehensive analysis shall be highlighted, and the role of the comparatively lower numerical variable will be weakened if the raw data is directly used for analysis. Accordingly, to ensure the reliability of the results, it was imperative for standardizing the table and simulation data to eliminate the dimensional influence between variables so that the data could be comparable. Data normalization involved the centralization and compression processing of the data simultaneously, the specific principles were as follows

$$x_{ij}^{*} = \frac{x_{ij} - \bar{x}}{s_j}, \qquad \begin{array}{l} i = 1,2,3,\ldots,n \\ j = 1,2,3,\ldots,p \end{array} \tag{3}$$

where, x_{ij} represents the value of the i-th row and the j-th column, x_{ij} * represents the normalized data of x_{ij}, i refers to the i-th row, and j refers to the j-th column, s_j represents the normalized parameter of column j.

2.2. Path Analysis

Path analysis was conducted to analyze the direct relationship between the impeller geometrical parameters and the pump performance, as well as the indirect coupling relationship between the parameters. Assuming that the p independent variable can be established, x_1, x_2, ... , x_p, the simple correlation coefficients between each of the two variables and the dependent variable y were capable of forming the normalized normal equation to solve the path coefficient:

$$\begin{array}{l} r_{11}\rho_1 + r_{12}\rho_2 + \ldots + r_{1p}\rho_p = r_{1y} \\ r_{21}\rho_1 + r_{22}\rho_2 + \ldots + r_{2p}\rho_p = r_{2y} \\ \ldots \quad \ldots \quad \ldots \quad \ldots \\ r_{p1}\rho_1 + r_{p2}\rho_2 + \ldots + r_{pp}\rho_p = r_{py} \end{array} \tag{4}$$

where, ρ_1, ρ_2, ... , ρ_p was the direct path coefficient. The direct path coefficient represented the direct effect size of the independent variable, while the indirect path coefficient suggested that the independent variable influences the dependent variable by impacting other independent variables. Besides, such coefficients could be calculated using the correlation coefficient r_{ij} and the direct path coefficient ρ_i. The direct path coefficient can be obtained by calculating the inverse matrix of the noted correlation matrix. With the assumption that B_{ij} was the inverse matrix of the correlation matrix r_{ij}, and then the direct path coefficient ρ_i (i = 1, 2, ... , p) was expressed as follows

$$\begin{bmatrix} \rho_1 \\ \rho_2 \\ \ldots \\ \rho_p \end{bmatrix} = \begin{bmatrix} B_{11} & B_{12} & B_{13} & \ldots & B_{1p} \\ B_{21} & B_{22} & B_{23} & \ldots & B_{2p} \\ \ldots & \ldots & \ldots & \ldots & \ldots \\ B_{p1} & B_{p2} & B_{p3} & \ldots & B_{pp} \end{bmatrix} \begin{bmatrix} r_{1y} \\ r_{2y} \\ \ldots \\ r_{py} \end{bmatrix} \tag{5}$$

The path coefficient ρ_{ye} of the remaining term is as expressed in Equation (6). In case the path coefficient ρ_{ye} of the remaining term was smaller, the impeller geometrical parameters and the performance would be well satisfied with the linear relation. Conversely, when the path coefficient ρ_{ye} of the remaining term was larger, it suggested that the test error was larger or other important factors were not introduced.

$$\rho_{ye} = \sqrt{1 - (\sum_{i=1}^{p} r_{iy}\rho_i)} \tag{6}$$

By path analyzing the impeller geometrical parameters and the pump head and the efficiency performance, the results were as listed in Table 2.

Table 2. Path analysis results between impeller geometric parameters and performance.

Factor	Direct Effect	Indirect Effect							
		$\gamma \to H$	$\beta_2 \to H$	$\varphi \to H$	$Z \to H$	$D_2 \to H$	$b_2 \to H$	$D_0 \to H$	$Y \to H$
γ	0.0076		0.0424	0.0000	−0.0502	−0.0748	−0.0365	−0.0031	0.1825
β_2	0.5091	0.0006		−0.1244	0.2510	0.0299	0.0122	0.0000	−0.0349
φ	−0.3731	0.0000	0.1697		0.1506	0.0299	−0.0487	0.0000	0.0295
Z	0.6025	−0.0006	0.2121	−0.1041		0.015	0.0122	0.0092	0.0322
D_2	0.1795	−0.0032	0.0849	−0.0622	0.0502		0.0487	0.0061	−0.3235
b_2	−0.1461	0.0019	−0.0424	−0.1244	−0.0502	−0.0598		0.0000	0.3087
D_0	−0.0368	0.0006	0.0000	0.0000	−0.1506	−0.0299	0.0000		0.1007
Y	0.4401	0.0032	−0.0404	−0.0250	0.0441	−0.1320	−0.1025	−0.0084	
Factor	**Direct Effect**	**Indirect Effect**							
		$\gamma \to \eta_1$	$\beta_2 \to \eta_1$	$\varphi \to \eta_1$	$Z \to \eta_1$	$D_2 \to \eta_1$	$b_2 \to \eta_1$	$D_0 \to \eta_1$	$Y \to \eta_1$
γ	−0.0528		0.0400	0.0000	0.0001	−0.2726	0.1070	0.0040	0.2509
β_2	0.4803	−0.0044		−0.1636	−0.0005	0.1090	−0.0269	0.0000	−0.048
φ	−0.4908	0.0000	0.1601		−0.0003	0.1090	0.1076	0.0000	0.0406
Z	−0.0011	0.0044	0.2001	−0.1227		0.0545	−0.0269	−0.0120	0.0443
D_2	0.6542	0.022	0.0800	−0.0818	−0.0001		−0.1076	−0.0080	−0.4445
b_2	0.3228	−0.0132	−0.0400	−0.1636	0.0001	−0.2181		0.0000	0.4242
D_0	0.0480	−0.0044	0.000	0.0000	0.0003	−0.1090	0.0000		0.1383
Y	0.6047	−0.0219	−0.0381	−0.0329	−0.0001	−0.4809	0.2265	0.0110	

3. Results

3.1. The Direct Impact Analysis of the Main Geometric Parameters of Nuclear Reactor Coolant Pump on Its Performance

As is evident from Table 2 the blade number, blade outlet angle, blade wrap angle, area ratio, impeller outlet diameter, and the blade outlet width in the eight impeller geometrical parameters had a large direct impact on the design point head of the nuclear reactor coolant pump, and the other two parameters had very little impact on the head. Among the six parameters with larger direct impact, the blade numbers were the largest; the blade outlet angle, blade wrap angle, area ratio, and impeller outlet diameter were ranked second; and the blade outlet width was the smallest. The direct path coefficient of the impeller blade number was 0.6025, suggesting that the blade number was the most critical in the geometrical parameters of the reactor coolant pump. When blade numbers were changed, the work efficiency of the impeller changed greatly, and its head also changed significantly. In a particular range, the head of the pump would definitely rise with the increase in the blade numbers. The direct path coefficients of the impeller blade outlet angle and area ratio were 0.5091 and 0.4401, respectively, which confirmed that the blade outlet angle and the area ratio on the head of the nuclear reactor coolant pump also had a major role. When the blade outlet angle was changed, the circumferential component of the absolute velocity at the impeller outlet was also changed. Thereafter, the circumferential component of the absolute velocity at the impeller outlet increased, while the head of the pump increased with the increase of blade outlet angle. As for area ratio the reaction of impeller and guide vane matching relationship, one of the important physical quantities, the performance of the pump was not unilaterally decided by the impeller but by the impeller, guide blade, and volute both (in this study, the volute remains the same, so it need not be considered). In the design process, the inlet area of the guide vane had hardly changed; hence the increase of area ratio in a certain scope was equivalent to the reduction in the impact loss and made the head increase. For the change of the blade wrap angle, the binding force of the fluid in the impeller channel was changed, and the relative velocity liquid angle of the impeller outlet was also changed. In a specific range, the restraint of the fluid in the impeller channel increased, while the blade wrap angle increased, however, the relative flow angle of the impeller outlet decreased, with the head. The direct path coefficient of the impeller outlet diameter was 0.1795, which indicated that in a certain range, the energy of the fluid increases, and the head also rises. The direct path coefficient of impeller outlet width minimum was −0.1461, which suggested that within a certain range, an increase of blade outlet width can make the impeller outlet edge overtilted, and a larger secondary flow would appear, resulting in the head decrease.

From Table 1, it can be observed that amongst the eight impeller geometric parameters, the blade outlet angle, blade wrap angle, blade outlet width, impeller outlet diameter, and area ratio had a greater direct impact on the efficiency performance of nuclear reactor coolant pump design point, while the other three parameters had a relatively smaller impact on its efficiency. Among the five parameters with larger direct impact, the impeller outlet diameter was the largest; the blade outlet angle, blade wrap angle, and area ratio ranked second; and blade outlet width was the smallest. The direct path coefficient of impeller outlet diameter reached 0.6542, which confirmed that impeller outlet diameter was the most critical of all the geometrical parameters of the nuclear reactor coolant pump.

The efficiency of nuclear reactor coolant pump varies with the impeller outlet diameter and the flow condition of the impeller outlet. This indicated that, within a certain range, with the increase of the impeller outlet diameter the impeller outlet speed is reduced, the impact loss between the blade wheel and guide vane decreases, and the hydraulic efficiency of the pump increases. The direct path coefficient of the area ratio between the impeller and the guide vane was 0.6047, which substantiated that increasing the impeller outlet area in a certain range makes the impeller and guide vane match better, reducing the impact loss and increasing the efficiency. The direct path coefficient of the impeller blade outlet angle and the blade wrap angle were 0.4803 and −0.4908, respectively, which confirmed that the blade angle and the blade wrap angle exert the main impact on the design point efficiency of the nuclear reactor coolant pump. Additionally, within a certain range, with the increase in the blade outlet angle, the circumferential component of the absolute speed at the impeller outlet increases, and the efficiency also increases. With the increase of the blade wrap angle, the fluid in the flow channel [22] gets restrained by the stronger blades, while the excessive flow channel increases the friction loss and decreases the efficiency of the pump. The direct path coefficient of the blade outlet width reached 0.3228, which suggested that in a particular range, with the increase of the blade outlet width, the pump can increase the efficiency.

3.2. *Indirect Effect Analysis of the Geometric Parameters of the Nuclear Reactor Coolant Pump on Its Performance*

Besides the direct impact on pump performance, the impeller geometry parameters have different degrees of mutual influence. The principle of path analysis is Correlation coefficient = direct path coefficient + indirect path coefficient, i.e., when a parameter changes, it not only has a direct impact on the performance, but also exerts an indirect impact on the performance by changing the other geometric parameters.

From the indirect path coefficient listed in Table 2, it is evident that for the head index, the indirect path coefficient of the outlet lean angle and the impeller inlet diameter were 0.0603 and −0.0792, respectively. This indicates that the outlet lean angle and the impeller inlet diameter exerted little indirect impact on the head by changing the other geometrical parameters, primarily by changing the area ratio ($\gamma \rightarrow Y \rightarrow H = 0.1825$, $D_0 \rightarrow Y \rightarrow H = 0.1007$). The indirect path coefficient of the blade outlet angle was 0.1344, indicating that blade outlet angle indirectly strengthens the head by changing the other geometrical parameters, and the outlet lean angle had the effect of reducing the head by changing the blade wrap angle ($\beta_2 \rightarrow \varphi \rightarrow H = -0.1244$). However, the outlet lean angle reinforced the head by blade numbers ($\beta_2 \rightarrow Z \rightarrow H = 0.251$). The indirect path coefficient of the blade wrap angle was 0.311, indicating that the blade wrap angle had an indirect strengthening effect on the head by changing the other geometrical parameters, and the blade wrap angle strengthens the head by changing the blade outlet angle and the blade numbers ($\varphi \rightarrow \beta_2 \rightarrow H = 0.1697$, $\varphi \rightarrow Z \rightarrow H = 0.1506$). The indirect path coefficient of the impeller blade numbers was 0.1868, suggesting that the impeller blade numbers exerted little indirect impact on the head by changing the other geometrical parameters, and the impeller blade numbers strengthened the head by changing the blade outlet angle ($Z \rightarrow \beta_2 \rightarrow H = 0.2121$), and yet the impeller blade numbers reduced the head by the blade wrap angle ($Z \rightarrow \varphi \rightarrow H$

$= -0.1041$). The indirect path coefficient of the blade outlet width was 0.0338, indicating that the blade outlet width had slightly enhanced the indirect impact on the head by changing the other geometrical parameters, and the blade outlet width reduced the head by changing blade wrap angle ($b_2 \to \varphi \to H = -0.1244$), and yet the blade outlet width had the effect of reinforcing the head by area ratio ($b_2 \to Y \to H = 0.3087$). The indirect path coefficient of the impeller outlet diameter was −0.199, which implied that the impeller outlet diameter had an abridged indirect effect on the head by changing the other geometrical parameters, and the impeller outlet diameter reduced the head by changing area ratio ($D_2 \to Y \to H = -0.3235$). The indirect path coefficient of the area ratio was −0.261, which suggests that the area ratio had an abridged effect on the head, and the area ratio reduced the head by changing the impeller outlet diameter and the blade outlet width ($Y \to D_2 \to H = -0.132$, $Y \to b_2 \to H = -0.3235$).

For the efficiency index, the indirect path coefficient of impeller outlet diameter was 0.0252, which proved that the impeller inlet diameter had a small indirect impact on the efficiency by changing the other geometrical parameters. The impeller inlet diameter impacted the efficiency by the impeller outlet diameter and the area ratio, while the impeller inlet diameter reduced the efficiency by changing the impeller outlet diameter ($D_0 \to D_2 \to \eta_1 = -0.109$), the impeller inlet diameter increased the efficiency by the area ratio ($D_0 \to Y \to \eta_1 = 0.1383$). The indirect path coefficient of the outlet lean angle was 0.1031, which indicated that the outlet lean angle had little indirect impact on the increase of the efficiency by changing the other geometrical parameters, the outlet lean angle increased the efficiency by the blade outlet width and the area ratio ($\gamma \to b_2 \to \eta_1 = 0.107$, $\gamma \to Y \to \eta_1 = 0.2509$), while it decreased the efficiency by area ratio ($\gamma \to D_2 \to \eta_1 = -0.2726$). The indirect path coefficient of the blade outlet angle was −0.1344, showing that blade outlet angle had an indirect impact on the decrease of the efficiency by changing the other geometrical parameters, and the blade outlet angle decreased the efficiency by the blade wrap angle ($\beta_2 \to \varphi \to \eta_1 = -0.1636$); the blade outlet angle decreased the efficiency by the impeller outlet diameter ($\beta_2 \to D_2 \to \eta_1 = 0.109$). The indirect path coefficient of the blade wrap angle was −0.417, hinting that the blade wrap angle had an indirect impact on the increase of the efficiency by changing the other geometrical parameters, and the blade wrap angle increased the efficiency by blade outlet angle, impeller outlet diameter, and the blade outlet width ($\varphi \to \beta_2 \to \eta_1 = 0.1601$, $\varphi \to b_2 \to \eta_1 = 0.109$, $\varphi \to b_2 \to \eta_1 = 0.1076$). The indirect path coefficient of the impeller blade numbers was 0.1417, which demonstrated that the impeller blade numbers had an indirect impact on the increase of the efficiency by changing the other geometrical parameters and the impeller blade numbers increased the efficiency by the blade outlet angle ($Z \to \beta_2 \to \eta_1 = 0.2001$), while the impeller blade numbers decreased the efficiency by the blade wrap angle ($Z \to \varphi \to \eta_1 = -0.1227$). The indirect path coefficient of the impeller outlet diameter was −0.54, denoting that the impeller outlet diameter had an indirect impact on the decrease of the efficiency by changing the other geometrical parameters, and then it decreased the efficiency by the blade outlet width and the area ratio ($D_2 \to b_2 \to \eta_1 = -0.1076$, $D_2 \to Y \to \eta_1 = -0.4445$). The indirect path coefficient of the blade outlet width was −0.0106, which implied that the blade outlet width had an indirect impact on the decrease of the efficiency by changing the other geometrical parameters, and the blade outlet width decreased the efficiency by the area ratio ($b_2 \to Y \to \eta_1 = 0.4242$), while it decreased the efficiency by the blade wrap angle and the impeller outlet diameter ($b_2 \to \varphi \to \eta_1 = -0.1636$, $b_2 \to D_2 \to \eta_1 = -0.2181$).

It was acquired by analyzing the indirect path coefficient between the different geometric parameters that the influence weight of each parameter was different when the different performances served as the index ($\beta_2 \to Z \to H = 0.251$, $\beta_2 \to Z \to \eta_1 = -0.0005$). Under the index of the same performance, the influence weight of each parameter was directional ($\beta_2 \to \varphi \to H = -0.1244$, $\varphi \to \beta_2 \to H = 0.1697$). Under the small indirect path coefficient, it did not mean that there was little interaction between the factor and other factors (the indirect path coefficient of the impeller outlet diameter reaches −0.0252, $D_0 \to D_2 \to \eta_1 = -0.109$, $D_0 \to Y \to \eta_1 = 0.1383$), which neutralized the indirect effect on each parameter.

3.3. Analysis of Residual Path Coefficient

The residual path coefficient is a critical value that determines the satisfaction of the linear relation between parameters and performances. In this paper, the determined path coefficients and the residual path coefficients between the eight geometrical parameters of the impeller and the performance of the nuclear reactor coolant pump are listed in Table 3.

Table 3. Residual path coefficient between impeller geometric parameters and performance.

Performance	Determine Path Coefficient	Residual Path Coefficient
H	0.8421	0.2909
η_1	0.6678	0.5541

It can be seen from the table that the determined path coefficient between the head and the performance reached 0.8421, which indicated that the eight parameters selected could calculate it accurately through the linear relationship, while the path coefficient of efficiency was determined as 0.6678. This proved that selecting the eight parameters and efficiency cannot satisfy the linear relationship, there may be larger errors, or the main parameter was not selected. However, according to the actual situation, as the flow of the components of nuclear reactor coolant pump, the impeller was employed to convert mechanical energy into potential energy, so that the head primarily becomes dependent on the impeller geometric parameters. For efficiency, the impeller was only a part of the nuclear main pump flow components, and the influence of the impeller geometry on the efficiency was greater than the selected seven. Thus, the determined path coefficient was normally not large, and hence the calculation process was accurate. The eight parameters selected could have been used as a representation of efficiency by linear. To visually represent the relationship between the efficiency index, the lift index, and the parameters, please refer the path diagram (shown in Figure 3).

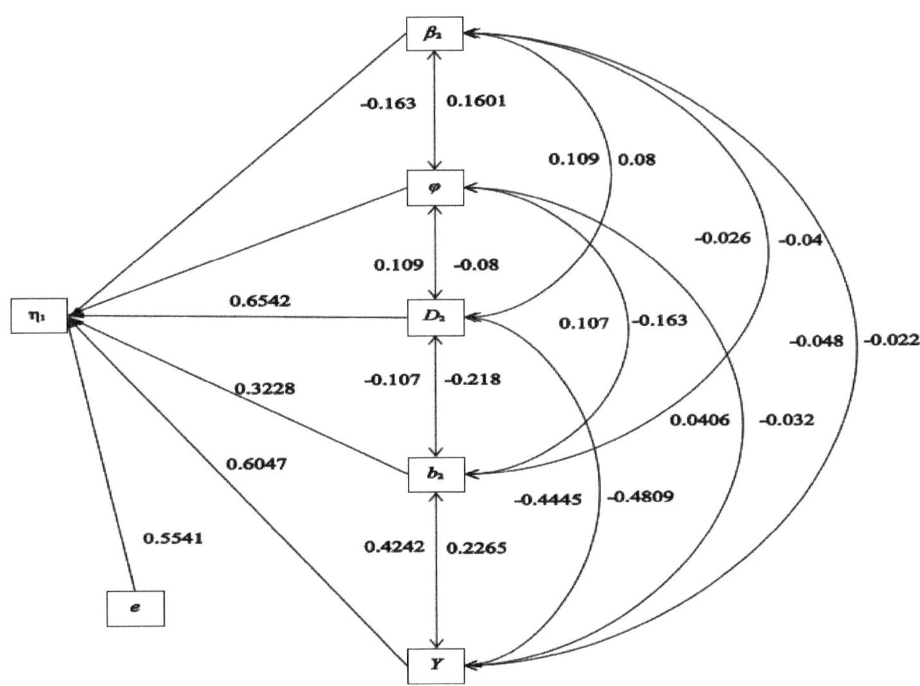

(**a**) The path diagram between the impeller geometrical parameters and the efficiency.

Figure 3. *Cont.*

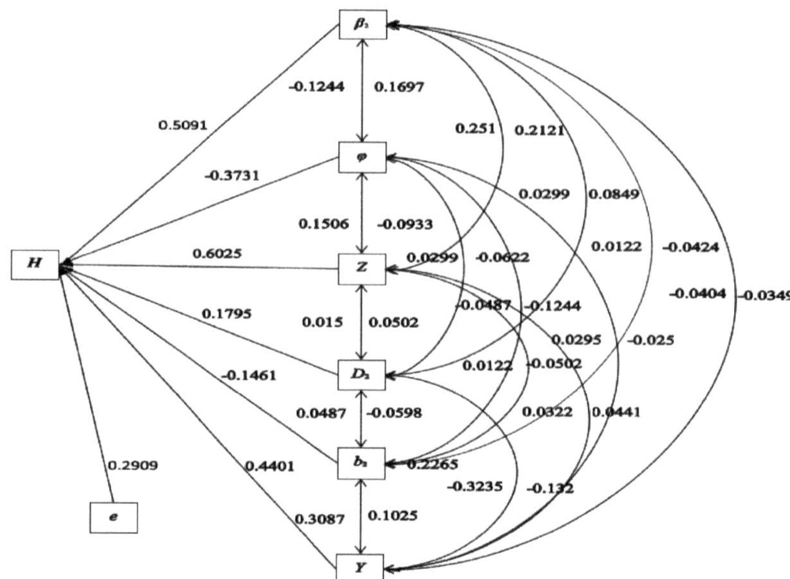

(b) The path diagram between the impeller geometrical parameters and the head

Figure 3. The path diagram between the impeller geometrical parameters and the performance.

3.4. Optimal Parameter Selection

By selecting the eight major parameters of the impeller, efficiency, and the head, respectively, as the performance evaluation indices, the geometrical parameters of the different indices have proved to have different effect sizes. When efficiency was the performance index, five geometrical parameters exerted the greatest influence on efficiency in line with the size of influence weight: blade outlet angle > impeller outlet diameter > blade wrap angle > area ratio > blade outlet width. When the head was the performance index, the six geometric parameters had the greatest impact on efficiency; according to influence weight: impeller blade numbers > blade outlet angle > blade wrap angle > area ratio > impeller outlet diameter > blade outlet width. This paper considers efficiency as the main performance index, the optimal parameters were selected in line with the results of partial correlation analysis and path analysis, and the results were as listed under Table 4.

Table 4. Optimal combination of impeller geometrical parameters.

Factor	γ	β_2	φ	Z	D_2	b_2	D_0	Y
Optimal results	23°	30°	115°	5	770 mm	200 mm	555 mm	1.002

4. Experiment Verification

To verify the effectiveness of this optimization method, the model pump, developed as per the specified parameters, was tested and verified. The model pump had the following specifications; design flow $Q_M = 104$ m^3/h, head $H_M = 3.6$ m, rotating speed $n = 1480$ r/min, and specific speed $n_s = 351$. The model pump and the reactor coolant pump size had a ratio of 5.56. The transient performance testbed for reactor coolant pump was presented in Figure 4. To complete the collection, the flow measurement used the LWGY-type turbine flow sensor instrument supplied by the Nanjing Ditai Electromechanical Equipment Co. Ltd (Nanjing, China), turbine flow meter diameter of DN125, output current signal of 4–20 mA, and acquisition accuracy of 0.5 grade. The instantaneous flow signal was acquired from Beijing Altai Technology Development Co., Ltd. We used a production USB3200 type data acquisition card with rotating speed function and a moment sensor: ZJ-type rotating speed and the moment sensor supporting WJCG dynamometer acquisition pump shaft speed, moment, and other data; its working principle was magnetoelectric conversion and electric phase difference, the measurement range of the moment was 0–50 NM, the number of teeth was 180, the precision was 0.2%, and the speed range was 0–5000 r/min. In the test process, the model pump was first, and then the outlet valve was adjusted after its operations were stabilized so that the pump could be shut down under the rated working condition. The start-up curves of the three groups with starting time of approximately 2 s, 4.5 s, and 8.5 s, respectively, were obtained through the coupling to connect the flywheels with different moments of inertia to reduce the starting acceleration of the unit. Through the similar conversion of the model pump, the starting characteristic curves of the three groups of different starting accelerations of the nuclear reactor coolant pump were determined as in Figure 5.

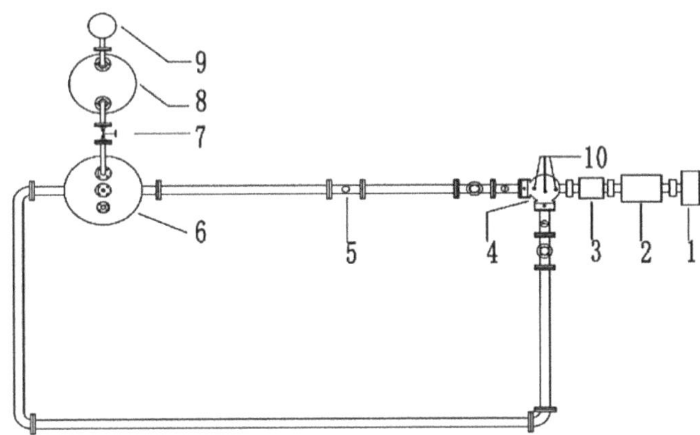

Figure 4. Test device for the coast-down transition process of the nuclear reactor coolant pump.

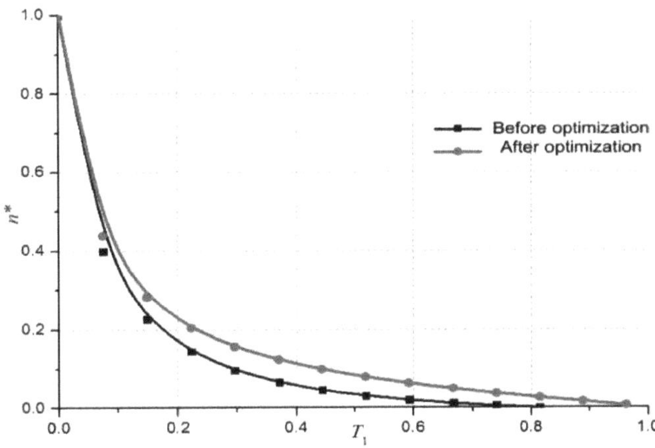

Figure 5. Change of rotational speed before and after the optimization of the nuclear reactor coolant pump during the coast-down transition process.

A dimensionless conversion was carried out with the optimized coast-down time of the pump as a cycle. From Figure 5, it is evident that the optimization had a great influence on the change of the rotation speed [23] in the coast-down transition of the reactor coolant pump. Because of the small moment of inertia of the impeller itself, the small changes in the impeller geometrical parameters and the amount of change in the moment of inertia can be neglected. Figure 3 illustrates that the curves of speed variations in the coast-down transition were different before and after the optimization. The preoptimization speed decreased to zero at nearly 0.8 T_1, and the optimized speed dropped to zero at nearly 1 T_1. Throughout the coast-down transition, the size of the speed before optimization was greater than the size of the speed of coast-down after optimization. This proved that almost the entire coast-down characteristics had been improved through the structural optimization of the nuclear reactor coolant pump.

5. Conclusions

The inertia moment energy stored in the rotor system was utilized to keep the nuclear reactor coolant pump running during coast-down transition. Considering the additional losses of the impeller were primarily caused by the nuclear reactor coolant pump under the off-design condition, the energy loss was reduced and the time of coast-down was delayed by optimizing the main structural parameters of the impeller.

(1) According to the energy conservation law, the calculation equation of coast-down time and the energy utilization of the inertia moment storage were listed, and the basis of coast-down optimization was given based on the reducing energy loss in the coast-down transition. Hydraulic optimization design of the reactor coolant pump impeller was carried out by combining orthogonal optimization test and CFD simulation software, while the hydraulic characteristics of the calculation model were also completed.

(2) The correlation between the impeller geometric parameters and efficiency, head, and different geometric parameters was computed, and the main parameters affecting efficiency and the pressure head were determined. By the path analysis on the results of hydraulic characteristics, the direct influence of geometric parameters on efficiency and head and the indirect influence of the geometric parameters on other parameters that changed the efficiency and head were ascertained.

(3) The efficiency of the pump was the target, the head was the constraint condition, and, combined with the calculation results of the partial correlation analysis and path analysis, the optimal parameters were selected as $\gamma = 23°$, $\beta_2 = 30°$, $\varphi = 115°$, $Z = 5$, $b_2 = 200$ mm, $D_2 = 770$ mm, $D_0 = 555$ mm, and $Y = 1.002$, respectively.

Author Contributions: Y.Z.: Experiments and simulations; X.S.: The writing and revision of the paper; X.W.: Ideas and fund support of the paper; R.Z., Q.F. and H.Z.: Experimental and simulated data processing.

References

1. Gao, H.; Gao, F.; Zhao, X.; Chen, J.; Gao, X. Analysis of reactor coolant pump transient performance in primary coolant system during start-up period. *Ann. Nucl. Energy* **2013**, *54*, 202–208. [CrossRef]
2. Tsukamoto, H.; Matsunaga, S.; Yoneda, H.; Hata, S. Transient characteristics of a centrifugal pump during starting/stopping periods: Analysis based on rotating circular cascade model and criterion for quasi-steady change. *Trans. Jpn. Soc. Mech. Eng. B* **1986**, *52*, 1291–1299. [CrossRef]
3. Wu, D.; Wang, L.; Hao, Z.; Li, Z.; Bao, Z. Experimental study on hydrodynamic performance of a cavitating centrifugal pump during transient operation. *J. Mech. Sci. Technol.* **2010**, *24*, 575–582. [CrossRef]
4. Li, Z.; Wu, D.; Wang, L.; Huang, B. Numerical simulation of the transient flow in a centrifugal pump during starting period. *J. Fluids Eng.* **2010**, *132*, 081102. [CrossRef]
5. Chalghoum, I.; Elaoud, S.; Akrout, M.; Taieb, E.H. Transient behavior of a centrifugal pump during starting period. *Appl. Acoust.* **2016**, *109*, 82–89. [CrossRef]
6. Elaoud, S.; Hadj-Taïeb, E. Influence of pump starting times on transient flows in pipes. *Nucl. Eng. Des.* **2011**, *241*, 3624–3631. [CrossRef]
7. Farhadi, K.; Bousbia-Salah, A.; D'Auria, F. A model for the analysis of pump start-up transients in Tehran Research Reactor. *Prog. Nucl. Energy* **2007**, *49*, 499–510. [CrossRef]
8. Farhadi, K. Transient behaviour of a parallel pump in nuclear research reactors. *Prog. Nucl. Energy* **2011**, *53*, 195–199. [CrossRef]
9. Farhadi, K. The effect of retarding torque during a flow transient for Tehran research reactor. *Ann. Nucl. Energy* **2011**, *38*, 175–184. [CrossRef]

10. Alatrash, Y.; Kang, H.O.; Yoon, H.G.; Seo, K.; Chi, D.Y.; Yoon, J. Experimental and analytical investigations of primary coolant pump coastdown phenomena for the Jordan research and training reactor. *Nucl. Eng. Des.* **2015**, *286*, 60–66. [CrossRef]
11. Lu, Y.; Zhu, R.; Fu, Q.; Wang, X.; An, C.; Chen, J. Research on the structure design of the LBE reactor coolant pump in the lead base heap. *Nucl. Eng. Technol.* **2018**, *51*, 546–555. [CrossRef]
12. Lu, Y.; Zhu, R.; Wang, X.; Fu, Q.; Li, M.; Si, X. Study on gas-liquid two-phase all-characteristics of CAP1400 nuclear main pump. *Nucl. Eng. Des.* **2017**, *319*, 140–148. [CrossRef]
13. Lu, Y.; Zhu, R.; Wang, X.; Fu, Q.; Ye, D. Study on the complete rotational characteristic of coolant pump in the gas-liquid two-phase operating condition. *Ann. Nucl. Energy* **2019**, *123*, 180–189.
14. Kim, J.H.; Ahn, H.J.; Kim, K.Y. High-efficiency design of a mixed-flow pump. *Sci. China Technol. Sci.* **2010**, *53*, 24–27. [CrossRef]
15. Varchola, M.; Hlbocan, P. Geometry design of a mixed flow pump using experimental results of on internal impeller flow. *Procedia Eng.* **2012**, *39*, 168–174. [CrossRef]
16. Long, Y.; Yin, J.L.; Wang, D.Z.; Li, T.B. Study on the effect of the impeller and diffuser blade number on reactor coolant pump performances. In Proceedings of the 7th International Conference on Pumps and Fans (ICPF 2015), Hangzhou, China, 18–21 October 2015.
17. Zhou, F.M.; Wang, X.F. The effects of blade stacking lean angle to 1400 MW canned nuclear coolant pump hydraulic performance. *Nucl. Eng. Des.* **2017**, *325*, 232–244. [CrossRef]
18. Zhu, R.S.; Xing, S.B.; Fu, Q.; Li, T.B.; Wang, X.L. Study on transient flow characteristics of AP1000 nuclear reactor coolant pump under exhaust transit condition. *At. Energy Sci. Technol.* **2016**, *50*, 1040–1046.
19. Gao, H.; Gao, F.; Zhao, X.; Chen, J.; Cao, X. Transient flow analysis in reactor coolant pump systems during flow coastdown period. *Nucl. Eng. Des.* **2011**, *241*, 509–514. [CrossRef]
20. Long, Y.; Zhu, R.; Wang, D.; Yin, J.; Li, T. Numerical and experimental investigation on the diffuser optimization of a reactor coolant pump with orthogonal test approach. *J. Mech. Sci. Technol.* **2016**, *30*, 4941–4948.
21. Kang, C.; Mao, N.; Zhang, W.; Gu, Y. The influence of blade configuration on cavitation performance of a condensate pump. *Ann. Nucl. Energy* **2017**, *110*, 789–797. [CrossRef]
22. Zhang, K.; Yuan, J.; Sun, W.; Si, Q. Internal flow characteristics of residual heat removal pump during different starting periods. *J. Drain. Irrig. Mach. Eng.* **2017**, *35*, 192–199.
23. Fu, Q.; Zhang, B.; Zhu, R.; Cao, L. Effect of rotating speed on cavitation performance of nuclear reactor coolant pump. *J. Drain. Irrig. Mach. Eng.* **2016**, *34*, 651–656.

8

Temporal Feature Selection for Multi-Step Ahead Reheater Temperature Prediction

Ning Gui [1], Jieli Lou [2], Zhifeng Qiu [3,*] and Weihua Gui [3]

[1] School of Comuputer Science and Engineering, Central South University, Changsha 410000, China
[2] School of Mechanical Engineering and Automation, Zhejiang Sci-Tech. University, Hangzhou 310000, China
[3] School of Automation, Central South University, Changsha 410000, China
* Correspondence: zhifeng.qiu@csu.edu.cn

Abstract: Accurately predicting the reheater steam temperature over both short and medium time periods is crucial for the efficiency and safety of operations. With regard to the diverse temporal effects of influential factors, the accurate identification of delay orders allows effective temperature predictions for the reheater system. In this paper, a deep neural network (DNN) and a genetic algorithm (GA)-based optimal multi-step temporal feature selection model for reheater temperature is proposed. In the proposed model, DNN is used to establish a steam temperature predictor for future time steps, and GA is used to find the optimal delay orders, while fully considering the balance between modeling accuracy and computational complexity. The experimental results for two ultra-super-critical 1000 MW power plants show that the optimal delay orders calculated using this method achieve high forecasting accuracy and low computational overhead. Moreover, it is argued that the similarities of the two reheater experiments reflect the common physical properties of different reheaters, so the proposed algorithms could be generalized to guide temporal feature selection for other reheaters.

Keywords: reheat steam temperature; temporal feature selection; delay order prediction; deep neural network; genetic algorithm

---

1. Introduction

Steam reheating plays an important role in power plants. It can increase thermal efficiency by 2% and it can also reduce steam humidity and improve the safety of the final stage's blade [1,2]. However, due to the complexity of the many influential factors, it is difficult to maintain the reheat steam temperature within a certain range [3]. For instance, the reheater steam temperature of two ultra-super-critical 1000 MW units investigated in this paper may fluctuate between 565 °C and 610 °C, while the normal reheater outlet steam temperature is 603 °C with tolerable fluctuation within the range of 503 to 608 °C [4] (the specific threshold may vary with the type of reheater). A temperature that is too high will cause damage to the metal material, while a temperature that is too low will reduce the thermal cycle efficiency [5]. Therefore, finding features that affect the modeling target and analyzing the extent of these features are crucial for the system's safety and efficiency.

A reheater system is a typical nonlinear hysteresis thermal system, which is highly coupled, complex, and impacted by many factors [6,7]. The selection of the most related features from a large variety of sensors is important for the realization of effective control [8]. Traditional feature selections are normally developed on the basis of mass balance, energy balance, and dynamic principles, which rely greatly on human expertise and normally require a long modeling time [9–11]. Recently, researchers have increasingly adopted the data-driven methodology that extracts features directly from huge amounts of accumulated process data [12–14]. Li et al. [15] analyzed operation parameters in power plants by correlation analysis to improve boiler efficiency. Wei et al. [16] used principle component analysis to transform higher-dimensional original data to lower-dimensional principle components, which were employed as the inputs to the NO_x emission model to reduce memory storage requirements and computational costs for data analytics. Buczyński et al. [17] judged whether features could exert substantial effects on a CFD (phase fluidised bed)-based model using sensitivity analysis to predict the performance of a domestic central-heating boiler fired with solid fuels. Pisica et al. [18] chose mutual information to assess the relevance of feature subsets in order to determine the operating states of power systems. Wang et al. [19] utilized the outputs of an improved random forest algorithm as inputs of a back propagation neural network to weight the importance of features and to improve the prediction accuracy of NO_x.

The above research works mainly focused on finding the most related features with respect to the modeling target, which only explores one dimension from all possible relationships. In practice, for the complex process, each feature may have a temporal effect on the modeling target [20]. For instance, some features might have a rapid impact on the target, while some other features might only display certain time-delay effects, i.e., effects after a certain period of time. In order to cover the temporal effects, multi-step features are often accumulated for data-driven modeling in the feature engineering process. Normally, the larger the delay order (number of steps selected) of a feature is, the more information it contains [21]. However, overly large delay orders of features may lead to overfitting, which may cause poor performance on unseen instances [22] and significantly increase memory storage and computational complexity for data analysis [23]. Therefore, it is necessary to find an optimal delay order set for each feature while maintaining a good balance between modeling accuracy and computational economy.

A few researchers have investigated the temporal feature selection problem. Lv et al. [21] used particle swarm optimization to determine delay orders and used a least square support vector machine (SVM) to predict the bed temperature of circulating fluidized bed boilers. However, this method suffered from computational complexity when modeling large-scale data sets. Shakil et al. [24] applied genetic algorithms to estimate the time delay of soft sensors for NO_x and O_2. Although these studies achieved good results on the delay order selection, their modeling targets were only for one particular future time instance, which has the potential not to include the features that impose too rapid or too slow impacts on the target. These approaches also provided little discussion on whether the generated delay order could be used to guide future modeling processes for similar equipment.

To address the optimal feature selection of delay orders for multi-step prediction, a method that combines a deep neural network (DNN) and a genetic algorithm (GA) is proposed. A prediction target with multiple future time steps is introduced to explore features that have rapid or slow effects. A DNN model is used to establish a steam temperature predictor [25] for the next 20 steps. A GA is proposed to find the optimal delay orders with the objective function of balancing modeling accuracy and computational complexity. The proposed method is tested in two 1000 MW coal-fired power plants, namely unit 3 and unit 4, which use more than two million records. The results of the two units display similar sets of delay orders for each feature, reflecting that the physical properties of reheater steam systems are similar to some extent.

The rest of this paper is organized as follows: Section 2 briefly describes the reheater system and proposes the problem statement. Section 3 establishes an objective function for model evaluation. The detailed introduction of the delay order selection mechanism is provided in Section 4. Section 5 presents experiments and discussions. Discussions and possible directions for future work are provided in the final section.

2. System Description and Problem Statement

2.1. Description of Reheater System

A reheater is a set of tubes located in a boiler, the main purpose of which is to avoid excess moisture in steam at the end of expansion to protect the turbine. The exhaust steam from the high-pressure turbines passes through these heated tubes to collect more energy before driving the intermediate- and then low-pressure turbines. The conceptual structure of the reheater unit is shown in Figure 1.

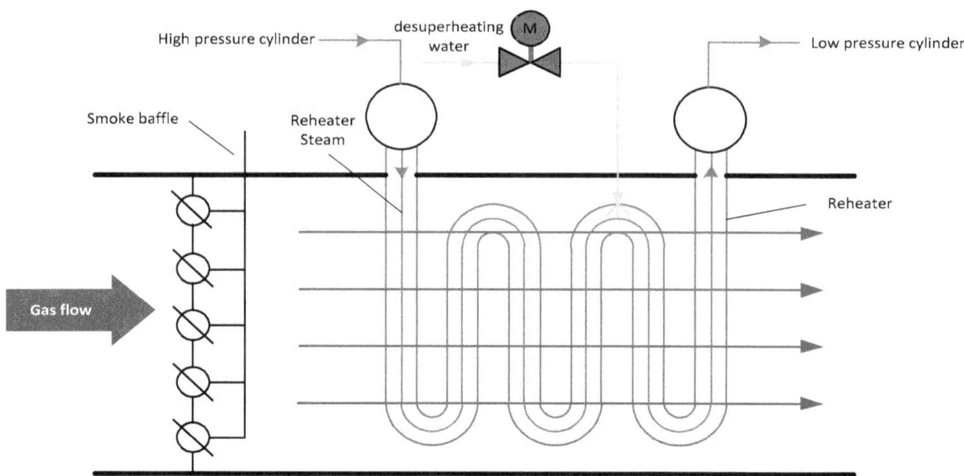

Figure 1. Reheater structure.

After the high-pressured turbine, the exhaust pressure and temperature at the inlet of the reheater are about 35–37 kg/cm^2 and 345–355 °C, respectively. A reheater is designed in the shape of a serpentine tube in order to increase the heated area. The hot smoke generated by the combustion of coal transfers heat to the reheater, meaning that the temperature of steam in the reheater rises. The steam temperature at the outlet of the reheater is kept around 603 °C. Reheater steam with high-temperature and high-pressure characteristics is collected into the high-temperature reheat steam container. A similar process is performed again in the low-pressure cylinder.

Table 1 denotes the influential features of our modeling target, which is the outlet steam temperature of the reheater. Many features affect the reheat steam temperature, such as the inlet steam temperature, inlet gas temperature, smoke baffle opening, etc. Also, these variables have different inertias toward the reheater outlet steam temperature. Therefore, these variables and their hysteresis times should be considered in the prediction model. Here, the previous values of the steam outlet temperature are also used in the modeling process and the multi-step steam temperatures are used as the outputs of the model. In order to simplify our discussion, the major factors are referred to by the notations shown in Table 1.

Table 1. Influential parameters for the temperature of the outlet steam.

Feature	Unit	Inertia	Not.
Inlet steam temperature	°C	small	$Steam^t$
Inlet steam pressure	Mpa	small	$Steam^p$
Inlet smoke temperature	°C	large	$Smoke^t$
Inlet smoke pressure	Kpa	small	$Smoke^p$
Smoke baffle opening	%	small	$Baffle^o$
Desuperheated water flow	t/h	large	D^{water}
Reheater steam temperature	°C	-	$Steam^o$

2.2. Problem Statement

One of the major control concerns of a reheater is the stability of $steam^o$. In respect to the reheater, some features are reheater-uncontrollable, e.g., smoke temperature and pressure. These features might influence the reheater wall temperature and then change the outlet steam temperature. $Steam^o$ has the characteristics of being non-linear and having a large inertia. Due to the change in operation conditions, it may deviate from the expected range. The normal operation changes the smoke flow toward the reheater by adjusting the smoke baffle opening degree. This operation exhibits a long delay before it imposes impacts on temperature. Another method is to spray the desuperheated water to the reheater steam. This method promptly lowers steam temperature, but also reduces the boiler's efficiency. Considering the economic benefits, the first method is always used. The second method is employed only in an emergency, such as when the steam temperature is too high or the working condition is changing.

Similar to the control variables mentioned above, other features also have impacts characterized by different inertias toward the steam temperature. One major concern is the complexity of accurately determining the impact inertia of different features, which highly depends on the physical nature laws of the reheater as well as the operational conditions of the reheater, e.g., combustion stability. One natural choice is to use long delay orders to compose the model inputs. However, the indiscriminate delay order settings make the feature dimension very high and introduce considerable overheads for both storage and computation. Thus, it is important to select the most cost-effective delay order for features while keeping the system model accurate enough.

3. Objective Function for Model Evaluation

3.1. Multi-Step Prediction

In order to predict the temperature trend of steam°, the nonlinear autoregressive exogenous model is presented. Differing from other approaches, the proposed model predicts values not for any given time, but for a set of future moments.

Since the reheater system displays different hysteresis characteristics toward different features, modeling the steam° with both short and long hysteresis parameters is important. A multi-step steam° prediction model, which generates a serial of predictions for the next $n + 1$ time steps, is given in Equation (1).

$$\begin{bmatrix} \hat{y}(t) \\ \hat{y}(t+1) \\ \cdots \\ \hat{y}(t+n) \end{bmatrix} = f(x_1(t-1),\cdots,x_1(t-\tau_1),\cdots,x_k(t-1),\cdots,x_k(t-\tau_k),y(t-1),\cdots,y(t-\tau_y)) \ , \quad (1)$$

where t is the current time, $t + n$ is the n-th future moment, x_k is the k-th independent variable, y is a dependent variable, τ_k represents the time delay order corresponding to x_k, and τ_y is the time delay order of dependent variable y.

3.2. Optimization Function

The prediction target increases the forecast performance for the next $n + 1$ time steps by selecting the most appropriate delay order. However, the total number of delay orders is proportional to the computational complexity and opposite to the model accuracy. Thus, the optimization goal defined is to strike a balance between the computational complexity and modeling accuracy. Accordingly, the objective function is used to minimize the total number of delay orders to minimize the computational complexity. Furthermore, the total number of delay orders is kept as high as possible but within a certain range in order to keep the prediction error low enough. Let ε be the maximum acceptable prediction error for the modeling target; thus, another optimization goal is transferred as one constraint, i.e., the prediction error is smaller than or equivalent to ε. Thus, a constrained optimization problem is formulated as Equation (2).

$$\begin{array}{rcl} \min \quad J & = & \tau_y + \sum_{k=1}^{K} \tau_k \\ \text{s.t.} \quad e & = & \frac{1}{m\cdot(n+1)}\|\hat{Y} - Y\|_1 \\ e & \leq & \varepsilon \\ e_{l+1} & \leq & e_l, \ \forall l = 1, 2, \cdots, L \\ 0 & \leq & \tau_y \\ \tau_k & \leq & C \end{array} \quad (2)$$

where K is the total delay orders of inputs, m is the total of test data, n is the n-th future moment, τ_k is the delay order of x_k, and τ_y is the delay order of the dependent variable. J is the total of delay orders. e is the error in total m samples and $n+1$ prediction numbers in the form of mean absolute error (MAE). e_l is the error generated by the l-th iteration. C is the max delay order. ϵ is the upper limit of MAE. $\hat{Y}$ is the prediction value vector and $\hat{Y} = [\hat{y}(t),\ \hat{y}(t+1), \ldots, \hat{y}(t+n)]^T$, Y is the actual value vector, and $Y = [y(t),\ y(t+1), \ldots, y(t+n)]^T$; $\hat{Y}$ and Y have m samples.

4. Delay Order Selection

In order to accurately select the temporal features, two parts—i.e., the DNN-based prediction model and the GA-based optimal feature selection algorithm—are designed. First of all, the GA generates the individuals of different delay order combinations, which are used as the inputs to the DNN. Then, the DNN outputs the multi-step predictions, which are evaluated by the test sets. The evaluated values are employed as fitness values, which are used in the GA.

4.1. Delay Order Optimization

Delay order optimization is performed by the GA algorithm. The schema of GA is shown in Figure 2. The algorithm starts from an initial population with 20 individuals and each individual has 28 genes. These randomly generated genes are divided into seven sections. Each section represents an input parameter and has 4 binary numbers which can delay the order range from 0 to 15. Then, the individuals are evaluated by the fitness function, which returns two fitness values (MAE and the total of orders). The different fitness values are assigned different fitness scores. The smaller the MAE value, the higher the fitness scores. In a case in which the MAE values are very close (the difference is below a certain threshold), the smaller the total number of delay orders, the higher the fitness scores. The fitness score determines the probability of being selected as a parent. The probability of being selected is according to the roulette wheel selection, shown in Equation (3).

$$p_i = \frac{f_i}{\sum_{j=1}^{N} f_j}, \quad (3)$$

where N is the number of individuals in the population, f_i is the fitness of individual i in the population, and p_i is the probability of individual i being selected in the population.

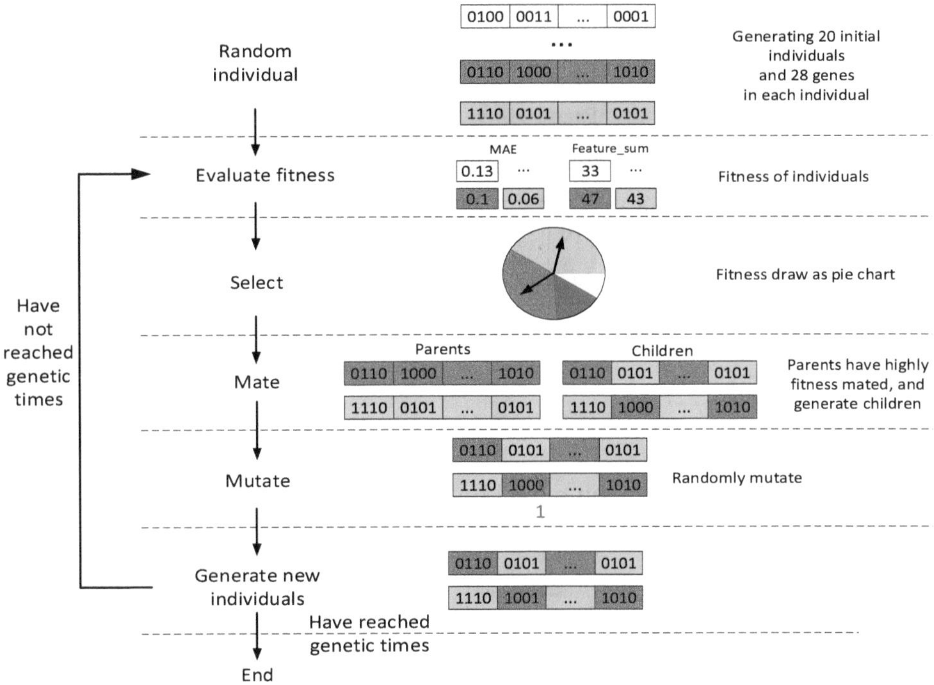

Figure 2. The schematic schema of the genetic algorithm (GA)-based optimized feature selection algorithm. MAE(mean absolute error).

Once the parents are selected, they have a certain probability (p_c) of being mated randomly and generating new individuals. If the parents are not mated, they become new individuals in the new population. Then, the new population has a certain probability (p_m) of deciding whether the individual is mutated. Mutating changes (0 changes to 1, or 1 to 0) randomly. The new individuals are evaluated, selected, mated, and mutated until the number of cycles is reached. At the end of the cycle, the GA obtains the best individuals [26,27].

4.2. Prediction Model

DNN is used to fit the correlation between the future steam° and the historical reheater inlet variables with the accumulated data sets. Figure 3 is the structure of the steam° trend prediction model. Let $m = \tau_1 + \tau_2 + \ldots + \tau_k$ be the total of input dimensions to DNN. The outputs of DNN are $n + 1$ values of steam°. DNN has one input layer, two hidden layers, one output layer, and a large number of neurons. The hypothesis function is shown in Equation (4).

$$h(X) = g(\Theta^3 \cdot g(\Theta^2 \cdot g(\Theta^1 \cdot X))), \tag{4}$$

where X is a vector with m dimensions and Θ^1, Θ^2, and Θ^3 are the weight matrixes between four layers, respectively. $g(\bullet)$ is the activation function.

The cost function of DNN is shown in Equation (5).

$$J(\theta) = \frac{1}{2m \cdot n} \sum_{i=1}^{m} \sum_{j=1}^{n} \left[h(X_j^i) - Y_j^i\right]^2 + \lambda \cdot L2, \tag{5}$$

where m is the total number of samples, n is the total number of output variables, l_k is the number of neurons in the k-th layer, and $h(X_j^i)$ is the prediction value in the i-th sample and the j-th predict value. Y_j^i is the prediction value in the i-th sample and the j-th actual value, λ is the regularization parameter, and $L2$ is the regularization term to limit over-fitting. The goal of the DNN is to minimize Equation (5) with the given sets of features and training samples.

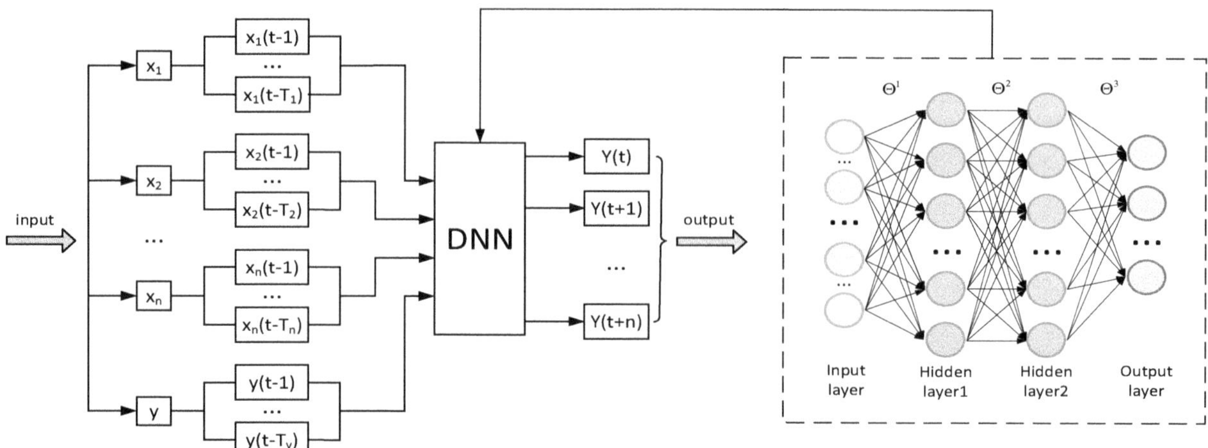

Figure 3. Multi-step prediction model for steam^{o}. DNN—deep neural network.

5. Experiments and Discussion

The data for modeling are collected every 3 s from unit 3 and unit 4 by the distributed control system (DCS). Unit 3 and unit 4 are two ultra-super-critical 1000 MW power plants with the same structure. In our experiment, in total, 7,084,800 records are used for evaluation, in which unit 3 and unit 4, respectively, have 3,542,400 records from 1 May 2016 to 31 August 2016.

5.1. Data Preprocessing

In the data preprocessing process, two steps are taken: Outlier removal and standardization.

Outlier removal: The outliers that violate the physical or technical limitations might affect the model's performance and should be removed before modeling. (1) The points out of the normal range of physical or technical are replaced with the average of adjacent points. For instance, for a certain period, the temperature of steam^{o} should be around 600 °C; thus, the points below 594 °C that violate the steady change characteristics of temperature should be replaced. (2) The errors of D^{water} control time should be modified. Under normal circumstances, the D^{water} control time (more than 0) takes a few minutes. For instance, if the collected data shows that the control time lasts for several hours, the abnormal control time will be modified to a maximum of 3 min.

Standardization: The different features might have different range of values. If these variables are used directly, the feature data with small values may be ignored, while the ones with large dimensions will be selected. Therefore, the Z-score standardization technique [28] is used to scale the data to the ones with a mean value of 0 and a standard deviation of 1, which will speed up the iteration rate of the optimization and convergence.

5.2. Experiment Settings

The parameters of DNN and GA are shown in Table 2. The DNN is a 2-hidden-layer neural network, and the learning rate is set to 0.001. MAE, which is the average absolute differences between predictions and actual observations, is used to evaluate the modeling error. Tanh is chosen as the activation function since it achieves the smallest average MAE compared to other activation functions (e.g., identity, logistic, relu) for the chosen data set.

The 4-month data for unit 3 and unit 4 are divided into 20 different sets. Each set consists of training data from 7 days (about 201,600 records) and test data from 1 day (about 28,800 records).

Table 2. The parameters of DNN and GA.

Neutral Network	Value	GA	Value
Number of hidden layers	2	Number of initial individuals	20
Number of first/second layer neurons	42/23	Mate rate	0.5
Number of outputs	20	Mutate rate	0.2
Activation function	tanh	Number of genes	0–15
Solver	sgd	Iterations	100
Learning_rate	0.001	E	0.14
Λ	0.0001	-	-

5.3. Results and Discussion

This proposed method is evaluated from three different perspectives: Firstly, a one-round simulation is performed with a set of data to demonstrate its capability for finding the optimal delay order for different features; secondly, the experiment is implemented on unit 3 and unit 4 at different times to demonstrate the adaptability of the presented method; finally, the delay order identified with data from the unit 3 is directly used in the modeling process for unit 4 to check its capability for generalization.

(1) Results of the one-round simulation

As for getting the preliminary delay order in unit 3, the data from ~23 July 2016–30 July 2016 is selected as the experiment data. The changes of MAE and the total number of selected orders during the iteration process are shown in Figure 4a. The accuracy level of MAE is set as 0.001. In the early iterations, MAE begins to decrease while the total delay order increases. Then, until MAE stabilizes at 0.13—i.e., the lower limit of MAE—the total delay order decreases. In the later iterations, these criteria remain constant, which indicates that the algorithm is converged. Figure 4b shows each feature's delay order. It can be seen that some features have a larger delay order, e.g., $smoke^p$, which indicates large hysteresis, while in contrast, the order of D^{water} shows timely but transient impacts.

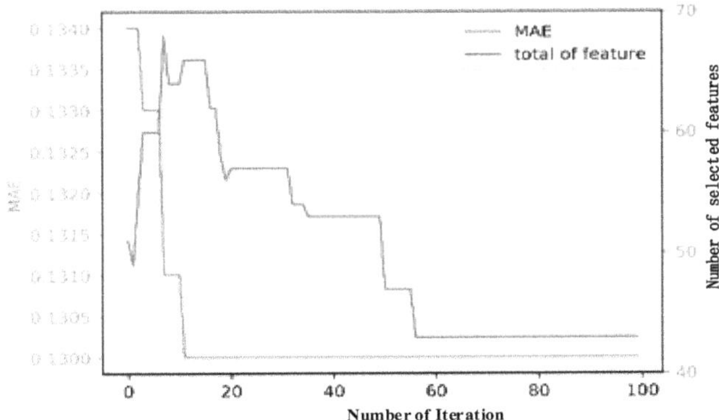

$steam^t$	$steam^p$	$smoke^t$
1	6	8
$smoke^p$	$baffle^o$	D^{water}
10	4	1
$steam^o$		
13		

(**a**) Fitness curve during the experiment (**b**) Optimal delay order

Figure 4. Results for the one-round simulation for unit 3 (~23 July 2016–30 July 2016).

In Figure 5, the forecasting errors in one-minute periods with 20 points in 30 July 2016 are plotted in a box plot which displays the distribution of five different metrics, i.e., minimum, first quartile, median, third quartile, and maximum. Figure 5 shows that MAE increases with the increase in the predicting time step. This is normal, as timely response factors, such as $steam^p$, $smoke^t$, and $baffle^o$, cannot be captured by predictor. However, the median MAE in one minute is less than 0.3 °C, and the average is near 0.1 °C. According to Figure 4b, the maximum delay order of the reheater steam temperature $steam^o$ is 13. This means that the historical data of $steam^o$ have major impacts on the accuracy of the model. It also shows that, in the current system, steam^{o} is not well controlled, as it should kept steady around 600 °C.

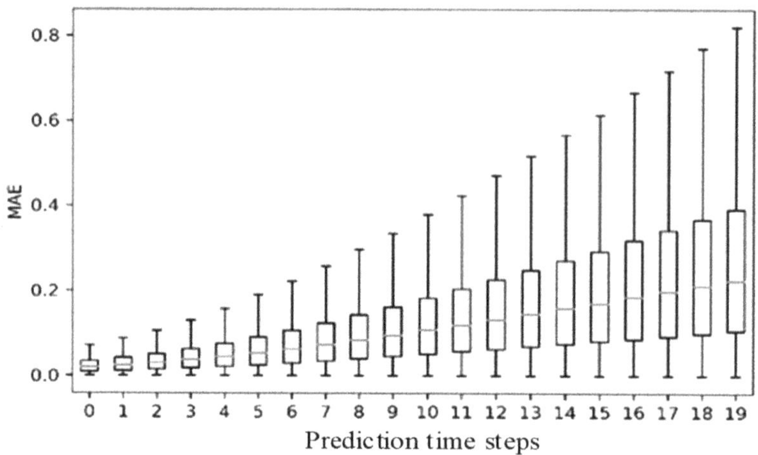

Figure 5. The box error curve.

(2) Comparisons of unit 3 and unit 4 from different perspectives

The feature selection method is tested for both unit 3 and unit 4 based on the operational data from 1 May 2016 to 31 August 2016. Since the records from some days contain too many abnormal data, the data from those days are not used for the model training. As shown in Table 3, the data periods are closed from the intra-comparisons within unit 3 or unit 4 or the inter-comparison between those two units.

Table 3 shows that the general range of the seven studied features has the corresponding length of delay orders with respect to their inertia toward steam^{o}. For all 20 tests, there is no significant deviation regarding MAE. This means that the designed DNN with the selected features as the inputs achieves good convergence. It also shows that the delay orders of smoke^{t} and smoke^{p} are larger than those of steam^{t} and steam^{p}, as the smoke has indirect impacts toward the steam^{o}. Thus, their delay orders are much larger than those of the feature of the inlet steam. D^{water} has a very small delay order due to the fast temporal response toward steam^{o}. For certain periods, the delay orders of D^{water} are zero, e.g., in tests 9, 10, 16, and 18. The zero value is due to the lack of training data for D^{water}. In those periods, the action of spraying de-superheated water is seldom performed. This is due to the insufficient training samples. At these stages, the numbers of sprays are, respectively, 31, 22, 26, and 18, while other tests have about 60 actions, owing to the comparable steam^{o} which is more stable. A similar phenomenon can also be observed for the optimal delay order for baffle^{o}. These results show the importance of the data coverage for the accuracy of feature selection.

Table 3. Results for both unit 3 and 4 (value before "/" is for unit 3 and after is for unit 4). MAE—Mean absolute error.

Test	Sample Date	Steam^{t}	Steam^{p}	Smoke^{t}	Smoke^{p}	Baffle^{o}	D^{water}	Steam^{o}	MAE
1/11	8 May–15 May/1 May–8 May	1/1	6/6	9/8	10/10	4/4	1/1	15/13	0.095/0.116
2/12	16 May–23 May/17 May–24 May	1/2	6/2	9/8	10/11	4/0	1/1	15/15	0.088/0.094
3/13	20 May–27 May/24 May–31 May	1/1	6/6	8/8	10/10	4/4	1/1	13/13	0.129/0.123
4/14	9 June–16 June/5 June –12 June	3/1	6/6	12/8	10/13	0/4	1/1	13/13	0.118/0.111
5/15	17 June–24 June/8 June–15 June	1/1	6/4	9/14	13/10	4/4	1/1	15/15	0.086/0.101
6/16	1 July–8 July/16 June–23 June	3/1	6/6	9/8	15/10	4/0	1/0	15/13	0.100/0.101
7/17	22 July–29 July/17 July–24 July	1/1	6/2	12/11	10/10	0/0	1/2	13/15	0.128/0.117
8/18	6 August–13 August/24 July–31 July	1/1	6/2	9/12	10/13	4/0	1/0	15/13	0.103/0.095
9/19	10 August–17 August/5 August –12 August	1/2	6/6	8/10	13/10	4/2	0/1	13/15	0.132/0.115
10/20	13 August–20 August/17 August–24 August	1/2	6/6	8/9	10/10	4/2	0/1	15/14	0.115/0.115

(3) Determination of delay order

For the purpose of controlling steam^o changes within the ideal range, properly finding a delay order is crucial to accurately describing the hysteresis of features for a prediction model. The variations of delay orders for each feature are shown in Figure 6; the shadow ranges from the maximum to minimum delay order. There is a large overlap between two units, which indicates the existence of common delay orders. The medians of overlap (2, 6, 10, 10, 2, 1, and 14) represent the general level of intervals and may serve as the references for delay orders regarding the steam^o system of ultra-super-critical 1000 MW power plants.

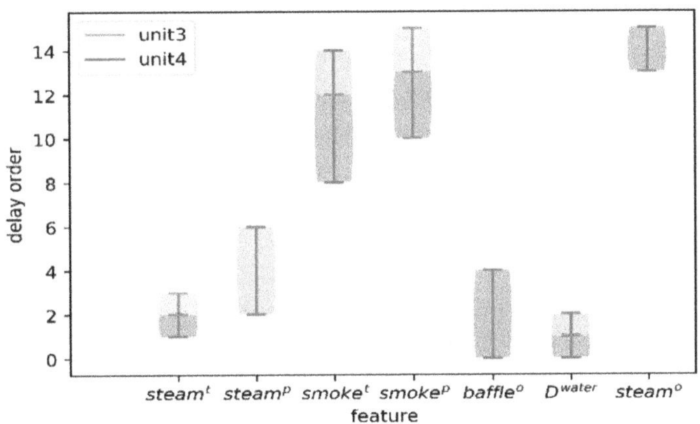

Figure 6. Delay order distribution of the seven features.

The features with delay orders of 2, 6, 10, 10, 2, 1, and 14 generated from the data from unit 3 are used as selected features for the reheater steam temperature prediction. We also adopt the same methods to find the optimal feature distributed for the unit 4. Then, those results are compared with the dataset of test 1 to test 20, which are from unit 4. The orange bars indicate the MAE with the identified delay order. The directly calculated optimal solution is shown by the blue bars. Figure 7 shows the comparisons, which obviously indicate that the MAEs of two cases are approximately equal. The maximum error is only 0.9% (on the 16th day), which means that it is almost the same as the results from the optimal solutions. This shows that the selected delay orders (2, 6, 10, 10, 2, 1, and 14) have good generalization capability, and can, it is argued, represent the physical characteristics of two reheaters.

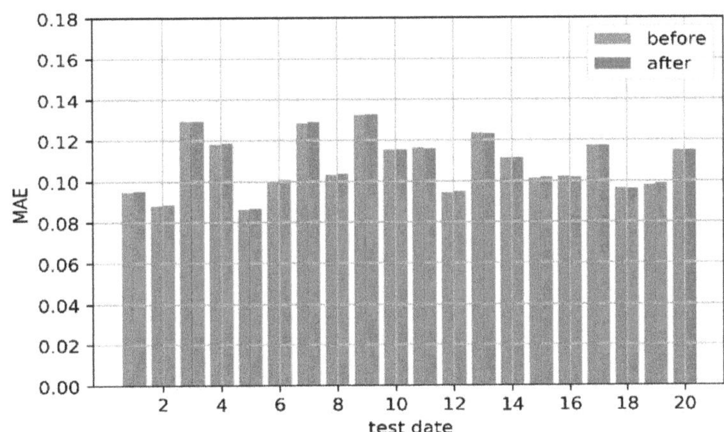

Figure 7. Comparisons between optimal and proposed methods. Blue bars are the MAE of the optimal solution and the orange bars indicate the MAE of the proposed method.

6. Conclusions

For many industrial processes, it is important to find the best feature delay orders as well as features that are most correlated with the prediction targets. In this paper, a delay order identification method based on GA and DNN is proposed. This method adopts the GA to generate candidate feature sets which try to find minimal numbers of features while keeping the MAE of the prediction model low enough. The DNN model is used for modeling processes that generate the multi-step predictions typically demanded in many industrial processes. This method is evaluated with experiments from different perspectives; data from two similar units are used to check whether the found time delays indeed demonstrate the physical characteristics of the underlying systems. The experimental results indicate that two units have similar delay orders and the delay order can be directly used for modeling similar devices with little loss of accuracy.

Of course, many interesting issues still need to be investigated. For instance, our solution limits the temporal feature selection. It is important for the delay order selection method to support both spatial and temporal feature selection. We are investigating the use of an attention mechanism to find the optimal solution for both dimensions. In addition, the GA demands considerable resources and computational costs. We are working to design more computationally efficient methods, e.g., filter-based feature selection for industrial feature processing.

Author Contributions: N.G. proposed the main idea of the method; J.L. and Z.Q. implemented the model and validated the field test. W.G. provided the funding.

Acknowledgments: This work is funded by the Nature Science Foundation of China 61403429, 61621062 and 61772473.

References

1. Zhang, L. *Principle of Boiler*; China Machine Press: Beijing, China, 2011.
2. Ge, Z.; Zhang, F.; Sun, S.; He, J.; Du, X. Energy Analysis of Cascade Heating with High Back-Pressure Large-Scale Steam Turbine. *Energies* **2018**, *11*, 119. [CrossRef]
3. Lee, K.Y.; Ma, L.; Boo, C.J.; Jung, W.H.; Kim, S.H. Intelligent modified predictive optimal control of reheater steam temperature in a large-scale boiler unit. In Proceedings of the 2009 IEEE Power & Energy Society General Meeting, Calgary, AB, Canada, 26–30 July 2009.
4. Fan, Q.G. *Principle of Boiler*; China Electric Power Press: Beijing, China, 2014.
5. Khartchenko, N.V.; Kharchenko, V.M. *Advanced Energy Systems*; CRC Press: Cleveland, OH, USA, 2013.
6. Hogg, B.W.; El-Rabaie, N.M. Multivariable generalized predictive control of a boiler system. *IEEE Trans. Energy Convers.* **1991**, *6*, 82–288. [CrossRef]
7. Liu, X.J.; Kong, X.B.; Hou, G.L.; Wang, J.H. Modeling of a 1000 MW power plant ultra super-critical boiler system using fuzzy-neural network methods. *Energy Convers. Manag.* **2013**, *65*, 518–527. [CrossRef]
8. Suryanarayana, G.; Lago, J.; Geysen, D.; Aleksiejuk, P.; Johansson, C. Thermal load forecasting in district heating networks using deep learning and advanced feature selection methods. *Energy* **2018**, *157*, 141–149. [CrossRef]
9. Staehelin, C.; Schultze, M.; Kondorosi, É.; Mellor, R.B.; Boiler, T.; Kondorosi, A. Structural modifications in Rhizobium meliloti Nod factors influence their stability against hydrolysis by root chitinases. *Plant J.* **1994**, *5*, 319–330. [CrossRef]
10. Gnanapragasam, N.V.; Reddy, B.V. Numerical modeling of bed-to-wall heat transfer in a circulating fluidized bed combustor based on cluster energy balance. *Int. J. Heat Mass Transf.* **2008**, *51*, 5260–5268. [CrossRef]
11. Black, S.; Szuhánszki, J.; Pranzitelli, A.; Ma, L.; Stanger, P.J.; Ingham, D.B.; Pourkashanian, M. Effects of firing coal and biomass under oxy-fuel conditions in a power plant boiler using CFD modelling. *Fuel* **2013**, *113*, 780–786. [CrossRef]
12. Guyon, I.; Elisseeff, A. An introduction to variable and feature selection. *J. Mach. Learn. Res.* **2003**, *3*, 1157–1182.
13. Saeys, Y.; Inza, I.; Larrañaga, P. A review of feature selection techniques in bioinformatics. *Bioinformatics* **2007**, *23*, 2507–2517. [CrossRef]

14. Chandrashekar, G.; Sahin, F. A survey on feature selection methods. *Comput. Electr. Eng.* **2014**, *40*, 16–28. [CrossRef]
15. Li, J.Q.; Gu, J.J.; Niu, C.L. The Operation Optimization based on Correlation Analysis of Operation Parameters in Power Plant. In Proceedings of the 2008 International Symposium on Computational Intelligence and Design, Wuhan, China, 17–18 October 2008.
16. Wei, Z.; Li, X.; Xu, L.; Cheng, Y. Comparative study of computational intelligence approaches for NOx reduction of coal-fired boiler. *Energy* **2013**, *55*, 683–692. [CrossRef]
17. Buczyński, R.; Weber, R.; Szlęk, A. Innovative design solutions for small-scale domestic boilers: Combustion improvements using a CFD-based mathematical model. *J. Energy Inst.* **2015**, *88*, 53–63. [CrossRef]
18. Pisica, I.; Taylor, G.; Lipan, L. Feature selection filter for classification of power system operating states. *Comput. Math. Appl.* **2013**, *66*, 1795–1807. [CrossRef]
19. Wang, F.; Ma, S.; Wang, H.; Li, Y.; Qin, Z.; Zhang, J. A hybrid model integrating improved flower pollination algorithm-based feature selection and improved random forest for NO_X emission estimation of coal-fired power plants. *Measurement* **2018**, *125*, 303–312. [CrossRef]
20. Sun, L.; Li, D.; Lee, K.Y. Enhanced decentralized PI control for fluidized bed combustor via advanced disturbance observer. *Control Eng. Pract.* **2015**, *42*, 128–139. [CrossRef]
21. Lv, Y.; Hong, F.; Yang, T.; Fang, F.; Liu, J. A dynamic model for the bed temperature prediction of circulating fluidized bed boilers based on least squares support vector machine with real operational data. *Energy* **2017**, *124* (Suppl. C), 284–294. [CrossRef]
22. Galicia, H.J.; He, Q.P.; Wang, J. A reduced order soft sensor approach and its application to a continuous digester. *J. Process Control* **2011**, *21*, 489–500. [CrossRef]
23. Souza, F.; Santos, P.; Araújo, R. Variable and delay selection using neural networks and mutual information for data-driven soft sensors. In Proceedings of the 2010 IEEE 15th Conference on Emerging Technologies & Factory Automation (ETFA 2010), Bilbao, Spain, 13–16 September 2010.
24. Shakil, M.; Elshafei, M.; Habib, M.A.; Maleki, F.A. Soft sensor for NOx and O_2 using dynamic neural networks. *Comput. Electr. Eng.* **2009**, *35*, 578–586. [CrossRef]
25. Xia, C.; Wang, J.; McMenemy, K. Short, medium and long term load forecasting model and virtual load forecaster based on radial basis function neural networks. *Int. J. Electr. Power Energy Syst.* **2010**, *32*, 743–750. [CrossRef]
26. Gosselin, L.; Tye-Gingras, M.; Mathieu-Potvin, F. Review of utilization of genetic algorithms in heat transfer problems. *Int. J. Heat Mass Transf.* **2009**, *52*, 2169–2188. [CrossRef]
27. Woodward, R.I.; Kelleher, E.J.R. Towards 'smart lasers': Self-optimisation of an ultrafast pulse source using a genetic algorithm. *Sci. Rep.* **2016**, *6*, 37616. [CrossRef] [PubMed]
28. Kreszig, E. *Advanced Engineering Mathematics*, 4th ed.; Wiley: Weinheim, Germany, 1979; p. 880.

9

Intelligent Energy Management for Plug-in Hybrid Electric Bus with Limited State Space

Hongqiang Guo *, Shangye Du, Fengrui Zhao, Qinghu Cui and Weilong Ren

School of Mechanical & Automotive Engineering, Liaocheng University, Liaocheng 252059, China; max_becker@163.com (S.D.); isukee@163.com (F.Z.); cui_qinghu@163.com (Q.C.); a1440107906@163.com (W.R.)

* Correspondence: guohongqiang@lcu.edu.cn

Abstract: Tabular Q-learning (QL) can be easily implemented into a controller to realize self-learning energy management control of a plug-in hybrid electric bus (PHEB). However, the "curse of dimensionality" problem is difficult to avoid, as the design space is huge. This paper proposes a QL-PMP algorithm (QL and Pontryagin minimum principle (PMP)) to address the problem. The main novelty is that the difference between the feedback SOC (state of charge) and the reference SOC is exclusively designed as state, and then a limited state space with 50 rows and 25 columns is proposed. The off-line training process shows that the limited state space is reasonable and adequate for the self-learning; the Hardware-in-Loop (HIL) simulation results show that the QL-PMP strategy can be implemented into a controller to realize real-time control, and can on average improve the fuel economy by 20.42%, compared to the charge depleting–charge sustaining (CDCS) strategy.

Keywords: plug-in hybrid electric bus; energy management; Q-learning; limited state space; Hardware-in-Loop (HIL) simulation

---

1. Introduction

Plug-in hybrid electric vehicle (PHEV) is a promising approach for energy-saving and lowering emissions, which would help the problems of energy shortage, global warming, and environment pollution [1]. Moreover, the advantage of the PHEV can be maximized by a well-designed energy management strategy (EMS) [2].

Rule-based control strategy is one of the most widely used methods in real-world situation, due to its easy implementation and real-time control performances [3]. Moreover, many investigations have demonstrated that the blended charge depletion (BCD) mode is the most efficient strategy, namely, the SOC can continuously decline to the expected value (such as 0.3) at the destination of route [4,5]. However, known driving conditions are usually indispensable for the BCD strategy, which brings great challenge to practical application.

Since the EMS can be taken as a nonlinearly constrained optimization control problem, the BCD mode can be well realized by optimal control methods [6]. It can be further classified into two categories: The optimization-based and the adaptive strategies [7]. The optimization-based strategies, such as dynamic programming (DP), Pontryagin's Minimum Principle (PMP), and Equivalent Consumption Minimization Strategy (ECMS), can obtain global optimization solutions [8–10]. However, driving conditions need to be known prior to use, which is the reason that they cannot be directly used in real-world situations and are usually only taken as the benchmark for other strategies. In contrast, the adaptive strategy has great potential in practical application [7]. For example, a model predictive control (MPC)-based EMS was proposed by combining Markov chain, DP, and a reference SOC plan method, based on the principle of receding horizon control [11]. An Adaptive PMP (A-PMP)-based-EMS was proposed by combining PI, PMP, and reference SOC plan method, based on the principle of feedback control [12]. An Adaptive ECMS (A-ECMS)-based-EMS was proposed by combining ECMS and particle swarm optimization (PSO) algorithms, where the equivalent factors (EFs) were firstly optimized, and then the actual EF was recognized based on a look-up table constituted by the optimized EFs [13]. A driving pattern recognition-based EMS was proposed based on PSO algorithm, where a series of typical reference cycles were firstly optimized off-line, then the corresponding optimal results were taken as the sampling sets for real-time control [14].

Reinforcement learning (RL) is an intelligent method that can be used in EMS, where Q-learning (QL) is the most popular method. It can be further classified into tabular and deep learning methods. The former can easily solve the self-learning problem, by employing a simplified Q-table. However, the state and the action should be discretized and the "curse of dimensionality" problem may be introduced once the state space is huge. In contrast, the Q-table will be substituted by neural network (NN) in the deep learning-based-EMS, and the control problem with continuous variable can also be solved. For the former, Ref. [15] proposed a Tabular QL-based EMS, where the required power, the velocity, and the SOC were taken as the states, meanwhile, the current of the battery and the shifting instruction of the automated mechanical transmission (AMT) were taken as the actions. Ref. [16] proposed a similar method by employing a Markov chain and a Kullback–Leibler (KL) divergence rate, where the power, the SOC and the state of voltage (SOC) were taken as the states, meanwhile, the current of the battery was taken as the action. Ref. [17] proposed a different Tabular QL-based EMS, where the SOC and the speed of the engine were taken as the states, and the throttle of the engine was taken as the action. For the latter, Ref. [18] proposed a deep QL (DQL)-based EMS based on NN, where the SOC, the engine power, the velocity and the acceleration were taken as the states, and the increment of the engine power was taken as the action. Ref. [19] proposed a similar DQL method, where the required torque and the SOC were taken as the states, and the engine torque was taken as the action. Ref. [20] proposed a different DQL-based EMS with AC (action–critic) framework, where the speed together with the torque of the wheel, the SOC together with the voltage of the battery, and the gear position of the AMT were taken as the states, meanwhile, the power of the motor was taken as the action. Ref. [21] proposed an interesting DQL-based EMS using a neural network, where the required power at the wheels, the SOC together with the distance to destination were taken as the states, and the power of the engine was taken as the action. Nevertheless, the existing tabular QL based-methods usually have huge state space, which is easy to result in the "curse of dimensionality" problem. On the other hand, the control performance of the DQL method may be greatly deteriorated once the fitting precision of NN is low. Moreover, the high computation burden of the deep learning may also restrict its application in currently used controllers.

This paper aims at solving the practical application problem of self-learning energy management for a single-parallel plug-in hybrid electric bus (PHEB). Since the shift instruction (discrete variable) and the throttle of the engine (continuous variable) can simultaneously influence the fuel economy of the vehicle, a mixed control variable with compact format is deployed into the control strategy. The main innovation of this paper is that a QL-PMP-based EMS is proposed, by combining the QL and the PMP algorithms, where the mixed control variables can be indirectly solved by the PMP using a self-learned co-state from the QL. More importantly, since the co-state is mainly dependent on the difference between the feedback SOC and the reference SOC, a limited state space with 50 rows and 23 columns is designed. Because the state space is greatly reduced, the QL-PMP algorithm can be directly implemented into controller.

The remainder of this paper is structured as follows. The configuration, parameters and models of the PHEB are described in Section 2. The QL-PMP algorithm is formulated in Section 3. The training process is discussed in Section 4. The Hardware-in-Loop (HIL) simulation results are detailed in Section 5, and the conclusions are drawn in Section 6.

2. The Configuration, Parameters, and Models of the PHEB

The single-parallel PHEB is shown in Figure 1, which includes an engine, a motor, an AMT, a clutch, and a battery pack.

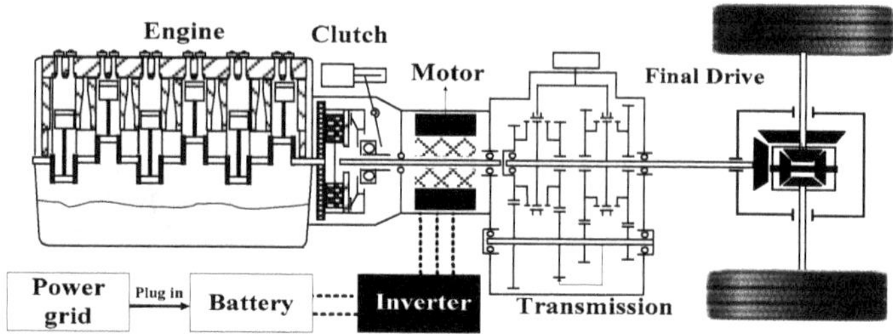

Figure 1. The configuration of the plug-in hybrid electric bus (PHEB).

The engine and the motor will work coordinately to provide the required power of the PHEB, and the battery pack is the energy source for the motor. The clutch can change the driving modes during the vehicle runs. In specific, the regenerative braking and the pure electric driving modes can be realized when the clutch is disengaged, otherwise, the hybrid driving with or without charging mode can be realized. In addition, the detailed parameters of the PHEB are shown in Table 1.

Table 1. The parameters of the PHEB. AMT: automated mechanical transmission.

Item	Description
Vehicle	Curb mass (kg): 8500
Passengers	Maximum number: 60; Passenger's mass (kg): 70
AMT	Speed ratios: 4.09:2.45:1.5:0.81
Final drive	Speed ratio: 5.571
Engine	Max torque (Nm): 639 Max power (kW): 120
Motor	Max torque (Nm): 604 Max power (kW): 94
Battery	Capacity (Ah): 35

2.1. Modeling the Engine

The modeling of the engine can be classified into theoretical and empirical methods. The former is formulated by combustion, fluid mechanics, and dynamic theories. As shown in Figure 2, the instantaneous fuel consumption of the engine can be interpolated by the brake specific fuel consumption (BSFC) map, based on Equation (1).

$$\dot{m}_e = \frac{T_e \cdot \omega_e \cdot b_e \cdot \Delta t}{3,600,000} \tag{1}$$

where $\dot{m}_e$ denotes the instantaneous fuel consumption of the engine; T_e denotes the torque of the engine; ω_e denotes the rotational speed of the engine; b_e denotes the fuel consumption rate of the engine, which can be obtained by the speed and torque of the engine; Δt denotes the sampling time.

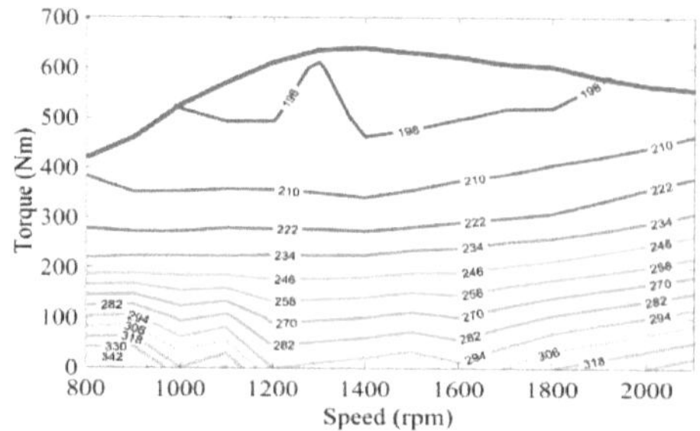

Figure 2. The BSFC map of the engine.

2.2. Modeling the Motor

Since only the working efficiency of the motor is need for the EMS, the empirical modeling method is adequate for the motor. In addition, because the motor can work in driving or regenerative mode, the motor model can be described as

$$P_m = \begin{cases} T_m \cdot \omega_m / \eta_m & T_m > 0 \\ T_m \cdot \omega_m \cdot \eta_g & T_m \leq 0 \end{cases} \tag{2}$$

where P_m denotes the power of the motor, T_m denotes the torque of the motor, ω_m denotes the rotational speed of the motor, and η_m and η_g denote the efficiency of the motor in driving mode and braking mode. When $T_m > 0$, motor works in driving mode, and when $T_m \leq 0$, motor works in braking mode. As shown in Figure 3, the power of the motor can be interpolated by the efficiency map of the motor, based on Equation (2).

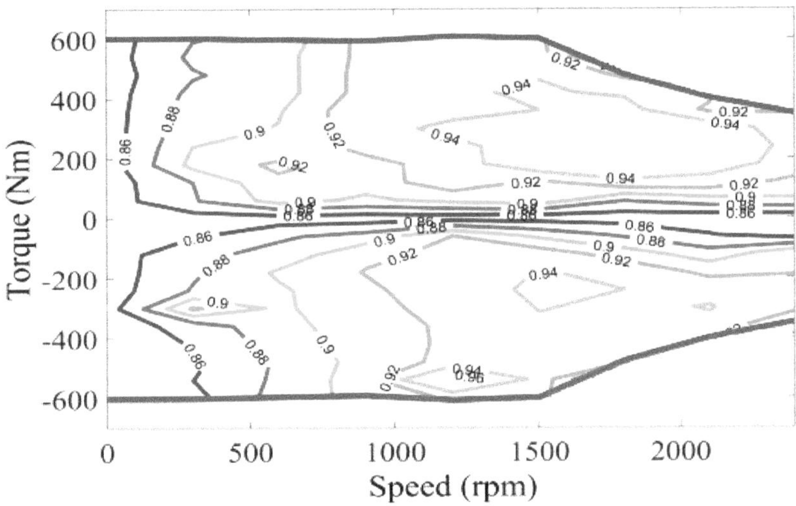

Figure 3. The efficiency map of the motor.

2.3. Modeling the Battery

For the energy management, the key issue of the battery is the estimation of the SOC, based on the voltage, the internal resistance and the current of the battery. Accordingly, a simplified battery model is deployed in Figure 4, and the power of the battery can be described as:

$$P_b = V_{oc} I_b - I_b^2 R_b \tag{3}$$

where P_b denotes the power of the battery, V_{oc} denotes the open-circuit voltage of the battery, I_b denotes the current of the battery and R_b denotes the internal resistance of the battery.

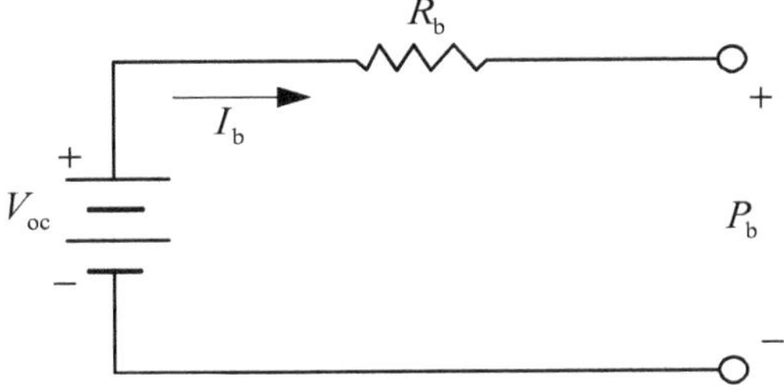

Figure 4. The simplified battery model.

2.4. The Dynamic Model of the Vehicle

Since only the required power is needed for the EMS, the model can be simplified to a lumped mass model, and only the longitudinal dynamic characteristic can be considered. Here, the driving and resistance forces can be described as:

$$F_t - F_R = \delta m \frac{dv}{dt} \tag{4}$$

where F_t denotes the driving force, δ denotes the rotating mass conversion factor, and m denotes the vehicle mass, which is constituted by the curb mass of the vehicle and the stochastic mass of the passengers, F_R denotes the resistance force, which is constituted by:

$$F_R = fmg\cos\alpha_s + mg\sin\alpha_s + \frac{1}{2}C_D A_a \rho_a v^2 \tag{5}$$

where f denotes the coefficient of the rolling resistance, g denotes the gravity acceleration, α_s denotes the road slope, C_D and A_a denote the coefficient of the air resistance and frontal area, respectively, ρ_a denotes the air density, and v denotes the velocity. The required power can be obtained by:

$$P_r = \frac{F_t \cdot v}{3600\eta_t} \tag{6}$$

where P_r denotes the required power and η_t denotes the efficiency of the transmission system.

3. The Formulation of the QL-PMP Algorithm

Different from the general traffic environment, the route of the PHEB is fixed and repeatable. Therefore, a series of historical driving cycles can be downloaded from remote monitoring system (RMS). In addition, many investigations have demonstrated that the factors of the velocity and the road slope have great effect on the EMS, which essentially points at the importance of the required power of the vehicle [2,7]. In this case, the factor of the stochastic vehicle mass is also a significant factor for EMS based on the Equations (4)–(6). Accordingly, a series of combined driving cycles constituted by the historical driving cycles, the road slope and the stochastic distributions of the vehicle mass are firstly designed. As shown in Figure 5, the training of the QL-PMP requires three steps.

Step 1 A series of co-states with respect to the combined driving cycles are firstly optimized, by an off-line PMP with Hooke–Jeeves algorithm [22]. Then, the average co-state is obtained and the corresponding optimal SOCs are extracted. Finally, a reference SOC model is established by taking the normalized distance as input and the optimal SOCs as output.

Step 2 Based on the known average co-state and the reference SOC model, the training of the QL-PMP is carried out as follows: firstly, the Q-table (denoted by Q_0), that is defined as zeros matrix, will be trained with the combined driving cycle 1; secondly, the trained Q-table (denoted by Q_1) will be taken as the initial Q-table for the second training with the combined driving cycle 2; and then this process will be continued until the Q-table is adequately trained.

Step 3 Verifying the adequately trained QL-PMP algorithm with HIL platform, using different combined driving cycles.

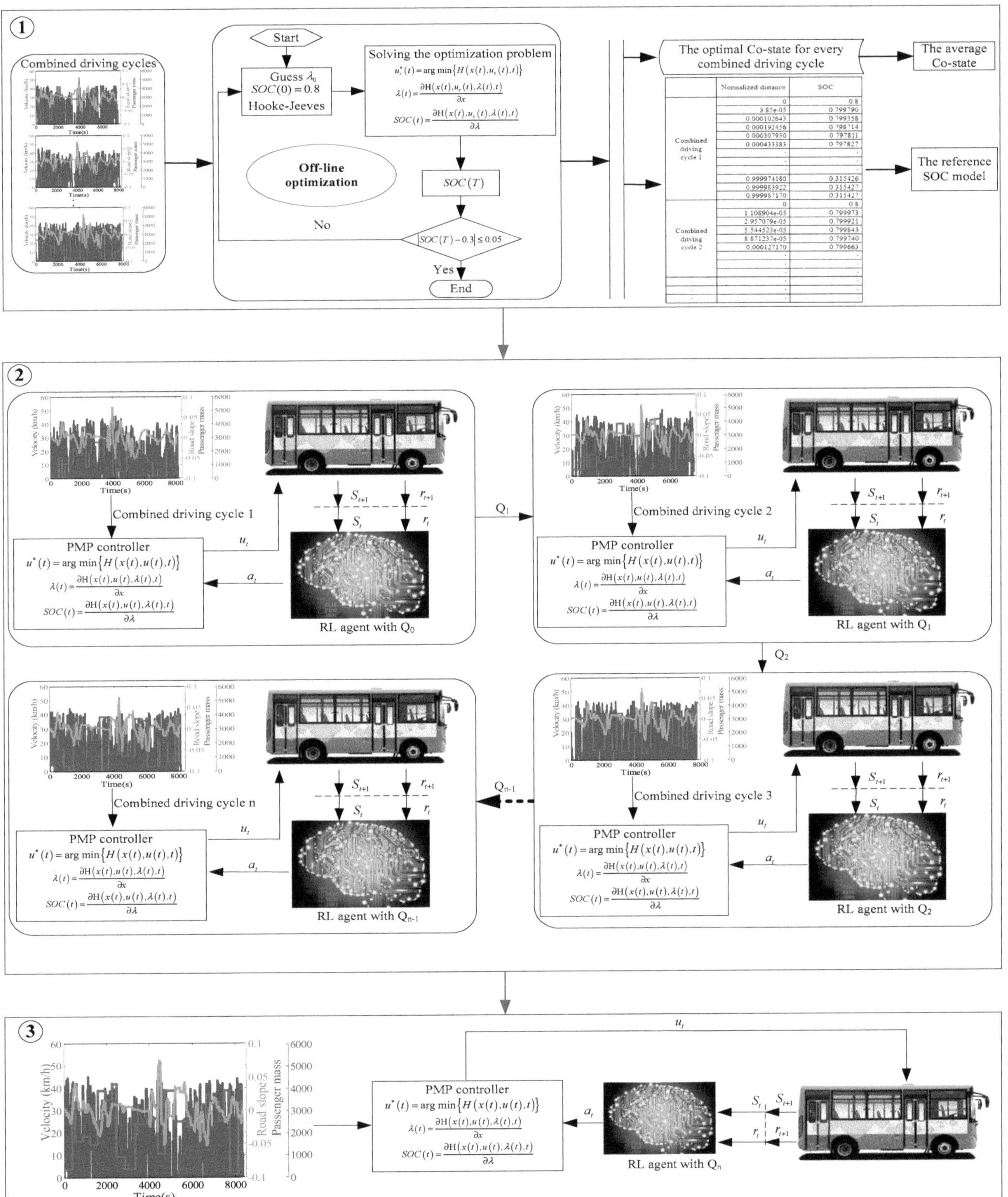

Figure 5. The QL-PMP algorithm (Q-learning and Pontryagin minimum principle (PMP)) algorithm.

3.1. The Combined Driving Cycle

In terms of the combined driving cycle, the historical driving cycle can be downloaded from the RMS, and the road slope can be obtained off-line based on the attitude of the road and the corresponding travelled distance. To simplify the EMS problem, the road slope is implemented into the controller in a prior process by designing a look-up table, through taking the travelled distance as input and the road slop as output. Similar to Ref. [23], the stochastic distributions of the vehicle mass were designed as follows:

Firstly, as shown in Figure 6, 25 road segments are defined based on the number of the neighbored bus stops. Because the distributions of the passenger in different road segments are stochastic, 25 factors with respect to the road segments were defined to describe the stochastic distribution of the vehicle mass over the bus route.

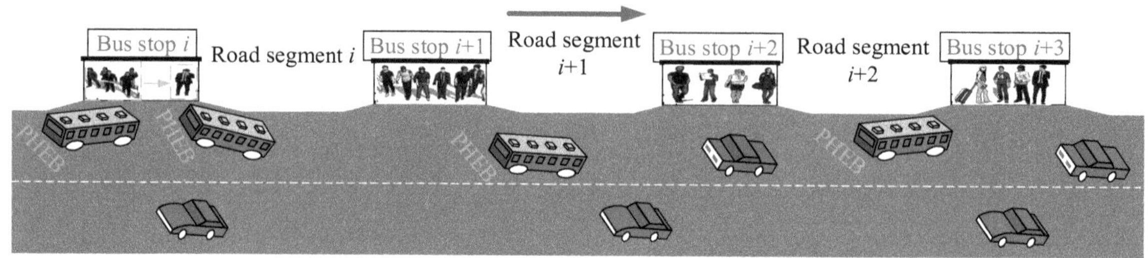

Figure 6. The route of the PHEB.

Secondly, assuming the stochastic distribution of the vehicle mass is reflected by the passenger mass, 60 levels are defined for each factor, based on the maximum passenger number. Moreover, to exhaustively probe the design space constituted by the 25 factors, the Optimal Latin hypercube design (Opt. LHD) is deployed, due to its good spatial filling and equalization performances. Finally, the combined driving cycles are constructed by stochastic matching of the historical driving cycles, the road slope and the stochastic distributions of passenger mass (kg) are generated. As shown in Figure 7, the blue line represents the velocity of the PHEB, the red line represents the total mass of passengers, and the green line represents the slope of the road. Besides, the stochastic vehicle mass is added by the curb vehicle mass and the stochastic distribution of passenger mass.

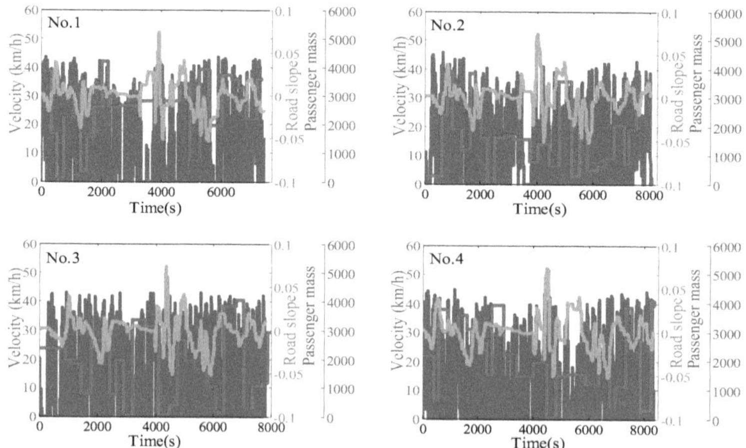

Figure 7. The combined driving cycles.

3.2. *The Off-Line PMP Algorithm*

Theoretically, EMS can be taken as an optimal control problem, which can make the vehicle reach the optimal state within a time domain through a series of discrete controls. Furthermore, since the bus route is fixed, and the BCD mode can be realized by the optimal control, and the terminal SOC can be easily controlled within the expected zone. This implies that the terminal SOC may be similar, despite of any combined driving cycle. Therefore, the electricity consumptions of the battery are similar for all of the combined driving cycles, and the minimum fuel consumption of the PHEB can be taken as one of the objectives. Moreover, the shift number should also be restricted to improve the control performance. In addition, only the state of the SOC is selected to simplify the control problem, and the control vector is reduced to a one-dimensional format constituted by the shift instruction of the AMT and the throttle of the engine, then the optimal control problem of the energy management can be described as:

$$\min J = \int_{t_0}^{t_f} \dot{m}_e(u(t)) + \beta|s(t)|dt$$
$$\text{S.t.}\begin{cases} \dot{SOC}(t) = -\frac{V_{oc}-\sqrt{V_{oc}^2-4R_{bat}P_{bat}}}{2R_{bat}Q_{nom}} \\ \omega_{e\_min} \le \omega_e(t) \le \omega_{e\_max} \\ \omega_{m\_min} \le \omega_m(t) \le \omega_{m\_max} \\ p_{e\_min} \le p_e(t) \le p_{e\_max} \\ p_{m\_min} \le p_m(t) \le p_{m\_max} \end{cases} \quad (7)$$

where J denotes the performance index function; $u(t)$ (Equation (8)) denotes the control vector, which is defined as a compacted format and is sampled by Opt. LHD. Here, the shift instruction is denoted by $s(t)$, and the throttle of the engine is denoted by $th(t)$. Moreover, $s(t)$ is defined as -1, 0 and 1, denoting the downshift, hold on and upshift, respectively; $th(t)$ is ranged from 0 to 1.

$$u(t) = \begin{bmatrix} s(t) \\ th(t) \end{bmatrix} \quad (8)$$

where the required power of the vehicle can be defined as P_r, $s(t)$ and $th(t)$ are changed with the state of P_r. If $P_r > 0$, it means that the vehicle runs in the driving state, the AMT can carry out the controls of the downshift, hold on and upshift, and the throttle of the engine can be ranged from 0 to 1; if $P_r = 0$, it means that the vehicle runs in the stopping state, the AMT will be downshifted to the lowest gear and the throttle of the engine is 0; if $P_r < 0$, the vehicle runs in the braking state, the gear of the AMT will be maintained and the throttle of the engine is 0.

In addition, $\dot{SOC}(t)$ denotes the state function, Q_{nom} denotes the battery capacity, $\omega_e(t)$ and $\omega_m(t)$ denote the rotate speeds of the engine and the motor, respectively, meanwhile, ω_{e_min}, ω_{e_max} and ω_{m_min}, ω_{m_max} denote the corresponding rotate speed boundaries of the $\omega_e(t)$ and $\omega_m(t)$, $p_e(t)$ and $p_m(t)$ denote the powers of the engine and the motor, meanwhile, p_{m_min}, p_{e_max} and p_{m_min}, p_{m_max} denote the corresponding power boundaries.

Finally, the Hamilton function can be descried as:

$$\begin{cases} u^*(t) = \text{argmin}\{H(x(t), u(t), t)\} \\ H(x(t), u(t), \lambda(t), t) = \dot{m}(u(t)) + \beta|s(t)| + \lambda(t) \cdot \dot{SOC} \end{cases} \quad (9)$$

where $u^*(t)$ denotes the optimal control vector, $H(x(t), u(t), t)$ denotes the Hamiltonian function and $\lambda(t)$ denotes the co-state.

As shown in Figure 5, the off-line optimization of the PMP can be formulated by

$$\begin{array}{ll} \min & F_{PMP} = f(\lambda)\frac{1}{2} \\ \text{S.t} & 2000 \le \lambda \le 3000 \\ & 0.27 \le SOC_f \le 0.33 \end{array} \quad (10)$$

where F_{PMP} denotes the objective function of the Hooke–Jeeves, and SOC_f denotes the terminal SOC, and is defined as a soft constraint to accelerate the convenience of the optimization. Moreover, the offline optimization of the PMP can be categorized into three steps.

Step 1 Taking the co-state as independent value of the Hooke–Jeeves and guessing an initial value of the co-state with a defined initial SOC value (0.8).

Step 2 Solving the dynamic optimal control problem by Equations (7)–(9), based on the given independent value from the Hooke–Jeeves.

Step 3 If the terminal SOC value satisfies the optimization objective, the co-state will be taken as the optimal co-state, otherwise, the iteration will be repeated from Step 1 to Step 3, till the objective is satisfied.

3.3. The Reference SOC Model

The most important innovation of this paper is the employment of a small state space for the Q-table, which is dependent on the difference between the feedback SOC and the reference SOC. Therefore, a reference SOC model should be designed. Since a series of optimal SOCs with respect to different combined driving cycles can be calculated off-line and the city bus route is fixed, the reference SOC model can be constructed by taking the normalized distance as input and the optimal SOC as output, based on partial least squares (PLS) method [23]. Here, the normalized distance can be described as

$$x(t) = \frac{d_{real}(t)}{d_{whole}(t)} \tag{11}$$

where $x(t)$ denotes the normalized distance, $d_{whole}(t)$ denotes the total distance and $d_{real}(t)$ denotes the travelled distance, which can be described as:

$$d_{real}(t) = \sum_{t=1}^{k}\left(v(t)\cdot\Delta t + \frac{1}{2}\cdot a(t)\cdot\Delta t^2\right) \tag{12}$$

The general form of the reference SOC model is described as:

$$SOC_r(t) = \frac{p_1+p_3x(t)+p_5x(t)^2+p_7x(t)^3+p_9x(t)^4+p_{11}x(t)^5+p_{13}x(t)^6+p_{15}x(t)^7+p_{17}x(t)^8+p_{19}x(t)^9}{1+p_2x(t)+p_4x(t)^2+p_6x(t)^3+p_8x(t)^4+p_{10}x(t)^5+p_{12}x(t)^6+p_{14}x(t)^7+p_{16}x(t)^8+p_{18}x(t)^9} \tag{13}$$

where $SOC_r(t)$ denotes the reference SOC; $p_i(i = 1,2,3\cdot\cdot 19)$ denotes the fitting coefficient. Based on the method in Ref. [23], the parameters of the reference SOC model can be obtained, and are shown in Table 2:

Table 2. The parameters of the reference state of charge (SOC) model.

p_1	p_2	p_3	p_4	p_5	p_6	p_7	p_8	p_9	p_{10}
0.7910	−8.0413	−6.3492	24.939	13.587	−53.294	15.148	143.58	−72.790	−162.76
p_{11}	p_{12}	p_{13}	p_{14}	p_{15}	p_{16}	p_{17}	p_{18}	p_{19}	
77.584	−73.520	−46.524	74.766	−23.603	90.256	73.1139	36.364	−8.6263	

As shown in Figure 8, a series of combined driving cycles from No.13 to No.20 are also designed to further verify the predictive precision of the reference SOC model.

As shown in Figure 9, the optimal SOC trajectories extracted from the off-line optimization fluctuate around the predicted SOC trajectory. As shown in Figure 10, the relative errors between the predicted SOC and the optimal SOCs are ranged from −0.089 to 0.0257, which implies that the reference SOC trajectory cannot be well predicted. Therefore, it is no requirement for the feedback SOC trajectory to strictly track the reference SOC trajectory. On the contrary, better fuel economy may be realized if it makes the feedback SOC fluctuate around the reference SOC, compared to the strictly tracking control.

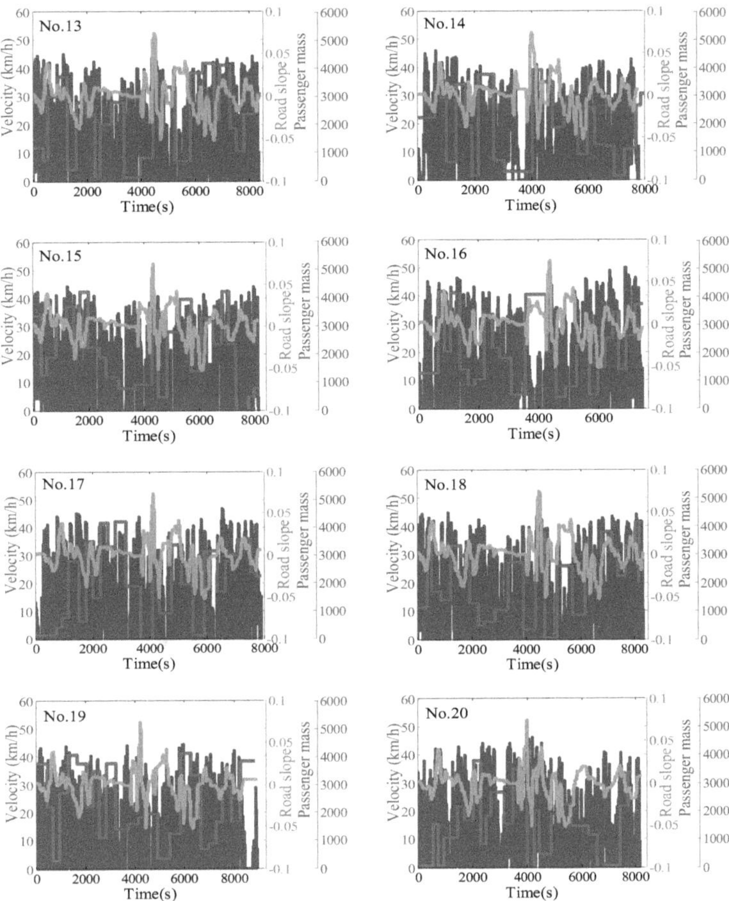

Figure 8. The combined driving cycles from No. 13 to No. 20.

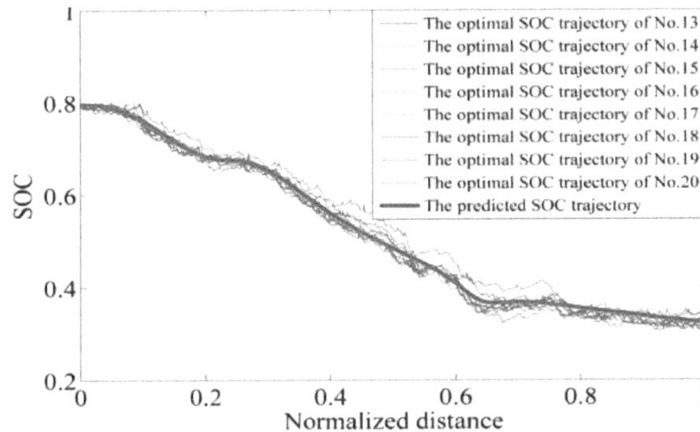

Figure 9. The comparison between the reference SOC and the optimal SOC trajectories.

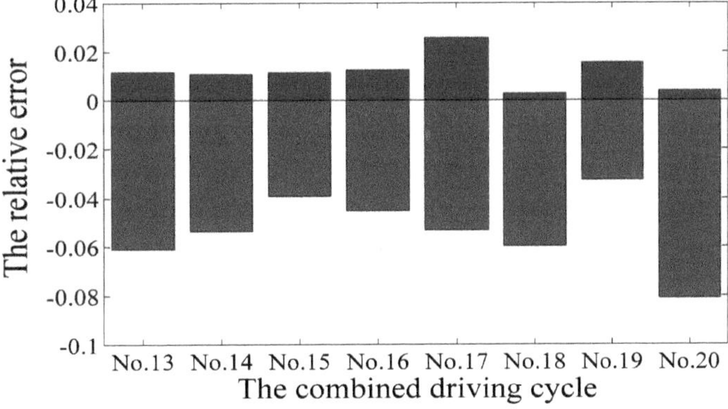

Figure 10. The relative errors between the reference and the optimal SOCs.

3.4. The QL-PMP Algorithm

QL is a value-based algorithm, which is also one of the most important temporal difference (TD) algorithms. It is defined by 4-tuple (S, A, R, γ), where S denotes the state space; A denotes the action space; R denotes the reward function; γ denotes the discount factor, which is defined as 0.8. Tabular QL is one of the most widely used methods, where the state-action function ($Q(s, a)$) is indispensable. The Q-table is the concrete manifestation of the $Q(s, a)$, which is designed to realize the mapping from the state to the action, and can be updated by:

$$Q(s_k, a_k) \leftarrow Q(s_k, a_k) + \alpha_l [r_{k+1} + \gamma Q(s_{k+1}, a_{k+1}) - Q(s_k, a_k)] \tag{14}$$

where $Q(s_k, a_k)$ denotes the old reward; α_l denotes the learning rate, which is defined as 1; $Q(s_{k+1}, a_{k+1})$ denote the maximum reward in future.

In the existing RL-based energy managements, the factors such as the current SOC, the required power, and the velocity are usually designed as the states, and should be discretized intensively to probe good action. This may lead to the "curse of dimensionality" problem, owing to the employment of a large state space in the Q-Table. In this paper, the difference between the reference SOC and the feedback SOC is exclusively designed as the state for the QL-PMP, and there are three reasons for this:

Firstly, since the reference SOC model can provide the corresponding reference SOC at any time step and the purpose of the QL-PMP is to make the feedback SOC track the reference SOC, the state of the difference between the reference SOC and the feedback SOC is a reasonable choice.

Secondly, the adaptive energy management control can be realized by making the feedback SOC fluctuate around the reference SOC, through adjusting the co-state. This implies that the difference of the SOC between the feedback SOC and the reference SOC can be taken as the sole state for the QL-PMP algorithm.

Thirdly, the range of the difference of the SOC is small, which means that the algorithm can find the optimal action with limited state space.

Based on the above discussion, the state of the difference of the SOC is ranged from −0.1 to 0.1, and are uniformly discretized to 50 points. The action is designed as the co-state, and is non-uniformly discretized by 23 points, namely, [−1000, −512, −256, −128, −64, −32, −16, −8, −4, −2, −1, 0, 1, 2, 4, 8, 16, 32, 64, 128, 256, 512, 1000]. The reward function is designed as:

$$r_{ss'}^{a} = \begin{cases} -\text{abs}(0.8 - SOC(k+1)) & \text{if } SOC(k+1) > 0.8 \\ \frac{1}{\text{abs}(SOC(k+1) - SOC_r(k+1))} & \text{if } SOC(k+1) <= 0.8 \ \& \ SOC(k+1) >= 0.3 \\ -\text{abs}(0.3 - SOC(k+1)) & \text{if } SOC(k+1) < 0.3 \end{cases} \tag{15}$$

where $r_{ss'}^{a}$ denotes the reward function. It means that if the $SOC(k+1)$ is bigger than 0.8, the punishment will be applied, and the farther from 0.8, the bigger punishment will be; if the $SOC(k+1) <= 0.8 \& SOC(k+1) >= 0.3$, the smaller of the difference of the SOC, the bigger reward will be, otherwise, if the $SOC(k+1) < 0.3$, the punishment will be applied, and the farther from 0.3, the bigger punishment will be.

As shown in Figure 5, at every time step, the RL agent will firstly receive the state (denoted by $\triangle SOC(k)$). Then, an action (denoted by $a(k)$) will be carried out, based on the current state and the greedy algorithm. In specific, if the random number is larger than the threshold value, the action will be selected by the random number, otherwise, it will be selected by the $Q(s_k, a_k)$, by finding the maximum value of the action-state value. Finally, the next state and the $Q(s_k, a_k)$ will be undated based on the selected action. It is worth mentioning that the co-state applied to the PMP algorithm is formulated by

$$\lambda_k = \lambda_{ave} + a_k \tag{16}$$

where λ_{ave} denotes the average co-state. The Algorithm 1 can be described as

Algorithm 1. The QL-PMP Algorithm

1: Initialize $Q(s,a) \leftarrow 0$ for all s
2: Initialize $R(s,a) \leftarrow 0$ for all s
3: Initialize $SOC(1) \leftarrow 0.8$
4: Initialize $\lambda \leftarrow \lambda_{ave} + a_1$
5: Repeat
6: Observe $s_k = \{\triangle SOC(k)\}$
7: Generate a random number $x \in [0,1]$
8: **if** $x > \varepsilon$
 select an action randomly $a_k = \{\lambda_k\}$
 else
 choose the optimal action by $a_k = \max(Q(s_k, a_k))$
 end
9: Calculate $\triangle SOC(k+1)$
10: Update the state-value function
 $Q(s_k, a_k) \leftarrow Q(s_k, a_k) + \alpha[r_{k+1} + \gamma Q(s_{k+1}, a_{k+1}) - Q(s_k, a_k)]$
11: $k \leftarrow k+1$
12: Until simulation stop

4. The Training Process

4.1. The Average Co-State

As shown in Figure 11, the co-states obtained from off-line optimizations are distributed evenly, and the values range from 1960 to 2280. The average co-state is 2202, which is calculated by the sum of the co-sates divided by the number of the combined driving cycles.

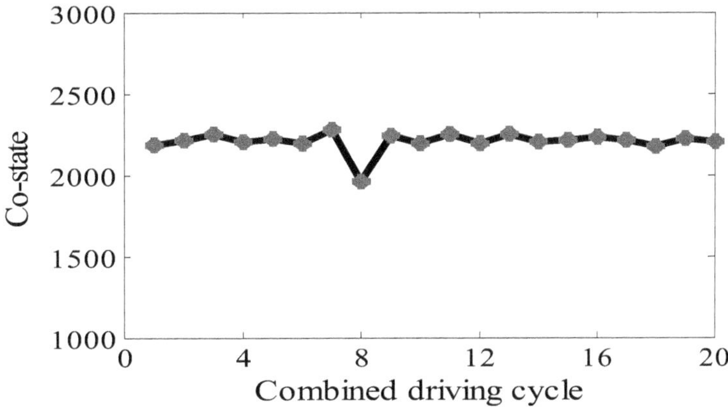

Figure 11. The co-state with different combined driving cycles.

4.2. The Off-Line Training

To train the Q-table of the QL-PMP, the combined driving cycles from No.13 to No.19 in Figure 8 are deployed. The complete training process is shown in Figures 12–17. The left figures denote the training results of the training episodes from 1 to 35, with the greedy factor of 0.3, the middle figures denote the training results of the training episodes from 36 to 45, with the greedy factor of 0.55, and the right figures denote the training results of the training episodes from 46 to 50, with the greedy factor of 0.9. The greedy factor 0.3 means that the QL-PMP tends to explore a new action to maximize the reward. The greedy factor 0.55 means that the QL-PMP tends to balance the chance between the exploration and the best action. The greedy factor 0.9 means that the QL-PMP tends to use the best action to realize the optimal control whilst providing a chance to find a better action.

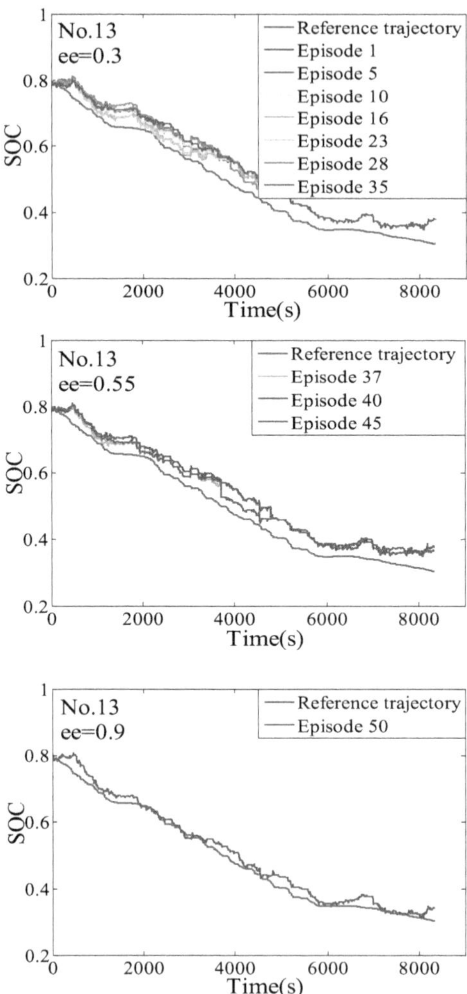

Figure 12. The training process of combined driving cycle 13.

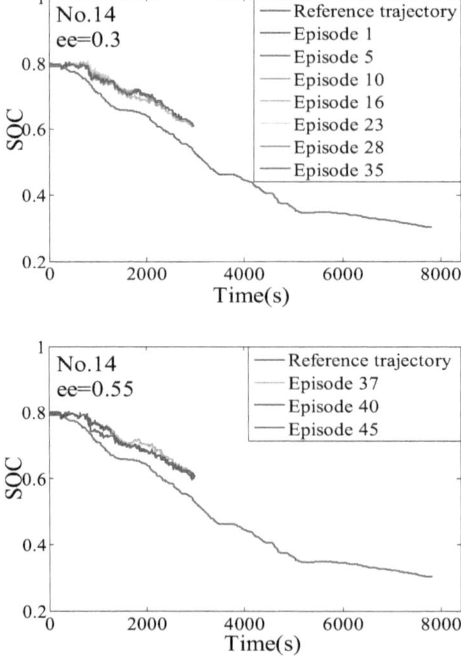

Figure 13. *Cont.*

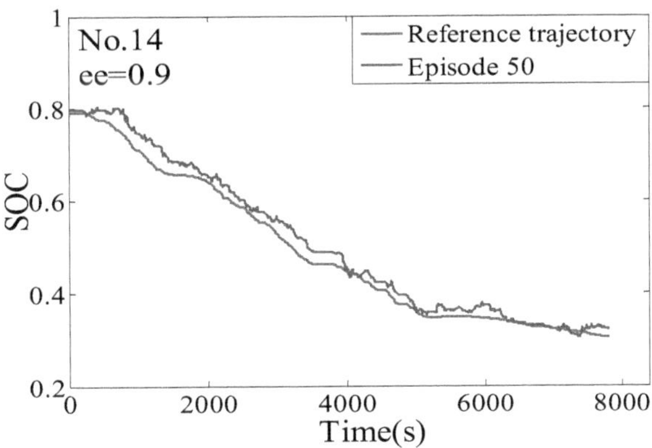

Figure 13. The training process of combined driving cycle 14.

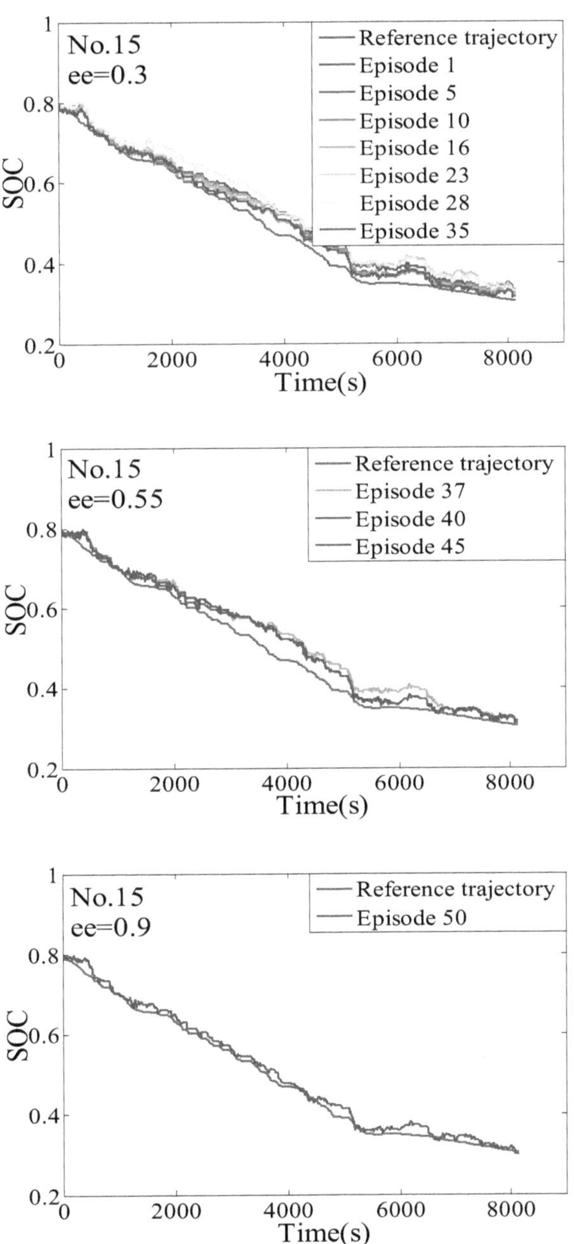

Figure 14. The training process of combined driving cycle 15.

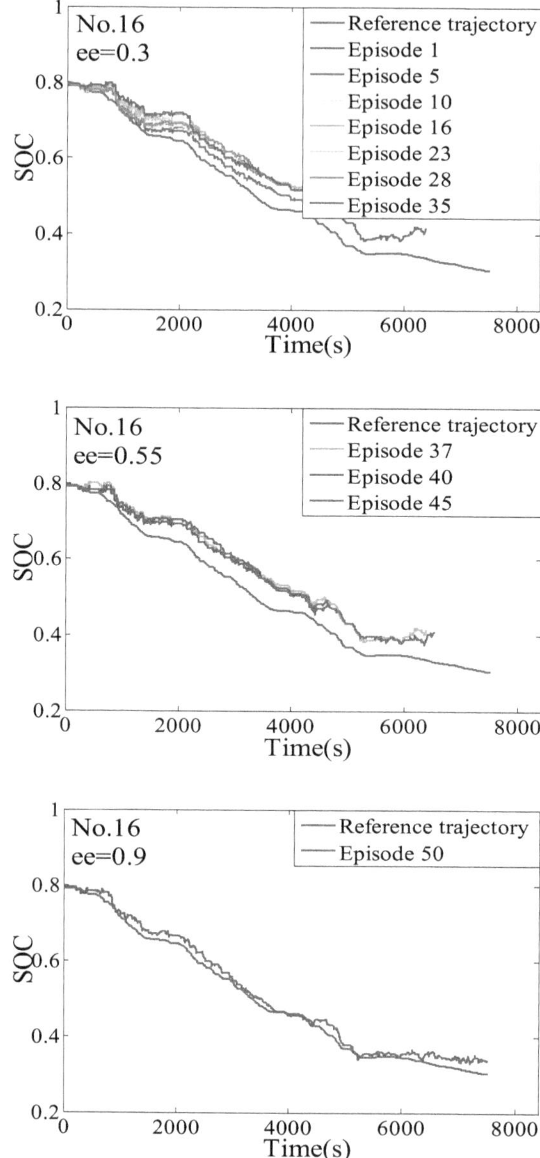

Figure 15. The training process of combined driving cycle 16.

As shown in Figure 12, for the training results of the combined driving cycle 13, the difference between the feedback SOC and the reference SOC is large when the greedy factor is 0.3, and the training fails all the time till the training reaches to episode 35. Moreover, although the feedback SOC trajectory is reasonable at the episode 35, the feedback SOC trajectory severely deviates from the reference trajectory. The QL-PMP becomes better and better with the continuous training, but is far from reliable When the greedy factor is 0.55, most of the feedback SOC trajectories can meet the requirements, but they are also far away from the reference SOC trajectory. When the greedy factor is 0.9, the feedback SOC trajectory of the episode 50 is reasonable and is close to the reference SOC trajectory. This implies that the trained Q-table is reasonable for the combined driving cycle 13.

As shown in Figure 13, for the training results of the combined driving cycle 14, the trainings are always fail when the episode is lower than 45 with the Q_1 (the trained Q-table with the combined driving cycle 13), which means that the Q_1 is not suitable for the combined driving cycle of 14, and the Q-table should be continually trained. Fortunately, the Q-table can be trained well enough through 45 trainings. It is worth noting that the failings of the feedback SOC trajectories do not mean that the Q_1 is not trained well, the failings are also attributed by the smaller greedy factor.

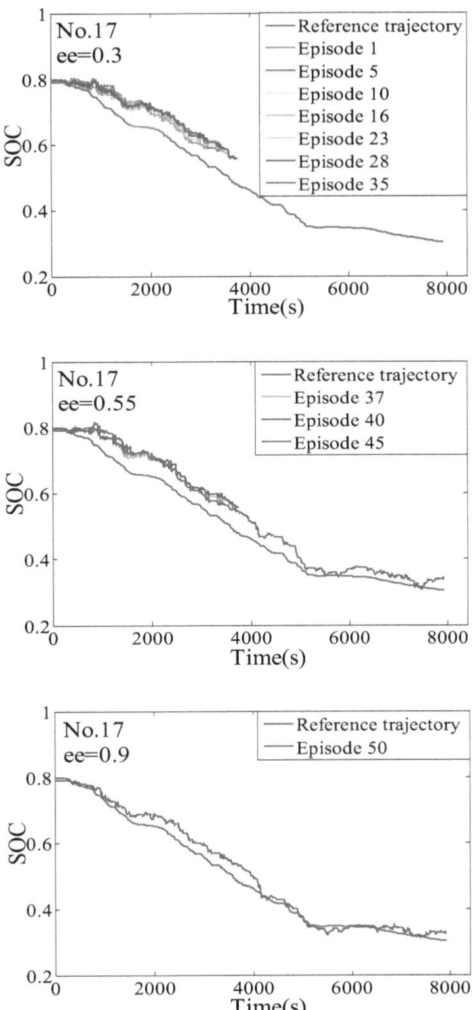

Figure 16. The training process of combined driving cycle 17.

As shown in Figure 14, for the training results of the combined driving cycle 15, the feedback SOC trajectories are always reasonable for any greedy factor and any episode. This implies that the Q_2 (the trained Q-table with the combined driving cycle 14) is suitable for this combined driving cycle. Moreover, the feedback SOC trajectory of the episode 50 with the greedy factor of 0.9 is closer to the reference SOC trajectory than others, which shows that the Q-table becomes more and more reliability through the trainings from episode of 1 to 45.

As shown in Figure 15, for the training results of the combined driving cycle 16, the feedback SOC trajectories always fail when the greedy factor is 0.3. However, the control performance is continuously improved with the increasing of training number. Similarly, the trainings still fail despite the greedy factor being 0.55. This implies that the driving conditions are different from the above combined driving cycles. Therefore, the Q-table should be continuously trained based on Q_3 (the trained Q-Table with the combined driving cycle 15). However, when the episode reaches to 50, the feedback SOC trajectory is reasonable and close to the reference SOC trajectory, which implies that the Q-table has been trained well.

As shown in Figure 16, for the training results of the combined driving cycle 17, although the Q_4 (the, trained Q-Table with the combined driving cycle 16) has been well trained, the training fails before episode 44. This implies that the Q_4 still not be suitable for all of the driving conditions. However, the Q_5 (the trained Q-Table with the combined driving cycle 17) can be suitable for the combined driving cycle 17 through a series of training.

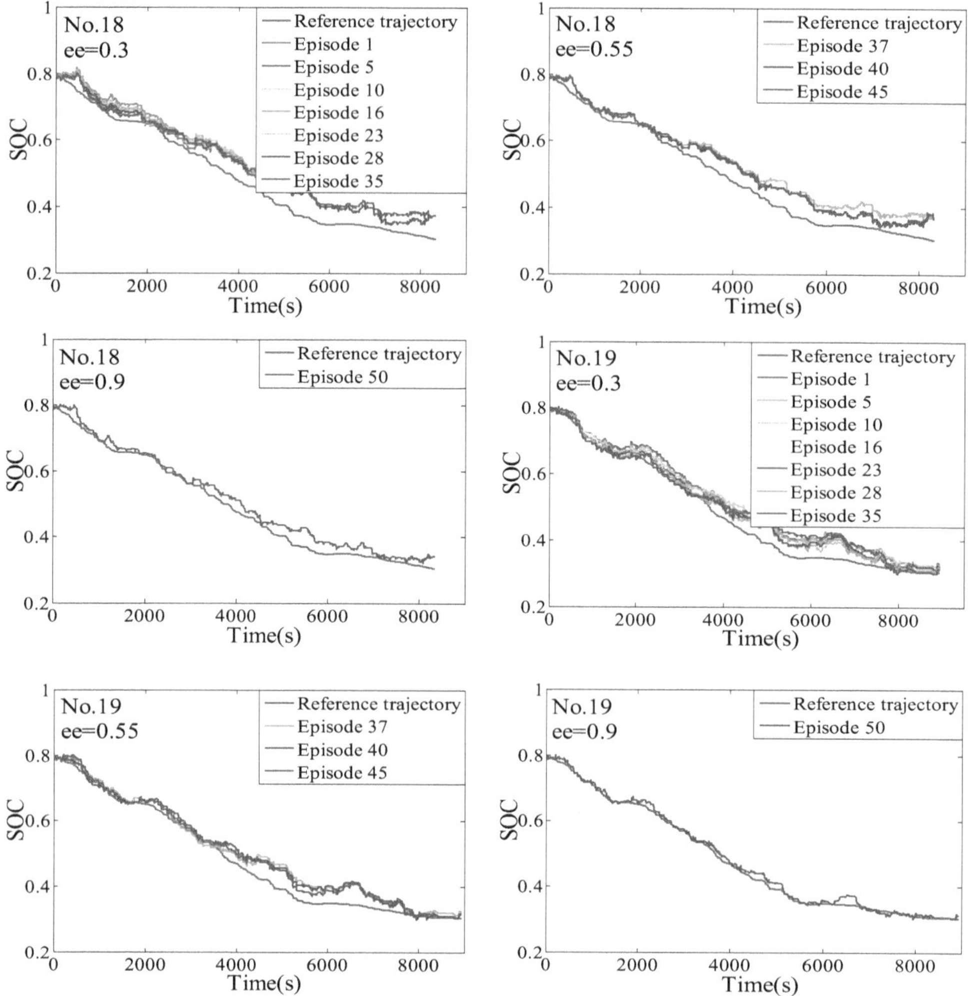

Figure 17. The training process of combined driving cycle 18 and 19.

As shown in Figure 17, the training results of the combined driving cycle 18 and 19 have demonstrated that Q_5 is well trained. The QL-PMP can be directly applied to the on-line control.

5. The Hardware-In-Loop Simulation

5.1. The Introduction of the HIL

To further verify the real-time and robust control performances of the EMS, a hardware-in-loop simulation (HIL) platform was also developed, based on the D2P rapid prototype control system. The D2P development platform mainly includes the MotoHawk development software and the MotoTune debugging software, where the former can bridge the gap between the control code and the MATLAB code; the latter is the debugging software for downloading and viewing the programs.

As shown in Figure 18, The Target Definition is the model selection and configuration module; the Main Power Relay is the main relay control module; the Trigger is the trigger module. In this paper, the controller type is ECM-5554-112-0904-Xd (DEV), the microprocessor is the MPC5554 of Freescale with 32-bit, and the frequency is 80 MHz. In terms of the EMS, the controlled models such as the engine, the motor and the battery are modeled as Simulink models, the algorithm of the QL-PMP will be implemented into the Hybrid Control Unit (HCU). Specifically, the QL-PMP algorithm will be firstly compiled and generated into the product-level code. Then, the code will be implemented into the HCU and the real-time communication between the Simulink models and the HCU will be built through the CAN bus.

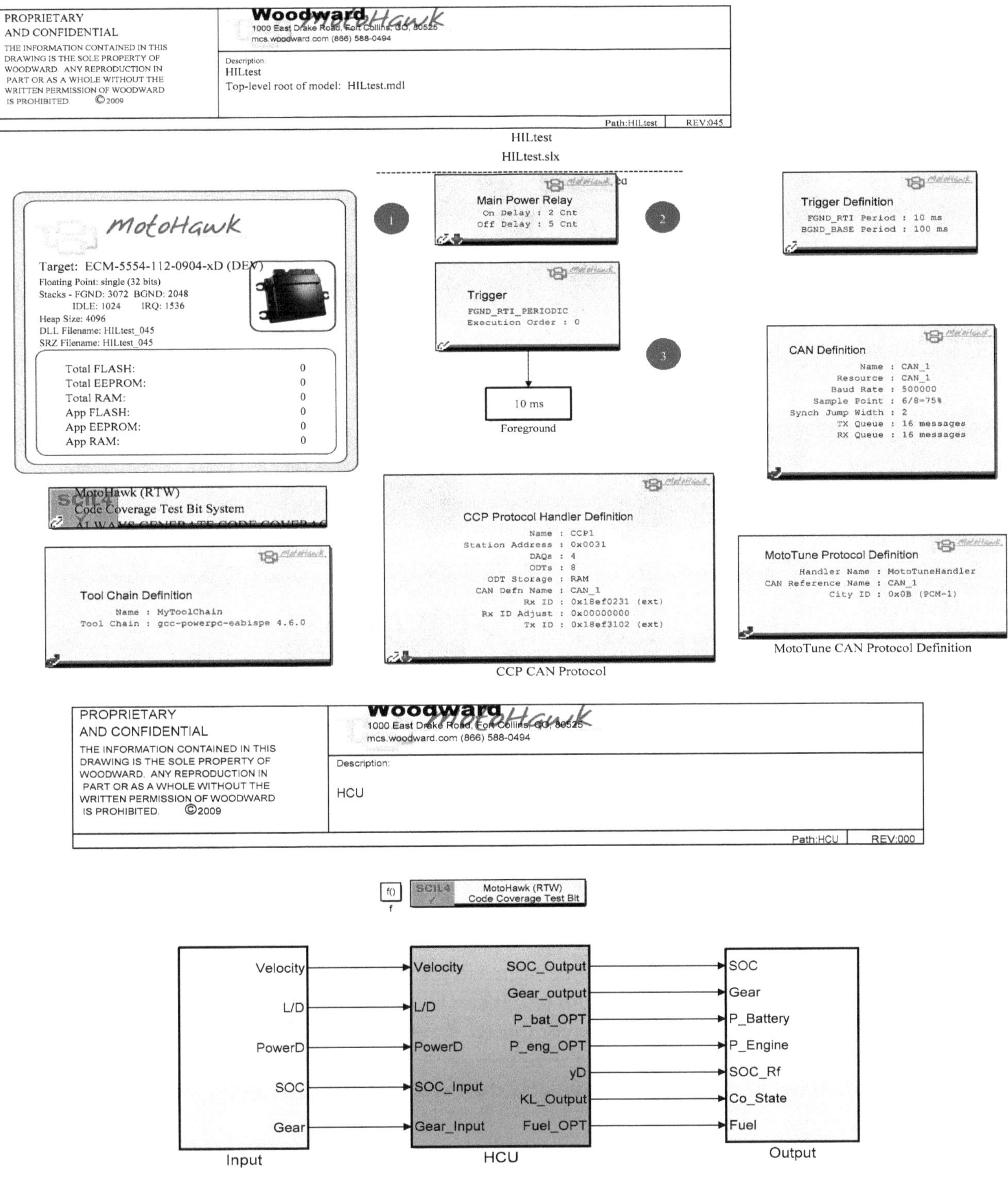

Figure 18. The MotoHawk-project interface.

5.2. *The Verification of the HIL*

As shown in Figure 19, the computer is the upper computer, which sends the signals of the Simulink models and monitors the signals transmitted from the HCU, through Kvaser. As shown in Figure 20, a series of combined driving cycles from No.21 to No.24 were designed to further verify the reliability of the Q-table and evaluate the fuel economy of the QL-PMP. In addition, a rule-based energy management (named by CDCS control strategy) which is characterized by CD followed by CS control strategy was also deployed to evaluate the fuel economy of the QL-PMP.

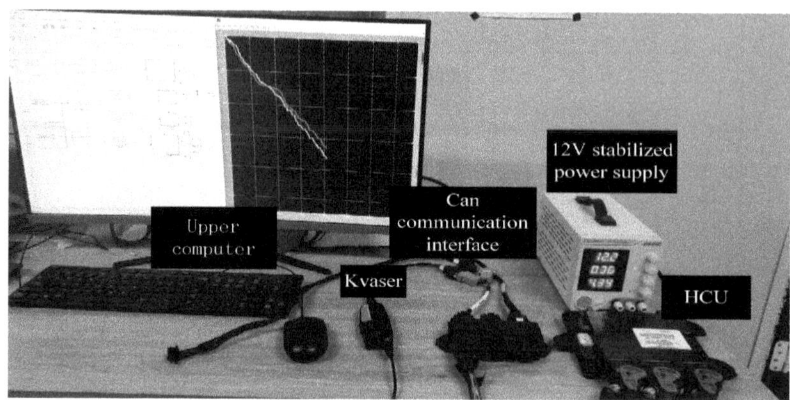

Figure 19. The Hardware-in-Loop (HIL) platform.

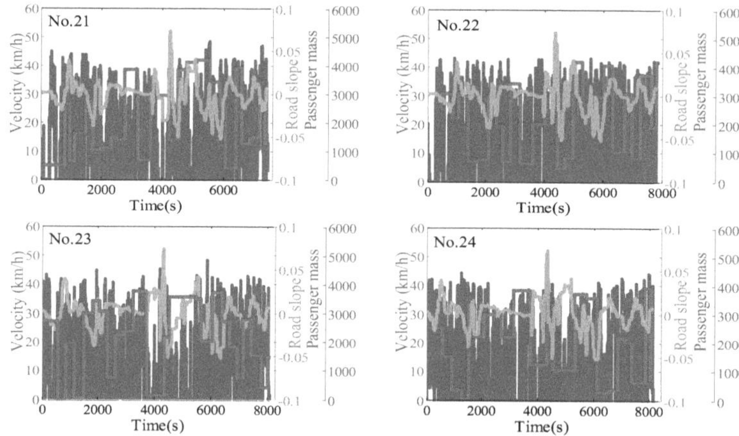

Figure 20. The combined driving cycles from No.21 to No.24.

As shown in Figures 21–24, (a) denotes the gears, (b) denotes the powers of the motor and the engine, (c) denotes the co-state, and (d) denotes the SOC trajectories. From (a), it can be seen that the gears of the AMT tended to maintain high gear to improve the fuel economy of the PHEB. From (b), it can be seen that the engine and the motor work coordinately to provide the required power of the vehicle, meanwhile, the motor works in driving or regenerative braking modes based on the driving conditions. From (c), it can be seen that the co-states are adjusted all the time to make the feedback SOC track the reference SOC trajectory. From (d), it can be seen that the feedback SOC can track the reference SOC trajectory well, despite of any combined driving cycle. Moreover, the SOC trajectory of the CDCS control strategy is far away from the reference SOC trajectory. At the same time, the CDCS control strategy firstly works in the CD mode, and then works in the CS mode.

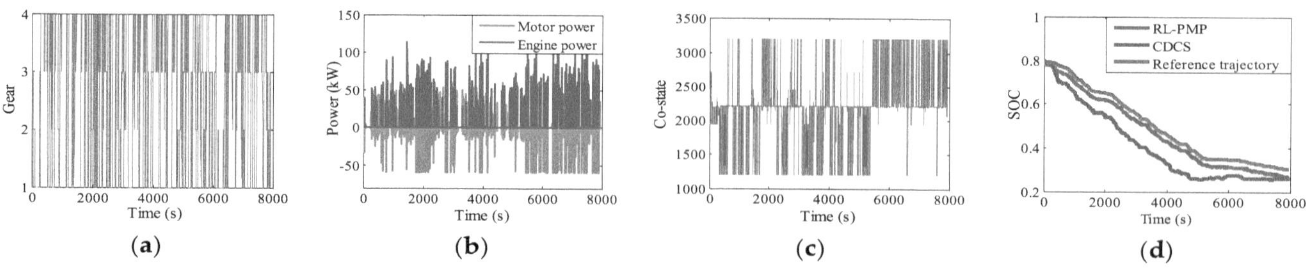

Figure 21. The HIL results of No.21. (**a**) denotes the gears, (**b**) denotes the powers of the motor and the engine, (**c**) denotes the co-state, and (**d**) denotes the SOC trajectories.

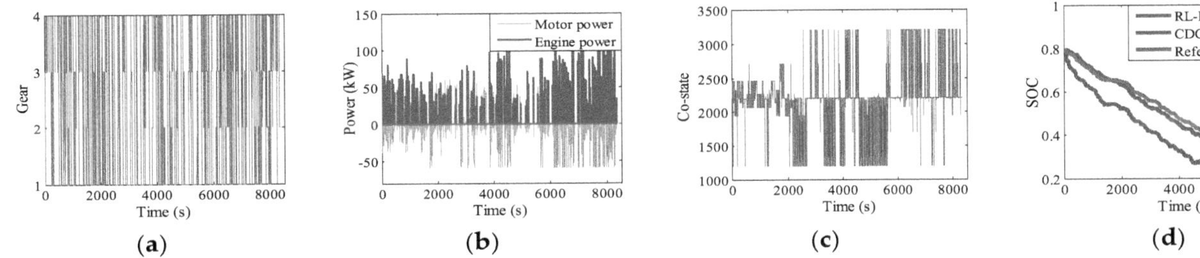

Figure 22. The HIL results of No.22. (**a**) denotes the gears, (**b**) denotes the powers of the motor and the engine, (**c**) denotes the co-state, and (**d**) denotes the SOC trajectories.

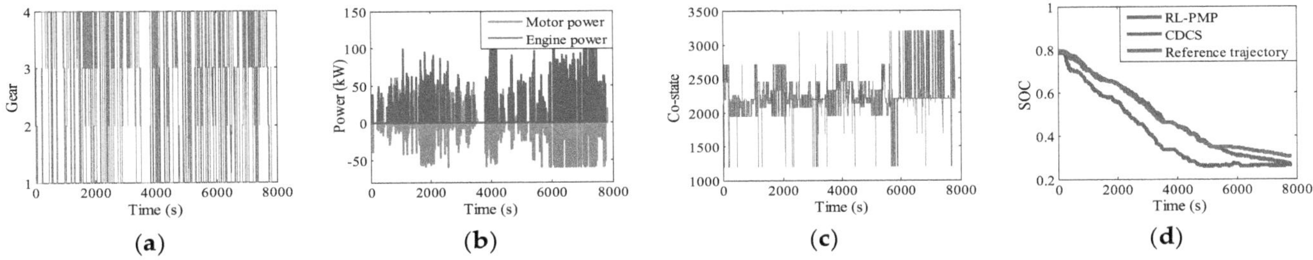

Figure 23. The HIL results of No.23. (**a**) denotes the gears, (**b**) denotes the powers of the motor and the engine, (**c**) denotes the co-state, and (**d**) denotes the SOC trajectories.

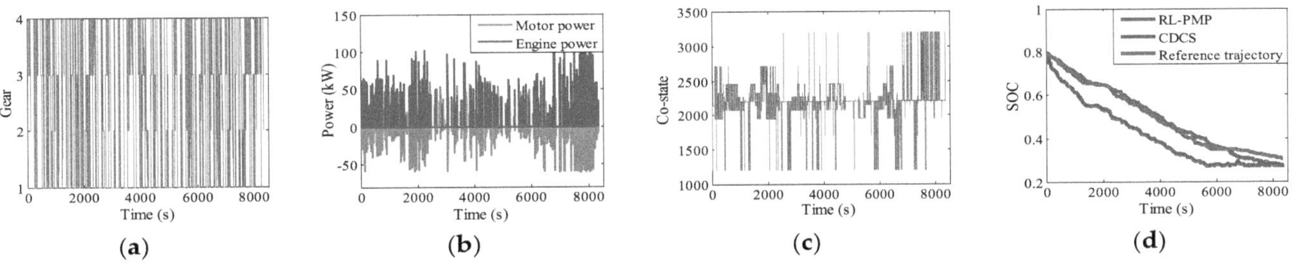

Figure 24. The HIL results of No.24. (**a**) denotes the gears, (**b**) denotes the powers of the motor and the engine, (**c**) denotes the co-state, and (**d**) denotes the SOC trajectories.

As shown in Figure 25, the fuel consumption of QL-PMP is significantly lower than the CDCS control strategy for any combined driving cycle, which further demonstrates the good economic performance of the QL-PMP algorithm. In specific, compared to the CDCS control strategy, the fuel economy can be improved by 23.01%, 23.57%, 14.13%, and 20.96% respectively. In a word, it can be improved by an average of 20.42%.

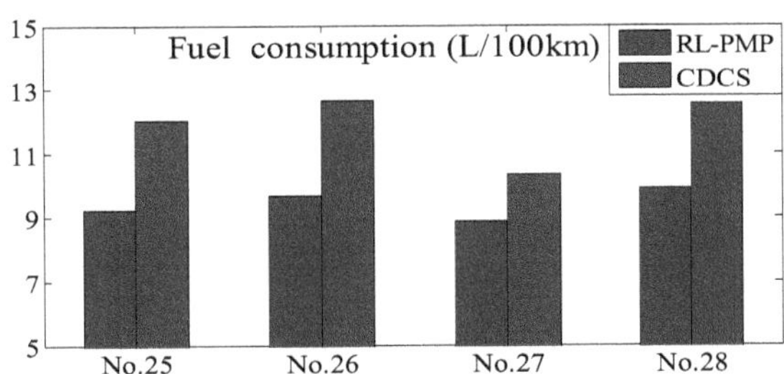

Figure 25. The fuel consumptions of the HIL verification.

6. Conclusions

This paper proposes a QL-PMP based- energy management with a small state space, which can realize real-time control in dynamically stochastic traffic circumstance. The conclusions are summarized as follows:

1. A self-learning energy management constituted by the QL and the PMP algorithms are proposed. The actual control variables of the throttle of the engine and the shift instruction of the AMT are exclusively solved by the PMP and the co-state in the PMP is designed as the action in the QL, which can effectively reduce the dependence of the action on the state and give a chance for the reduced state space design.
2. A limited state space constituted by the difference between the feedback SOC and the reference SOC is proposed. Since the co-state is designed as the action of the QL-PMP and has weak dependence on the state, only a Q-matrix with 50 rows and 25 columns are adequate for the Q-Table. This also gives a chance for the real-time energy management control in real-world.
3. The complete training process of the QL-PMP algorithm and the HIL verification are presented. The training process show that the QL-PMP can realize the self-learning energy management control for any combined driving cycle, where the control performance can be continuously strengthened. The verification of HIL demonstrates that the completely trained QL-PMP has high reliability, and can realize the real-time control, despite of any combined driving cycle. Moreover, the fuel economy can be improved by 20.42%, compared to CDCS control strategy.

Author Contributions: H.G. put forward the idea of intelligent energy management strategy, and analyzed the context of the whole paper. S.D. was mainly responsible for the writing and editing. F.Z., Q.C. and W.R. were mainly responsible for the data collecting, processing and analysis.

Nomenclature

$\dot{m}_e$	the instantaneous fuel consumption of the engine
T_e	the torque of the engine
b_e	the fuel consumption rate of the engine
Δt	the sampling time
λ_{ave}	the average co-states
T_m	the torque of the motor
γ	the discount factor
η_m	the efficiency of the motor in driving mode
η_g	denote the efficiency of the motor in braking mode
P_b	the power of the battery
V_{oc}	the open-circuit voltage of the battery
I_b	the current of the battery
R_b	the internal resistance of the battery
F_t	the driving force
δ	the rotating mass conversion factor
m	the vehicle mass
F_R	the resistance force
f	the coefficient of the rolling resistance
g	the gravity acceleration
α_s	the road slope
C_D	the coefficient of the air resistance
A_a	the frontal area
ρ_a	the air density
v	the velocity
P_r	the required power
η_t	the efficiency of the transmission system
H	the Hamiltonian function
$\dot{SOC}$	the state function
Q_{nom}	the battery capacity

ω_e	the rotate speed of the engine
ω_m	the rotate speed of the motor
$\omega_{\mathrm{m\_min}}$	the lower boundary of the ω_m
$\omega_{\mathrm{m\_max}}$	the higher boundary of the ω_m
ω_{e_min}	the lower boundary of the ω_e
ω_{e_max}	the higher boundary of the ω_e
p_e	the power of the engine
p_m	the power of the motor
p_{m_min}	the lower boundary of the p_m
p_{m_max}	the higher boundary of the p_m
p_{e_max}	the higher boundary of the p_e
p_{e_min}	the lower boundary of the p_e
u^*	the optimal control vector
λ	the co-state
F_{PMP}	the objective function
SOC_f	the terminal SOC
x	the normalized distance
d_{whole}	the total distance
d_{real}	the travelled distance
SOC_r	the reference SOC
p_i	the fitting coefficient
$Q(s_k, a_k)$	the old reward
α_l	the learning rate
$Q(s_{k+1}, a_{k+1})$	the maximum reward in future
$r^a_{ss'}$	the reward function
s	the shift instruction
S	the state space
A	the action space

References

1. Xie, S.; Hu, X.; Xin, Z.; Brighton, J. Pontryagin's Minimum Principle based model predictive control of energy management for a plug-in hybrid electric bus. *Appl. Energy* **2019**, *236*, 893–905. [CrossRef]
2. Huang, Y.; Wang, H.; Khajepour, A.; Li, B.; Ji, J.; Zhao, K.; Hu, C. A review of power management strategies and component sizing methods for hybrid vehicles. *Renew. Sustain. Energy Rev.* **2018**, *96*, 132–144. [CrossRef]
3. Yan, M.; Li, M.; He, H.; Peng, J.; Sun, C. Rule-based energy management for dual-source electric buses extracted by wavelet transform. *J. Clean. Prod.* **2018**, *189*, 116–127. [CrossRef]
4. Padmarajan, B.V.; McGordon, A.; Jennings, P.A. Blended rule-based energy management for PHEV: System structure and strategy. *IEEE Trans. Veh. Technol.* **2016**, *65*, 8757–8762. [CrossRef]
5. Pi, J.M.; Bak, Y.S.; You, Y.K.; Park, D.H.; Kim, H.S. Development of route information based driving control algorithm for a range-extended electric vehicle. *Int. J. Automot. Technol.* **2016**, *17*, 1101–1111. [CrossRef]
6. Sabri, M.F.M.; Danapalasingam, K.A.; Rahmat, M.F. A review on hybrid electric vehicles architecture and energy management strategies. *Renew. Sustain. Energy Rev.* **2016**, *53*, 1433–1442. [CrossRef]
7. Zhou, Y.; Ravey, A.; Péra, M.C. A survey on driving prediction techniques for predictive energy management of plug-in hybrid electric vehicles. *J. Power Source* **2019**, *412*, 480–495. [CrossRef]
8. Tie, S.F.; Tan, C.W. A review of energy sources and energy management system in electric vehicles. *Renew. Sustain. Energy Rev.* **2013**, *20*, 82–102. [CrossRef]
9. Hou, C.; Ouyang, M.; Xu, L.; Wang, H. Approximate Pontryagin's minimum principle applied to the energy management of plug-in hybrid electric vehicles. *Appl. Energy* **2014**, *115*, 174–189. [CrossRef]
10. Tulpule, P.; Marano, V.; Rizzoni, G. Energy management for plug-in hybrid electric vehicles using equivalent consumption minimization strategy. *Int. J. Electr. Hybrid Veh.* **2010**, *2*, 329–350. [CrossRef]
11. Li, G.; Zhang, J.; He, H. Battery SOC constraint comparison for predictive energy management of plug-in hybrid electric bus. *Appl. Energy* **2017**, *194*, 578–587. [CrossRef]

12. Onori, S.; Tribioli, L. Adaptive Pontryagin's Minimum Principle supervisory controller design for the plug-in hybrid GM Chevrolet Volt. *Appl. Energy* **2015**, *147*, 224–234. [CrossRef]
13. Yang, C.; Du, S.; Li, L.; You, S.; Yang, Y.; Zhao, Y. Adaptive real-time optimal energy management strategy based on equivalent factors optimization for plug-in hybrid electric vehicle. *Appl. Energy* **2017**, *203*, 883–896. [CrossRef]
14. Lei, Z.; Qin, D.; Liu, Y.; Peng, Z.; Lu, L. Dynamic energy management for a novel hybrid electric system based on driving pattern recognition. *Appl. Math. Model.* **2017**, *45*, 940–954. [CrossRef]
15. Lin, X.; Wang, Y.; Bogdan, P.; Chang, N.; Pedram, M. Reinforcement learning based power management for hybrid electric vehicles. In Proceedings of the 2014 IEEE/ACM International Conference on Computer-Aided Design, San Jose, CA, USA, 3–6 November 2014; IEEE Press: Piscataway, NJ, USA, 2014; pp. 32–38.
16. Xiong, R.; Cao, J.; Yu, Q. Reinforcement learning-based real-time power management for hybrid energy storage system in the plug-in hybrid electric vehicle. *Appl. Energy* **2018**, *211*, 538–548. [CrossRef]
17. Liu, T.; Wang, B.; Yang, C. Online Markov Chain-based energy management for a hybrid tracked vehicle with speedy Q-learning. *Energy* **2018**, *160*, 544–555. [CrossRef]
18. Wu, J.; He, H.; Peng, J.; Li, Y.; Li, Z. Continuous reinforcement learning of energy management with deep Q network for a power split hybrid electric bus. *Appl. Energy* **2018**, *222*, 799–811. [CrossRef]
19. Hu, Y.; Li, W.; Xu, K.; Zahid, T.; Qin, F.; Li, C. Energy management strategy for a hybrid electric vehicle based on deep reinforcement learning. *Appl. Sci.* **2018**, *8*, 187. [CrossRef]
20. Liessner, R.; Schroer, C.; Dietermann, A.M.; Bäker, B. Deep Reinforcement Learning for Advanced Energy Management of Hybrid Electric Vehicles. *ICAART* **2018**, *2*, 61–72.
21. Qi, X.; Luo, Y.; Wu, G.; Boriboonsomsin, K.; Barth, M. Deep reinforcement learning enabled self-learning control for energy efficient driving. *Transp. Res. Part C Emerg. Technol.* **2019**, *99*, 67–81. [CrossRef]
22. Moser, I.; Chiong, R. A hooke-jeeves based memetic algorithm for solving dynamic optimisation problems. *Hybrid Artif. Intell. Syst.* **2009**, *5572*, 301–309.
23. Guo, H.; Lu, S.; Hui, H.; Bao, C.; Shangguan, J. Receding horizon control-based energy management for plug-in hybrid electric buses using a predictive model of terminal SOC constraint in consideration of stochastic vehicle mass. *Energy* **2019**, *176*, 292–308. [CrossRef]

10

Salp Swarm Optimization Algorithm-Based Controller for Dynamic Response and Power Quality Enhancement of an Islanded Microgrid

Touqeer Ahmed Jumani [1,2,*], Mohd. Wazir Mustafa [1], Madihah Md. Rasid [1], Waqas Anjum [1,3] and Sara Ayub [1,4]

1 School of Electrical Engineering, Universiti Teknologi Malaysia, Johor Bahru 81310, Malaysia; wazir@utm.my (M.W.M.); madihahmdrasid@utm.my (M.M.R.); Waqas.anjum@iub.edu.pk (W.A.); sara.ayub@buitms.edu.pk (S.A.)

2 Department of Electrical Engineering, Mehran University of Engineering and Technology SZAB Campus Khairpur Mirs, Khairpur 66020, Pakistan

3 Department of Electronic Engineering, The Islamia University of Bahawalpur, Bahawalpur 63100, Pakistan

4 Department of Electrical Engineering, Balochistan University of Information Technology, Engineering and Management Sciences, Quetta 87300, Pakistan

* Correspondence: atouqeer2@graduate.utm.my

Abstract: The islanded mode of the microgrid (MG) operation faces more power quality challenges as compared to grid-tied mode. Unlike the grid-tied MG operation, where the voltage magnitude and frequency of the power system are regulated by the utility grid, islanded mode does not share any connection with the utility grid. Hence, a proper control architecture of islanded MG is essential to control the voltage and frequency, including the power quality and optimal transient response during different operating conditions. Therefore, this study proposes an intelligent and robust controller for islanded MG, which can accomplish the above-mentioned tasks with the optimal transient response and power quality. The proposed controller utilizes the droop control in addition to the back to back proportional plus integral (PI) regulator-based voltage and current controllers in order to accomplish the mentioned control objectives efficiently. Furthermore, the intelligence of the one of the most modern soft computational optimization algorithms called salp swarm optimization algorithm (SSA) is utilized to select the best combination of the PI gains (k_p and k_i) and dc side capacitance (C), which in turn ensures optimal transient response during the distributed generator (DG) insertion and load change conditions. Finally, to evaluate the effectiveness of the proposed control approach, its outcomes are compared with that of the previous approaches used in recent literature on basis of transient response measures, quality of solution and power quality. The results prove the superiority of the proposed control scheme over that of the particle swarm optimization (PSO) and grasshopper optimization algorithm (GOA) based MG controllers for the same operating conditions and system configuration.

Keywords: microgrid; optimization; voltage and frequency regulation; dynamic response enhancement; salp swarm optimization algorithm; power quality

1. Introduction

Due to the recent developments in power electronics and artificial intelligence technology, the flexible deployment and performance of distributed generators (DGs) have been improved significantly. This is necessary in order to cope up with the increasing power demand along with improvement in power quality and reliability of the power system. Furthermore, these DGs combine together with a group of loads to form a small entity of an electrical network called Microgrid (MG) [1]. MG can be interconnected with the utility grid through a static switch at point of common coupling (PCC) or operated isolated. In the grid-tied mode, the power can be imported or exported between the MG and main grid as needed while in the islanded mode the MG operates to support the load at its own without any external support from the utility grid. Furthermore, in grid-tied mode, the voltage magnitude and frequency are dictated by the giant power system, while power flow between DGs and the main grid is the major control concern. Contrary to that, one of the very basic tasks of any control architecture in islanded mode is to regulate the voltage magnitude and frequency at their rated values. A general structure of the MG is shown in Figure 1.

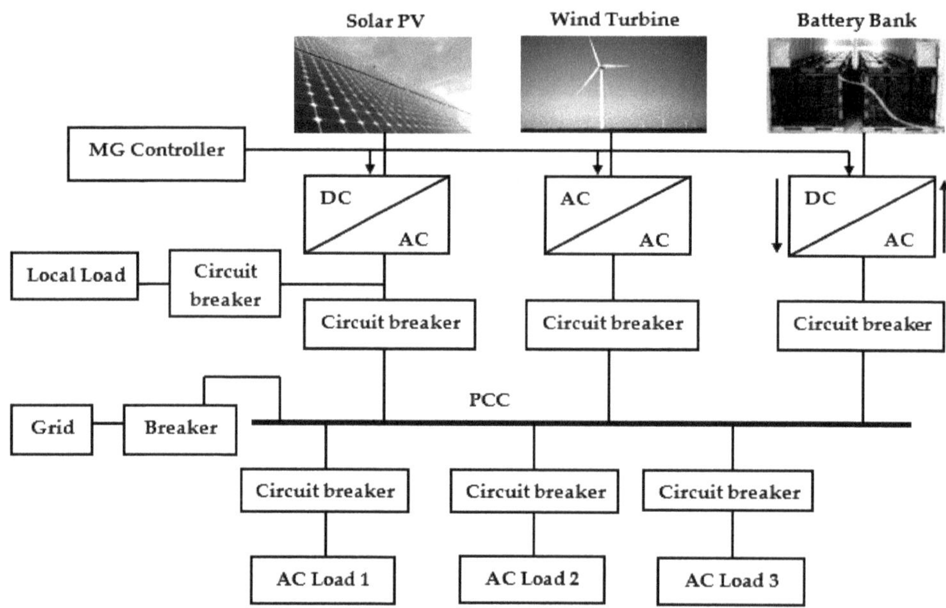

Figure 1. A basic structure of Microgrid [2].

Figure 1 portrays a very general structure of an MG where the DGs (wind and solar in this case), battery banks and loads are connected to the point of common coupling through three-phase circuit breakers. To inject a controlled amount of power at rated voltage and frequency, generally, a controlled non-linear power electronic device is utilized. These devices are non-linear in nature and hence need to be properly controlled in order to achieve a higher power quality along with the proper accomplishment of control objectives such as voltage, frequency and power flow regulation. In grid-connected mode, the sole objective of the MG controller is to ensure proper sharing of power among DGs and between MG and utility grid while in islanded mode the prime control objective is to regulate voltage and frequency at their rated values along with controlled power-sharing among DGs and power quality. Since MG faces more challenges in the islanded mode as compared to that of the grid-connected mode, this research is purely dedicated to the islanded mode of MG operation.

The study aims to improve the transient response of an islanded MG under different operating conditions. The mentioned task is accomplished by using the intelligence of the salp swarm optimization algorithm (SSA) to obtain the optimal combination of Proportional plus Integral (PI) controller parameters and dc-link side capacitance value provides the least settling time and overshoot during DG injection and load switching conditions. SSA is one of the most intelligent optimization algorithms introduced by Mirjalili et al. [3]. As compared to Genetic Algorithm (GA) and Particle Swarm Optimization (PSO), SSA is a more evolved algorithm for solving the different optimization problems [3]. It has been applied and found better than its competitor algorithms in solving many engineering problems like parameter identification of Photo-Voltaic (PV) modules [4], optimal allocation and capacity of DGs [5], extracting optimal parameters of fuel cells [6], training ANN for pattern recognition [7] and load-frequency control [8]. The mentioned studies have duly verified the effectiveness of SSA in solving the studied optimization problems better than conventional optimization techniques. This research work utilizes the intelligence of SSA for obtaining the optimal combination of PI gains and dc-link capacitance value to improve the dynamic response of studied islanded MG and regulating voltage and frequency under the DG insertion and load change conditions. The controller utilizes the droop control in the control structure as a power-sharing controller along with the voltage and current control loops. To authenticate the effectiveness of the proposed control technique, its performance is compared with that of the PSO and GOA based controllers for the identical operating conditions.

The subsequent sections are described as follows; Section 2 discusses the modern MG control architectures and a brief literature review of the available techniques used to tackle the presented problem. Section 3 describes the considered islanded MG along with its complete mathematical model and control architecture. In Section 4 the proposed methodology for obtaining an optimal transient response of studied MG system and basics of SSA along with the justification for considered fitness function has been provided. Finally, in Section 5, the obtained results are presented.

2. Modern Microgrid Control Architectures

Recently, several control architectures have been developed for the islanded MGs using proportional plus integral (PI) regulator-based power, voltage and current control loops. The generalized control architecture of the MG based on the above-mentioned control loops is shown in Figure 2.

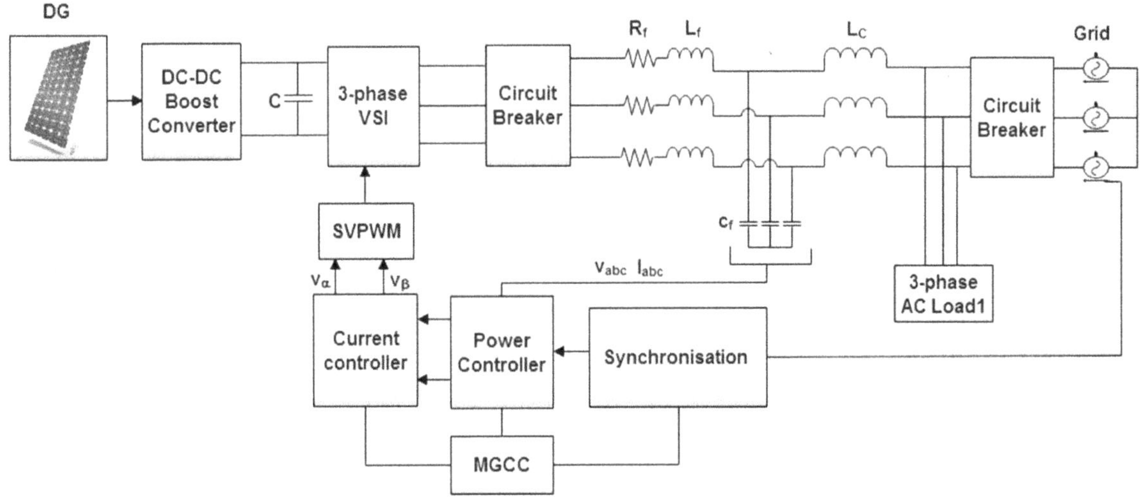

Figure 2. A generalized control structure for MGs.

For the grid-tied mode, there is no requirement of the voltage controller as it is dictated by the main grid while the power controller may be replaced by a voltage controller in case of the autonomous mode. Due to the robust structure and simple control action, PI controllers are found to be the most widely used regulators in these control loops. The main drawback of using PI regulators in modern MG control architectures is that their performance is dictated by their proportional and integral gains (k_p and k_i). Hence, selection of these gains decides the overall performance of the controller and consequently, the studied system. Over the last years, these gains were selected by using the "trial and error" or Ziegler-Nicolas (Z-N) method. These conventional methods of PI tuning suffer from many disadvantages, such as excessive time consumption, uncertainty in gain selection and complex calculations which restricted PI controller application in the latest MG control architectures up to a large extent. Most recently, with the development in the area of artificial intelligence (AI) and its applications in optimization field, several AI methods such as fuzzy logic (FL), genetic algorithm (GA) and particle swarm optimization (PSO) were explored to optimize PI controller parameters for dynamic response enhancement of islanded ac MGs. The results presented in the mentioned research papers clearly show the importance of the AI techniques in obtaining optimal PI parameters, which led to the enhanced transient response of the studied MG systems.

Many researchers around the world have worked on solving the mentioned optimization problem using different AI techniques. Al-Saedi et al. [9] developed an MG controller for controlling voltage and frequency of an autonomous MG. The PSO was utilized to optimize the system parameters and hence dynamic response. The developed controller successfully regulated the voltage magnitude and managed to keep it within standard limits (±5% of the rated value). However, the frequency level was not up to the standard (±1% of the rated value) during transient conditions. An attempt was made to solve the mentioned problem in reference [10] where the PSO based controller was developed for regulating voltage and frequency of an islanded MG. The designed controller successfully regulated the frequency and kept in well within the standard limits despite the huge load and source variations. However, the study did not consider the voltage profile, which is one of the very important parameters in the islanded mode of MG. Authors in the references [11,12] developed an AI-based controller for the voltage regulation of an islanded MG using Pareto based BB-BC and hybrid big-bang big-crunch (BB-BC) algorithms, respectively. In both case studies, the frequency declined to 59.7 Hz from its rated value (60 Hz) during DG insertion and load change conditions. Most recently, Qazi et al. [13] used the whale optimization algorithm (WOA) for regulating frequency and voltage independently by formulating two different fitness functions in an islanded MG. However, it may be noted that the voltage and frequency are two interrelated parameters [12], and therefore, it is not practically possible to independently optimize these two inter-dependent parameters. Most recently, in November 2018, the authors in reference [14] explored the grasshopper optimization algorithm (GOA) in order to regulate voltage and frequency along with the optimal transient response of an islanded MG. The authors achieved a better dynamic response as compared to previous quoted literature; however, as compared to WOA and PSO, no significant improvement was observed in overshoot and settling time during load change conditions. To address the stated limitation, for the very first time as per the best of authors' knowledge, an additional parameter that is dc side capacitance has been adopted as optimization variable along with four gains of two PI regulators in islanded MG controls. In almost all previous MG control architectures the dc side capacitance value has been adopted based on "trial and error" method whose disadvantages are already discussed earlier.

3. Proposed Islanded MG Architecture

In this section, a detailed insight of the islanded MG model along with the proposed SSA based voltage-frequency (v-f) controller is provided. A comprehensive block diagram of the power and control circuit of the studied islanded infinite bus MG system is shown in Figure 3.

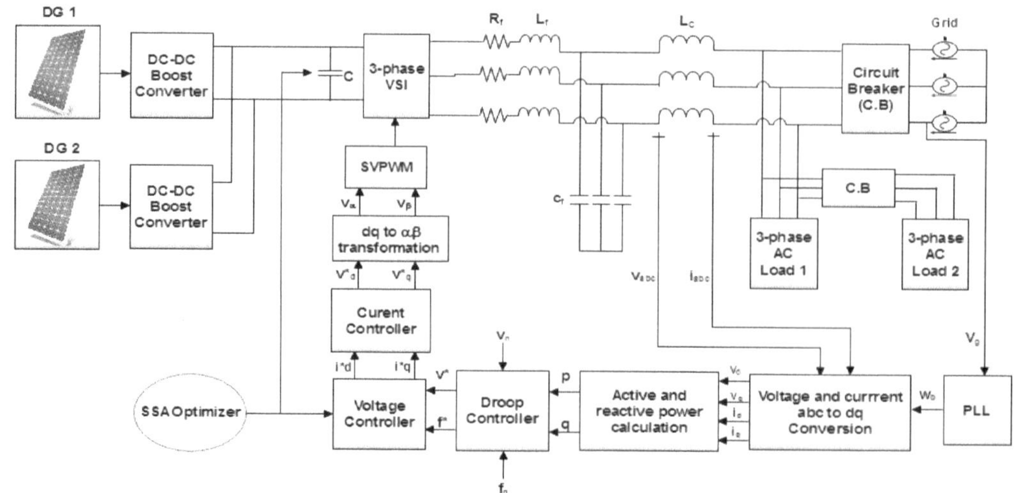

Figure 3. Studied islanded MG with power and control architecture.

The power circuit contains two solar photovoltaic (PV) modules, a three-phase voltage source inverter (VSI), RLC filter, coupling inductor and two RL loads. Since the studied MG is operated in an islanded mode, the utility grid disconnected from the rest of the network using a three-phase circuit breaker. In order to control the voltage and frequency of the studied MG system, the voltage and current signals are converted from *abc* to *dq* reference frame by using well-known Park's transformation as given in Equations (1) and (2), respectively:

$$\begin{bmatrix} v_d \\ v_q \\ v_o \end{bmatrix} = \sqrt{\frac{2}{3}} \begin{bmatrix} \cos\theta & \cos\left(\theta - \frac{2\pi}{3}\right) & \cos\left(\theta + \frac{2\pi}{3}\right) \\ -\sin\theta & -\sin\left(\theta - \frac{2\pi}{3}\right) & -\sin\left(\theta + \frac{2\pi}{3}\right) \\ \frac{1}{\sqrt{2}} & \frac{1}{\sqrt{2}} & \frac{1}{\sqrt{2}} \end{bmatrix} \begin{bmatrix} v_a \\ v_b \\ v_c \end{bmatrix} \tag{1}$$

$$\begin{bmatrix} i_d \\ i_q \\ i_o \end{bmatrix} = \sqrt{\frac{2}{3}} \begin{bmatrix} \cos\theta & \cos\left(\theta - \frac{2\pi}{3}\right) & \cos\left(\theta + \frac{2\pi}{3}\right) \\ -\sin\theta & -\sin\left(\theta - \frac{2\pi}{3}\right) & -\sin\left(\theta + \frac{2\pi}{3}\right) \\ \frac{1}{\sqrt{2}} & \frac{1}{\sqrt{2}} & \frac{1}{\sqrt{2}} \end{bmatrix} \begin{bmatrix} i_a \\ i_b \\ i_c \end{bmatrix} \tag{2}$$

The basic aim of this transformation is to achieve an equivalent two-phase orthogonal stationary reference frame from three-phase rotating signals. This conversion is necessary because the PI controllers do not work properly on sinusoidal reference signals. A more detailed version of the proposed control architecture is depicted in Figure 4.

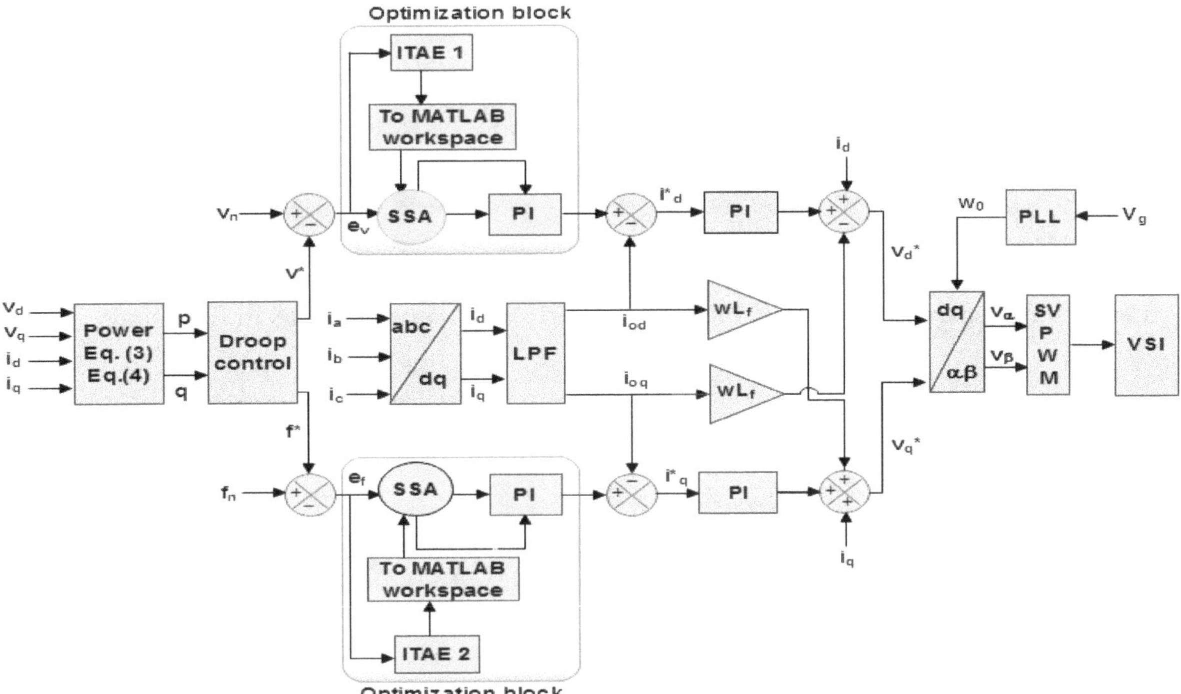

Figure 4. Detailed diagram of the salp swarm optimization algorithm based controller for islanded MG.

After obtaining the voltage and current signals in *dq* reference frame, the active and reactive power is measured using Equations (3) and (4), respectively:

$$p = v_d i_d + v_d i_q \tag{3}$$

$$q = v_d i_q - v_q i_q \tag{4}$$

In order to remove the harmonic contents from the fundamental active and reactive power curves, a low pass filter (LPC) with the following transfer function models is used:

$$\frac{P}{p} = \frac{w_c}{s + w_c} \tag{5}$$

$$\frac{Q}{q} = \frac{w_c}{s + w_c} \tag{6}$$

It may be noted that in conventional synchronous generator-based power systems, the voltage and frequency of the system are regulated by an auto-voltage regulator (AVR) and governor system respectively. However, these methods do not work for DGs such as solar PV, where the output power from panels is dc and fluctuating in nature due to unpredictable weather conditions and continuous load switching. Hence, in this case, a droop controller is utilized to compensate for the voltage and frequency sags and swells during any source and load change conditions. Droop control can instantly compensate for the voltage and frequency drop without any communication link with the central MG controller. It does so by continuously measuring the power output and accordingly changing the reference signals (v^* and f^*) for the voltage controller. The dynamics for the droop control are given in Equations (7) and (8):

$$f^* = f_n - k_w P \tag{7}$$

$$v^* = v_n - k_v Q \tag{8}$$

The values for the v_n and f_n are taken as 220 V and 50 Hz, respectively. The two output signals of the droop controller, i.e., which v^* and f^*, are then compared with their respective nominal values using a comparator and errors (e_v and e_f) are fed to two PI controllers whose gains are optimized by SSA optimizer. The sole aim of optimizing PI gains is to achieve an optimal dynamic response of the studied MG system with minimum overshoot and settling time in voltage and frequency curves. In order to precisely track the current reference signals (i_d^* and i_q^*), two more PI controllers are used in the current controller block of the proposed controller. Since the gains for the PI controller used in voltage controller block were optimized using SSA, optimizing gains for PI regulators in the current controller block will increase the complexity and duration of the overall optimization process and may result in un-optimal results at the end due to the excessive number of optimization variables. The transfer function equations for the current controller as derived from Figure 4 are as in Equations (9) and (10):

$$v_d^* = k_{pv} + \frac{k_{iv}}{s}\left(i_{d_{ref}} - i_{od}\right) - w * L_f * i_{oq} + v_d \tag{9}$$

$$v_q^* = k_{pf} + \frac{k_{if}}{s}\left(i_{d_{ref}} - i_{od}\right) + w * L_f * i_{od} + v_q \tag{10}$$

Finally, the space vector pulse width modulation (SVPWM) technique is used to generate the firing pulses for VSI so that a regulated quantity of power with high power quality can be injected to feed the connected load.

4. Proposed Methodology

This section is further divided into two subsections. The first section discusses the mathematical formulation of SSA for optimizing MG control parameters, while the fitness function to be minimized by the SSA is presented in the second section.

4.1. Salp Swarm Algorithm and Its Implementation

In this section, the motivation behind the proposed optimization algorithm along with its complete mathematical model is discussed in detail. Salps are the family member of the Salpidae group. They possess the translucent barrel-shaped figure and move similar to that of jellyfish. To model their movement mathematically, their initial positions are initialized randomly as portrayed in Equation (11) [3]:

$$K_1^{1:n} = rand\,(\ldots) * (ub_j - lb_j) + lb_j, \quad \forall j \in no.\,of\ variables \tag{11}$$

where $K_1^{1:n}$ shows the initial positions of the salps, ub_j and lb_j represents the upper and lower limit, respectively, and *rand* (...) is the command for generating random numbers between 0–1.

Afterward, two groups are formed: leader and followers. All the salps except those at the rear are labeled as followers while the front ones are called leaders. The leader salps guide the whole swarm while the followers follow them continuously in order to search for a food source labeled as M in this case. Similar to other metaheuristic optimization techniques, the location of each salp is defined in a defined search space having n number of dimensions that are equal to the number of controlled parameters in an optimization problem. Furthermore, the location of the entire number of salps is kept in a matrix K, which is defined in Equation (12) [3]:

$$K_j^1 = \begin{cases} M_i + c_1\big((ub_j - lb_j)c_2 + lb_j\big)\ c_3 \geq 0 \\ M_i - c_1\big((ub_j - lb_j)c_2 + lb_j\big)\ c_3 < 0 \end{cases} \tag{12}$$

where the symbol K_j^1 symbolizes the position of leader salp in the jth dimension, M_i shows the location of the food source in the jth dimension and c_1, c_2 and c_3 denote the random numbers. It is significant to note that, unlike conventional optimization methods such as GA and PSO, the SSA effectively manages to avoid trapping into local minimum due to its adaptive optimization mechanism. SSA updates the position of follower salps with respect to each other and allows them to move gradually towards the leading salp, which prevents the algorithm from stagnating into the local optima. Thus, the algorithm produces an optimal or near-optimal solution precisely during an optimization process [3]. Furthermore, SSA has better exploration versus exploitation balancing capability, which is the fundamental requirement for reaching the best available solution for an optimization problem. As can be seen from Equation (12), the leader salp upgrades its location with reference to the food source only. The coefficient c_1 in Equation (12) is one of the very important parameters in SSA since it helps in balancing the exploitation and exploration characteristics of SSA and is defined in Equation (13):

$$c_1 = 2e^{-\left(\frac{4l}{L}\right)^2} \tag{13}$$

where l represents the current number of iteration, while L denotes the number of maximum iterations. The symbols c_2 and c_3 used in Equation (12) represents the random numbers between 0 and 1. In order to upgrade the location of the follower salps, a similar equation to that of Newton's law of motion is used:

$$K_j^i = \frac{1}{2}\,at^2 + v_0\,t \tag{14}$$

where $i \geq 2$ and K_j^i is the positions of ith follower salp in jth dimension, t denotes the time and v_0 is the symbol used for the velocity at the start of the optimization process, which is generally taken as 0. As the time in the optimization procedure can be replaced by the iterations and the variance between two successive iterations cannot be in a fractional number, hence by assuming $v_0 = 0$, the Equation (14) can be re-written as given underneath:

$$K_j^i = \frac{K_j^i + K_j^{i-1}}{2} \tag{15}$$

where $i \geq 2$ and K_j^i denotes the position of ith follower salp in jth dimension. It is important to note that the salp chain possesses the potential of moving towards the continuously changing global optimum (food source) in order to find a better solution by exploring and exploiting the defined search space. Some of the very important features of SSA which make it different from the conventional optimization algorithms are listed as follows [3];

1. The algorithm keeps the best-obtained solution after each iteration and assigns it to the global optimum (food source) variable. Hence, it can never be wiped out even if the whole population deteriorates.
2. SSA updates the position of the leading salp with respect to the food source only, which is the best solution obtained thus far; therefore, the leader salp always explores and exploits the space around it for a better solution.
3. SSA updates the position of follower salps with respect to each other in order to let them move towards the leading salp gradually.
4. Gradual movements of follower salps prevent the SSA from being easily stagnating into local optima.
5. Parameter c_1 is decreased adaptively over the course of iterations, which helps the algorithm to explore the search space at starting and exploits it at the ending phase.
6. SSA has only one main controlling parameter (c_1), which reduces the complexity and makes it easy to implement.

The above-mentioned merits of SSA make it potentially able to solve the optimization problems better than conventional optimization methods and hence became the motivation for the current research work. In addition, the adaptive mechanism of SSA allows this algorithm to find an accurate estimation of the best solution by continuously avoiding the trapping into local solutions.

Further details of SSA along with its pseudo-code and benchmarking against the conventional optimization methods can be found in reference [3]. It may be noted that for tuning the characteristics of the SSA for solving a specific problem, two parameters may be adjusted, namely the number of searching salps and the total iterations' number. The complete methodology of obtaining optimal PI parameters and dc-link capacitance for enhancing the dynamic response of the grid-tied MG system using SSA is depicted in the flow-chart shown in Figure 5.

Similar to other metaheuristic optimization methods, SSA spread a specified number of random search agents within searching space in the first stage. Based on the operating mechanism of SSA as given in Figure 5, these search agents are permitted to change their position in the restricted search space to minimize the stated Fitness Function (FF) subjected to a defined set of constraints. It is important to note that the mathematical model for simulating salp chains cannot be directly employed to solve optimization problems. In other words, there is a need to tweak the model according to the studied system configuration in order to make it applicable to solve the specific optimization problem. The ultimate goal of any optimization algorithm is to determine the global optimum which is generally carried out by minimization or maximization process of a defined FF. In SSA, the mentioned objective is accomplished by making the leading salp to move towards the food source and the follower salps follow the leading salp as per the governing equations of the algorithm. In the context of the optimization process, the food source is treated as the global optimum; therefore, the salp chain automatically moves towards it. It is assumed that the best solution obtained so-far is the global optimum and the food source to be chased by the salp chain until the termination criterion is reached. In the current case study, the completion of the maximum number of iterations is set as the termination criteria of the optimization process. Hence, once the pre-decided maximum number of iterations is reached, the obtained optimized PI gains and capacitance value are collected from the MATLAB workspace and are exported to the PI controller and dc side capacitor block in MG Simulink model. After that, the MG Simulink model is executed for 0.5 s to obtain the optimized power, voltage and current curves.

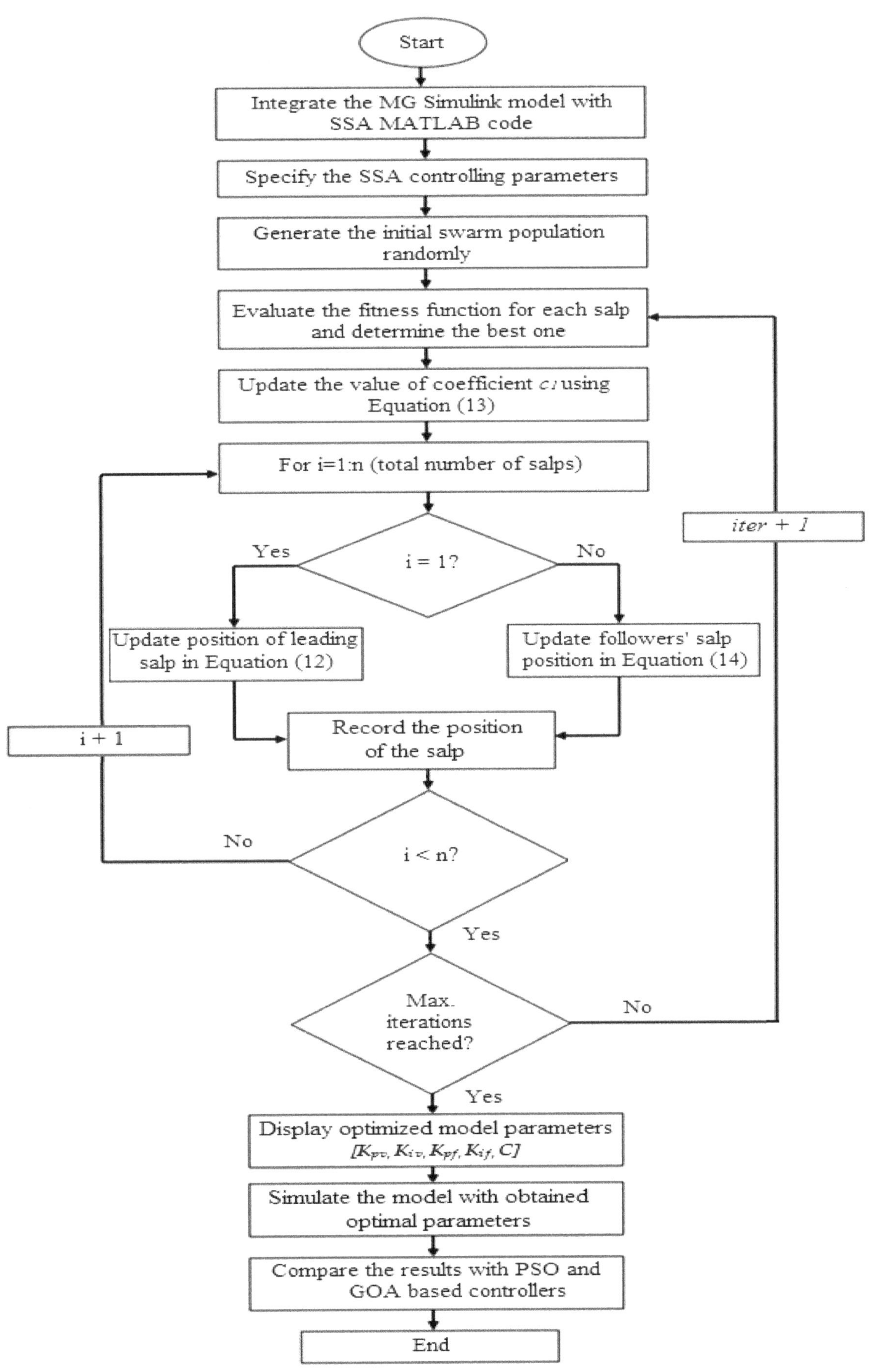

Figure 5. Proposed SSA based optimal plus integral (PI) parameter selection in studied islanded MG.

4.2. Fitness Function Formulation

The operating point of the power system fluctuates continuously and abruptly due to continuous and sudden load variations. Furthermore, in grid-connected MG power systems, the sudden insertion and desertion of DGs into the power system may push it into an unstable region of operation due to large power fluctuations and overshoots. One of the solutions to avoid such a situation is to use optimized controller parameters that can provide an optimum transient response of the system. Keeping in view the mentioned issues in the power system, the two gains of each PI controller (K_{pv}, K_{iv}, K_{pf}, K_{if}) placed in the first control loop of the studied MG controller along with the dc link capacitance (C) are optimized by SSA. In fact, for all metaheuristic and evolutionary optimization methods, the presence of the FF is a compulsory requirement that evaluates the fitness of each and every search agent and selects the best one for further comparison during the next iteration. In the context of the current study, where the optimal tuning of PI controller is required, four FF criterions that are generally considered in the literature are Integral Square Error (ISE), Integral Absolute Error (IAE), Integral Time Square Error (ITSE) and Integral Time Absolute Error (ITAE). However, ITAE is the most widely used FF criterion than its compilators due to the easy implementation, realistic error indexing and better outcomes [15,16]. The ITSE and ISE used to square the error which produces large perturbation in results even for a very small change in error signal, and hence generates impractical results. In addition, due to the continuous-time multiplication with the absolute value of error, the ITAE produces more realistic error indexing as compared to IAE. Hence looking at the prominent features of the ITAE criterion, it is adopted as the FF to be minimized in this study. Mathematically, the ITAE is expressed as provided in Equations (16) and (17):

$$ITAE1 = \int_0^\infty t|e_v|\, dt \tag{16}$$

$$ITAE2 = \int_0^\infty t\left|e_f\right|\, dt \tag{17}$$

where t refers to the total time of simulation, $|e_v|$ and $\left|e_f\right|$ are the absolute values errors in active and reactive power measurement, respectively, which are figured out by arithmetically deducting the measured power values from the setpoint power values. The overall FF is formed by mathematically adding two measured ITAE values as given in Equation (18):

$$FF = Min\left\{ \int_0^\infty t * |e_v|dt + \int_0^\infty t * \left|e_f\right|dt \right\} \tag{18}$$

It is important to understand that the magnitude of the FF in Equation (18) needs to be minimized by SSA in order to acquire the optimal PI parameter values that provide an optimum transient response of the MG system. The islanded MG modeling along with its controller and formulation of FF is carried out in MATLAB/SIMULINK version 2018a, while the coding of SSA along with its parameters like the total number of salps, iterations and optimization variables are depicted in the MATLAB editor window. Furthermore, the Simulink model is made to run for the pre-set maximum number of iterations using the "sim ( )" command from the editor window. Finally, once the simulation is finished, the optimized PI gains are placed into the corresponding PI block in MATLAB/SIMULINK model to achieve the optimal dynamic behavior of the designed MG power system.

5. Results and Discussion

The evaluation of the dynamic response for the developed islanded MG system is studied by using MATLAB/SIMULINK version 2018a. In order to make a justified evaluation of PSO, GOA and SSA the identical system parameters are used for both simulations and are depicted in Table 1.

Table 1. Studied MG system parameters.

Parameter	Symbol	Value
Solar PV rating	P_s	150 kW
Filter capacitance	C_f	2.5 mF
Filter inductance	L_f	95 mH
Switching frequency	f_{sw}	10 kHz
Sampling frequency	f_s	500 kHz
Load 1	P_1, Q_1	50 kW, 30 kVAR
Load 2	P_2, Q_2	40 kW, 20 kVAR
Load 3	P_3, Q_3	40 kW, 20 kVAR

The optimized parameters obtained at the end of the optimization process are provided in Table 2.

Table 2. The optimized parameters obtained at the end of the simulation.

Optimization	K_{pv}	K_{iv}	K_{pf}	K_{if}	C (mF)
PSO	0.2571093	25.6392019	0.9374905	9.3847852	23.817
GOA	0.9441557	12.8365850	26.768654	1.2474575	17.458
SSA	1.5485963	0.87302975	2.1385992	15.583932	19.954

The considered MG is purely operated in islanded mode, and hence, no grid-connection is made throughout its operation. Furthermore, the voltage magnitude and frequency curves are measured at the load connection point. In addition, since the operating conditions, the number of iterations and simulation time is pre-defined for studied islanded MG model, the optimized variables obtained through PSO, GOA and SSA at the end of the simulation ensure optimal dynamic response of the studied MG system for its complete operation. The obtained optimized parameters are inserted into the islanded MG SIMULINK model and the developed model is then evaluated for the three cases which are thoroughly discussed in the subsequent subsection.

5.1. Voltage and Frequency Regulation during DG Insertion and Load Change

As discussed earlier, the regulation of voltage and frequency is one of the major control concerns in any islanded MG due to the absence of support from the main grid. Achieving a stable rated voltage and frequency after the DG insertion and abrupt load changes with minimum overshoot and settling time needs fine-tuning of the system parameters. Hence, three different metaheuristic techniques, i.e., PSO, GOA and SSA, were utilized separately to obtain the optimal set of PI parameters and dc-link capacitance in order to minimize the overshoot and settling time after a disturbance in the MG system. Once the simulation started from the MATLAB editor, the SSA initiated its searching process for the most suitable PI gains and capacitance combination that delivers the least value of FF in order to achieve optimal dynamic response with least possible overshoot and settling time. It simulates the developed SIMULINK MG model for the pre-set number of iterations. Finally, at the end of the simulation, optimal values of four PI gains along with the capacitance value were obtained that provided minimum error integrating the FF value, which in turn guarantees the optimal dynamic behavior of the developed MG model. At exactly 0.05 s, the solar PV modules were switched on through a three-phase CB. Once the DGs were injected into the power system, they caused overshoots in the system voltage, as depicted in Figure 6.

It may be noted from Figure 6 that the system voltage observed an overshoot during the DG insertion at 0.05 s of simulation. The magnitude of the overshoot depended on the DG rating as well as the selection of controller parameters. In order to have a fair comparison between the three optimal parameter selection methods, i.e., PSO, GOA and SSA, the DG rating and other system parameters were taken identical for all three cases. At 0.25 s of the simulation, a load of 40 kW, 20 kVAR was added to the system, and as a result of that, the system voltage observed a dip in voltage. Similarly, at 0.55 s, the load-3 was disconnected from the system. The corresponding response shows a swell in voltage at the mentioned simulation time. A zoomed version of Figure 6 during DG injection, load injection and load detachment is shown in Figure 7a–c, respectively.

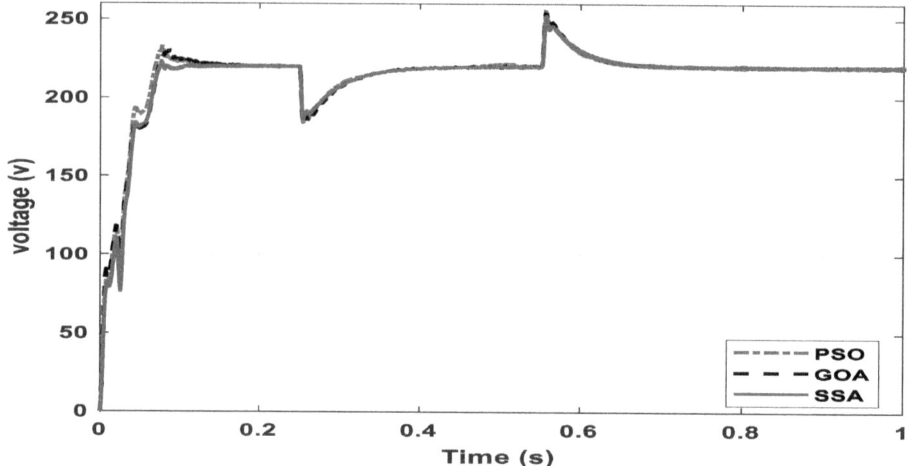

Figure 6. The voltage response of the system during DG insertion and load change.

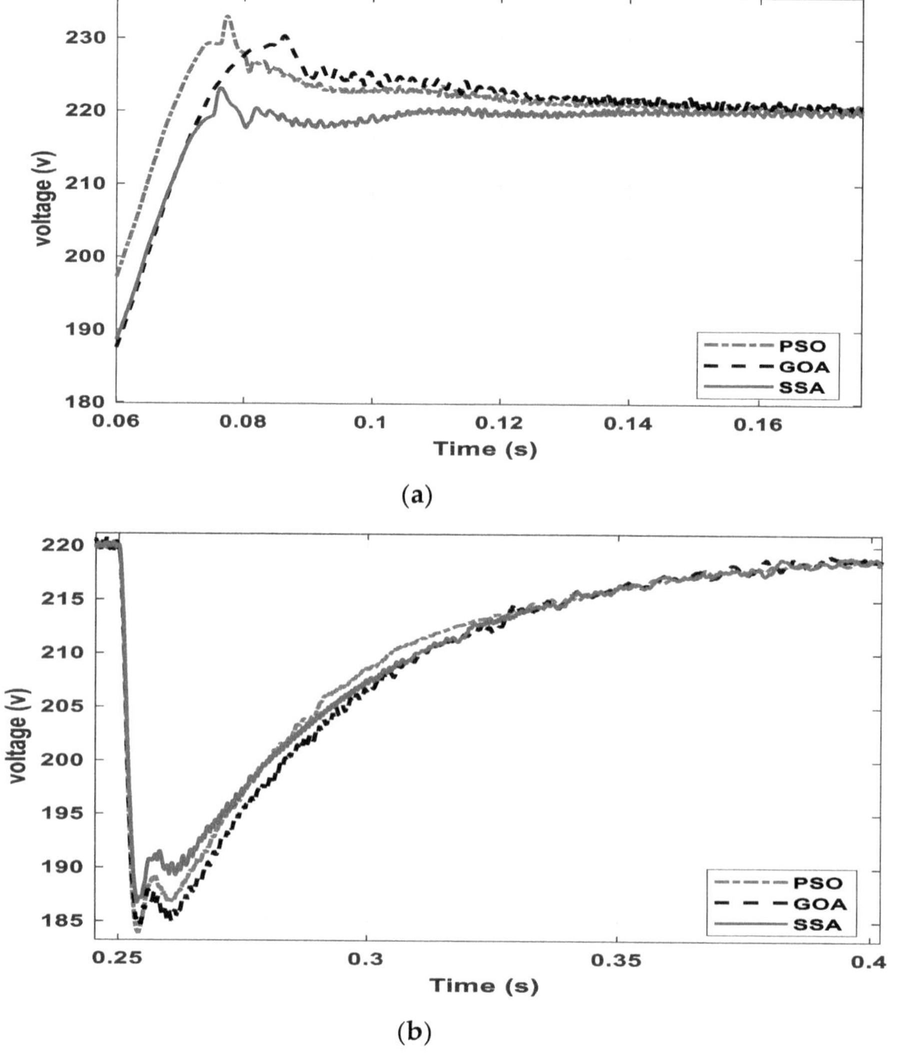

Figure 7. *Cont.*

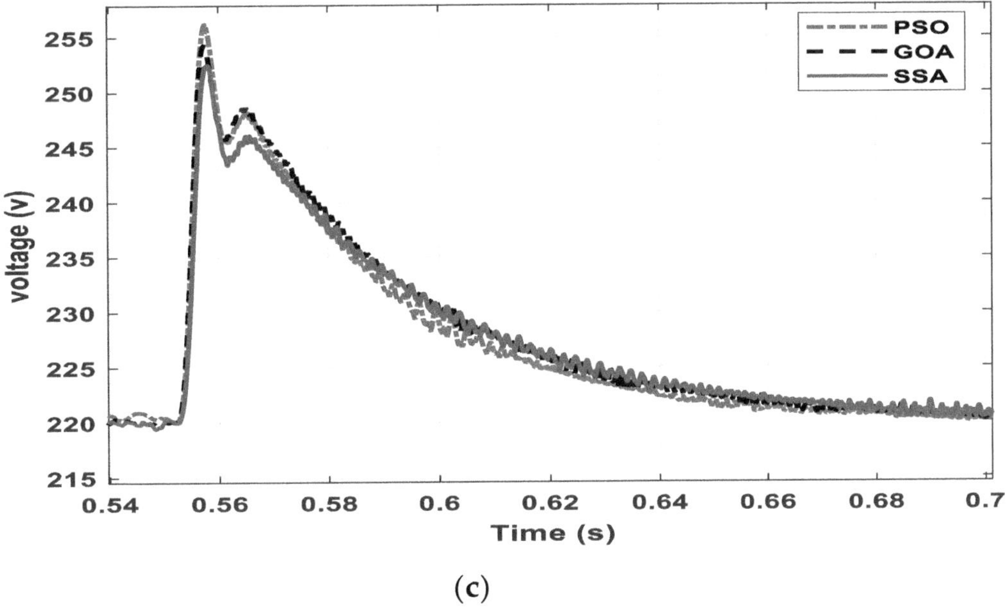

(c)

Figure 7. Voltage profile at (**a**) DG insertion (**b**) abrupt Load increment (**c**) abrupt load decrement.

It is obvious from Figure 7a–c that the optimal parameters obtained by the SSA optimization method provide better results compared to PSO and GOA in terms of overshoot and settling time for all three studied conditions. Another parameter that needs to be regulated during the islanded mode of MG operation is the frequency of the system. The frequency response of the system for PSO, GOA and SSA based MG system is shown in Figure 8.

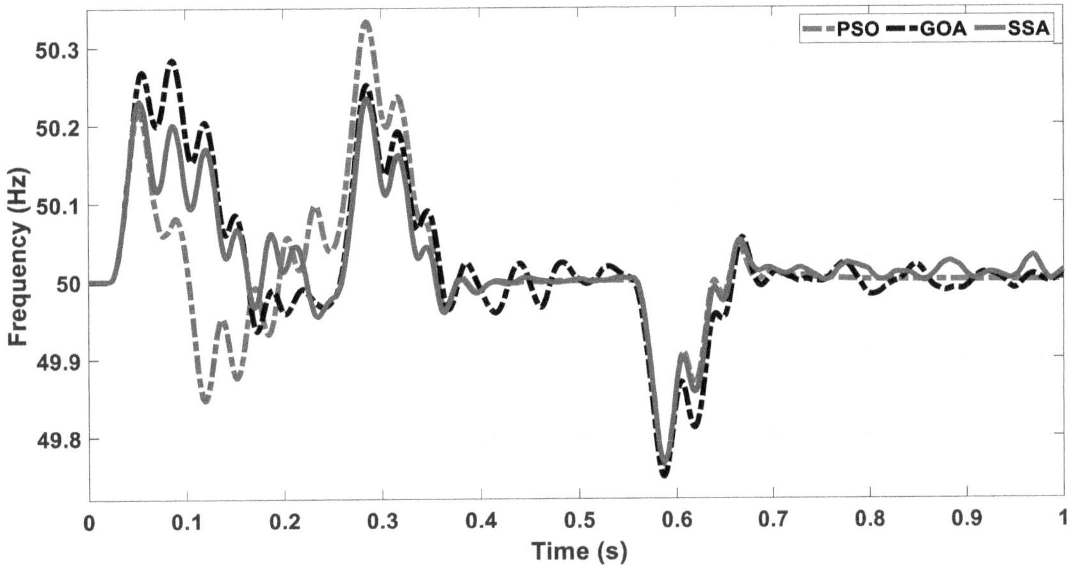

Figure 8. The frequency response of the MG system for the studied optimization methods.

Figure 8 shows the frequency response of the MG system for all three optimization methods. The result shows that all the metaheuristic methods provide a stable system frequency response within the allowable deviation range (±1%). However, the SSA based MG provides a better dynamic response as compared to its compotators. Table 3 shows a comparative analysis of all the studied methods for the voltage and frequency regulation of the studied islanded MG.

Table 3. Dynamic response evaluation of the proposed controller for voltage and frequency regulation.

	Studied Condition	Method	Maximum Overshoot/Undershoot (%)	Peak Time (ms)	Settling Time (ms)
Voltage	MG insertion	PSO	5.86	27.2	37.7
		GOA	4.68	36.3	64.5
		SSA	1.45	26.2	26.36
	Load injection	PSO	16.45	4.00	94.21
		GOA	16.00	4.70	94.20
		SSA	15.04	3.90	94.19
	Load detachment	PSO	16.41	7.70	73.50
		GOA	15.59	7.50	78.50
		SSA	14.77	7.80	77.40
Frequency	MG injection	PSO	0.44	2.05	-
		GOA	0.54	5.58	-
		SSA	0.46	2.30	-
	Load injection	PSO	0.66	35.2	-
		GOA	0.50	34.8	-
		SSA	0.46	35.0	-
	Load detachment	PSO	0.50	36.4	-
		GOA	0.48	36.7	-
		SSA	0.46	36.8	-

It may be noted from Table 3 that the SSA based controller provided better results for the most important dynamic response indicators as compared to PSO and GOA based controllers for the same operating conditions. Furthermore, it provided the most stable operation of the studied MG system and maintained the voltage within ±5% and frequency within ±1% of their nominal values, and hence, satisfied the IEEE standards. It is important to note that the settling time for the frequency is not provided. This is due to the reason that the frequency curve did not cross the ±2% of the rated value, and hence, settling time calculation is not applicable in this case.

5.2. Performance Evaluation of Studied Optimization Algorithms

In this section, the results obtained from the performance evaluation of the studied optimization algorithms are presented. Three different optimization algorithms, namely PSO, GOA and SSA, were tested to minimize the stated fitness function under identical operating conditions and system parameters. Furthermore, in order to carry out a fair comparison among the mentioned algorithms, all algorithms were tested for an identical number of iterations, i.e., 50 iterations, and number of search agents, i.e., 50 number of search agents. Since all the metaheuristic algorithms were initiated by spreading the search agents randomly in the bounded search area, the current study was tested for 20 simulation runs for each algorithm and the best (minimum) fitness function values were adopted for comparison. The convergence curve for the tested algorithms is shown in Figure 9.

It can be seen from Figure 9 that the SSA achieves the least value of fitness function (0.5840618) in the 17th iteration, whereas for the PSO and GOA the least magnitude obtained was recorded as 0.9211586 and 0.8748774 in the 21st and 25th iteration, respectively. Hence, the SSA converges faster and provides a higher quality of solution as compared to its competitors.

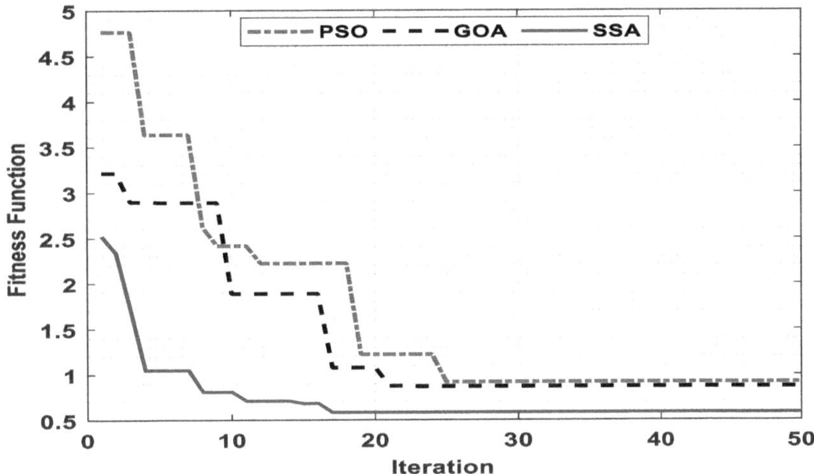

Figure 9. Convergence profile for PSO, GOA and SSA.

5.3. Power Quality Analysis

Due to the presence of non-linear power electronic devices like VSI and absence of the utility grid, maintaining the power quality and hence sinusoidality of the supply voltage and current is a very challenging task. A proper control architecture with optimal parameters can ensure pure sinusoidal voltage and current waveforms along with the high-power quality in such cases. The three-phase output current waveform for the studied system during all studied conditions is shown in Figure 10a, while its zoomed version from 0.1 to 0.22 s of simulation run is given in Figure 10b.

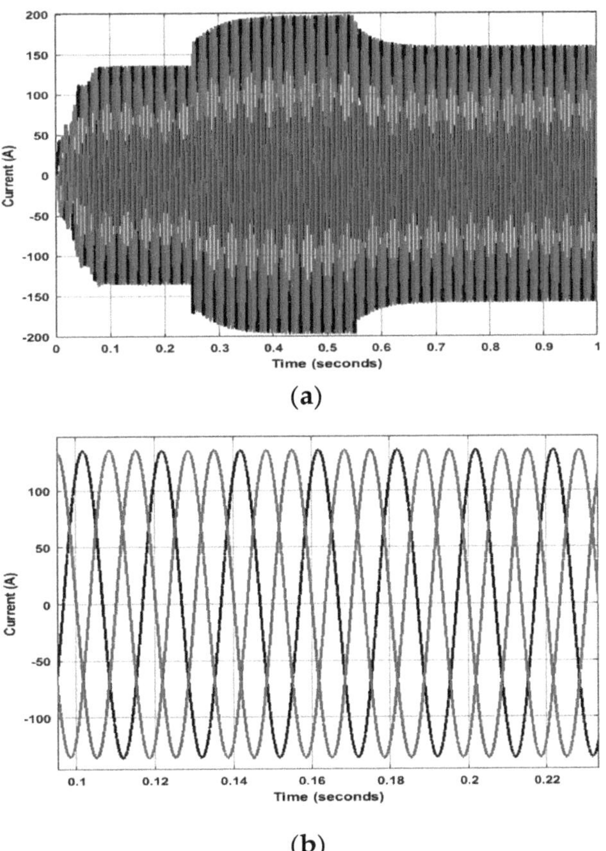

Figure 10. Three-phase sinusoidal current waveform (**a**) complete operation (**b**) zoomed version from 0.1–0.22 s.

It can be seen from Figure 10 that the solar PV dc output current is inverted into almost pure sinusoidal waveform by the proposed controller with the least possible distortion. Furthermore, to analyze the harmonic contents present in the obtained current waveform, the Fast Fourier Transform (FFT) analysis has been carried out for all three operating conditions of the system and the results are depicted in Figure 11.

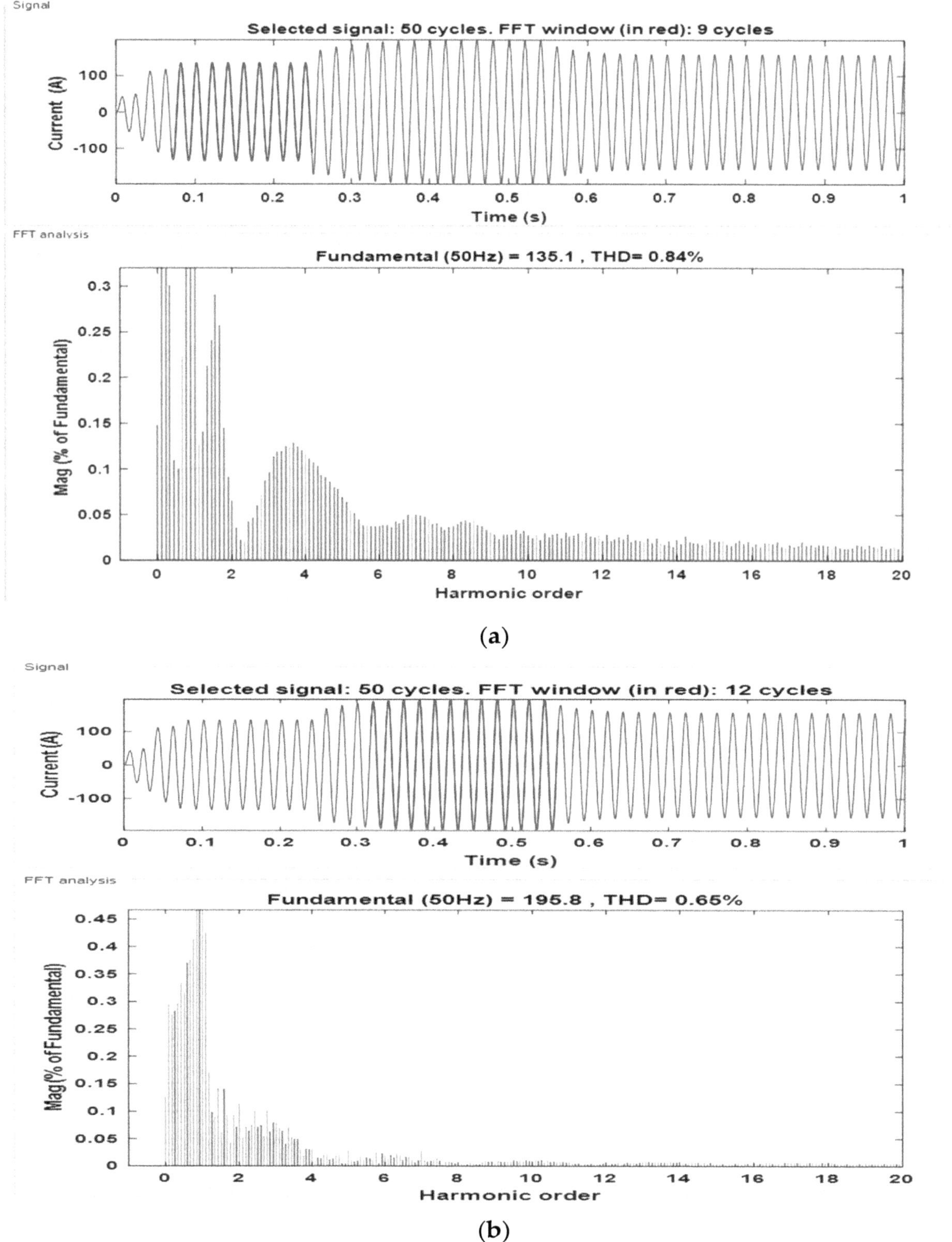

Figure 11. *Cont.*

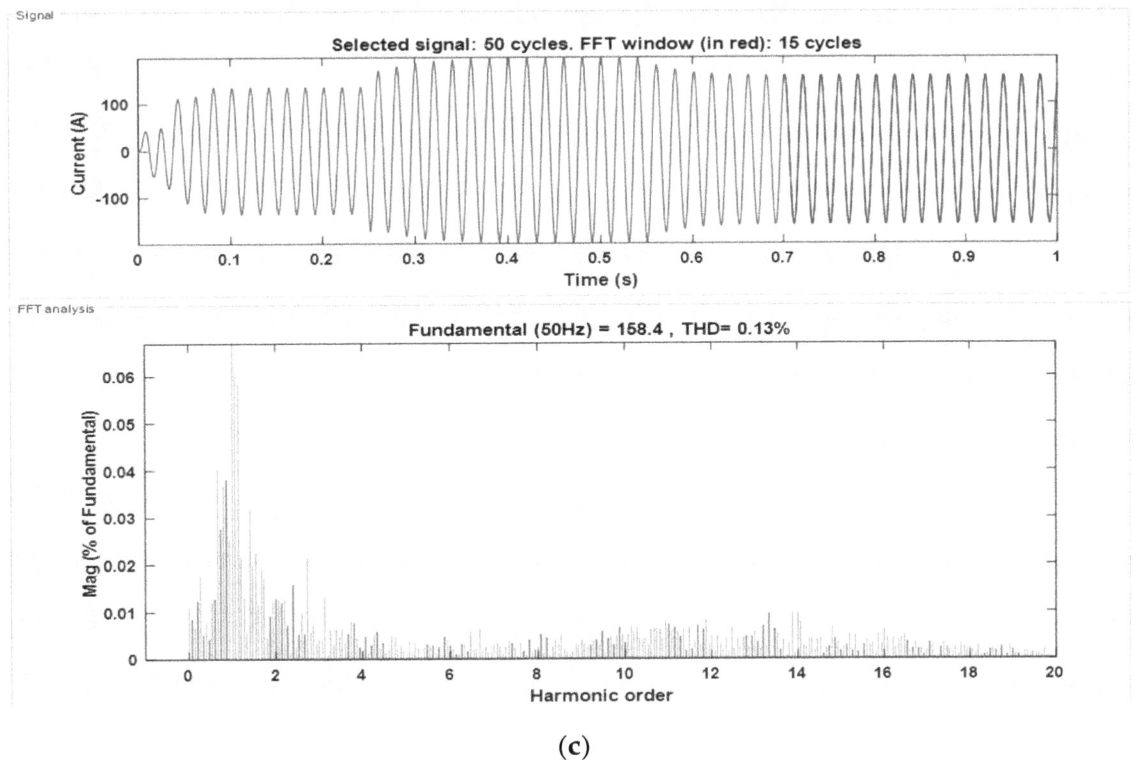

(c)

Figure 11. FFT analysis of studied MG system during (**a**) DG injection, (**b**) load injection and (**c**) load detachment.

It is clear from the FFT analysis of the studied power system that the SSA based controller duly satisfies the power quality standards set by IEEE 1547–2003 [17]. The results from Figure 11 are explained in tabular form in Table 4.

Table 4. FFT analysis understudied operating conditions.

Operating Condition	Percentage Harmonics (%)
MG injection	0.84
Load injection	0.65
Load detachment	0.13

It is obvious from Figure 11 and Table 4 that the proposed controller provides a high quality of power supply under all studied operating conditions and provides pure sinusoidal waveform, which shows the effectiveness of the proposed controller in maintaining optimal transient response with high power quality.

6. Conclusions

In this paper, an optimal controller for islanded MG has been successfully developed using the intelligence of SSA. The developed controller has successfully maintained the voltage and frequency at their nominal values with minimum possible overshoot and settling time during MG injection and load change conditions. The convergence behavior of the studied optimization algorithms proved that the SSA offers a higher quality of solution and faster optimization capabilities as compared to its counterpart optimization algorithms. Furthermore, the results from the power quality analysis show that the developed controller achieves high power quality and maintained almost perfect sinusoidal waveform of the voltage and current. The outcomes of the proposed controller were compared with that of the PSO and GOA based controller for the same operating conditions and system configuration.

The results show that the SSA based parameter selection provides an optimal dynamic response during all studied conditions and has superior performance as compared to its competitors.

Author Contributions: Each author has contributed significantly. The first and the corresponding author has initiated the idea and carried out the simulations along with preparation of the initial draft. The second author (M.W.M.) supervised the whole project and made significant corrections in the initial version of the article. M.M.R. has reviewed the article and helped in validation process. W.A. and S.A. helped in software and writing the revised version of the article respectively.

List of Symbols

Symbol	Name
C_f	Low pass filter capacitance
$K_1^{1:n}$	initial positions of the salps
K_j^1	Position of leader salp
L_f	Low pass filter inductance
M_i	Location of the food source in the jth dimension
R_f	Low pass filter resistance
V_g	Grid voltage
e_f	Frequency error
e_v	Voltage error
f^*	Reactive frequency
f_n	Nominal frequency
i_{abc}	Three-phase current
i_d^*	Direct reference current
i_q^*	Quadrature reference current
k_v	Droop constant for voltage
k_w	Droop constant for frequency
lb_j	Lower bound of search boundary
ub_j	Upper bound of search boundary
v^*	Reference voltage
v_{abc}	Three-phase voltage
v_d^*	Direct reference voltage
v_n	Nominal or rated voltage
v_q^*	Quadrature reference voltage
v_α, v_β	Reference voltage in $\alpha\beta$ frame
a	Acceleration of the leading salp
i	Salp number
C	dc-link capacitance
c_1, c_2, c_3	Random numbers
K_{pf}, K_{if}	Gains for the lower arm PI controller
K_{pv}, K_{iv}	Gains for the upper arm PI controller
l	Number of iterations
L	Number of maximum iterations
M	Food source position
θ	Reference angel
p	Active power
q	Reactive power
$rand$	Random number
t	Total simulation time
ω	Angular frequency
ω_c	Filter cut-off frequency
v_0	Initial velocity of leading salp

References

1. Hatziargyriou, N. *Microgrids: Architectures and Control*; John Wiley & Sons: Chichester, UK, 2014.
2. Jumani, T.A.; Mustafa, M.W.; Rasid, M.M.; Mirjat, N.H.; Baloch, M.H.; Salisu, S. Optimal Power Flow Controller for Grid-Connected Microgrids using Grasshopper Optimization Algorithm. *Electronics* **2019**, *8*, 111. [CrossRef]
3. Mirjalili, S.; Gandomi, A.H.; Mirjalili, S.Z.; Saremi, S.; Faris, H.; Mirjalili, S.M. Salp Swarm Algorithm: A bio-inspired optimizer for engineering design problems. *Adv. Eng. Softw.* **2017**, *114*, 163–191. [CrossRef]
4. Abbassi, R.; Abbassi, A.; Heidari, A.A.; Mirjalili, S. An efficient salp swarm-inspired algorithm for parameters identification of photovoltaic cell models. *Energy Convers. Manag.* **2019**, *179*, 362–372. [CrossRef]
5. Tolba, M.; Rezk, H.; Diab, A.; Al-Dhaifallah, M. A novel robust methodology based Salp swarm algorithm for allocation and capacity of renewable distributed generators on distribution grids. *Energies* **2018**, *11*, 2556. [CrossRef]
6. El-Fergany, A.A. Extracting optimal parameters of PEM fuel cells using Salp Swarm Optimizer. *Renew. Energy* **2018**, *119*, 641–648. [CrossRef]
7. Abusnaina, A.A.; Ahmad, S.; Jarrar, R.; Mafarja, M. Training neural networks using salp swarm algorithm for pattern classification. In Proceedings of the 2nd International Conference on Future Networks and Distributed Systems, Amman, Jordan, 26–27 June 2018; p. 17.
8. Kumari, S.; Shankar, G. A Novel Application of Salp Swarm Algorithm in Load Frequency Control of Multi-Area Power System. In Proceedings of the 2018 IEEE International Conference on Power Electronics, Drives and Energy Systems (PEDES), Chennai, India, 18–21 December 2018; pp. 1–5.
9. Al-Saedi, W.; Lachowicz, S.W.; Habibi, D.; Bass, O. Voltage and frequency regulation based DG unit in an autonomous microgrid operation using Particle Swarm Optimization. *Int. J. Electr. Power Energy Syst.* **2013**, *53*, 742–751. [CrossRef]
10. Vinayagam, A.; Alqumsan, A.A.; Swarna, K.; Khoo, S.Y.; Stojcevski, A. Intelligent control strategy in the islanded network of a solar PV microgrid. *Electr. Power Syst. Res.* **2018**, *155*, 93–103. [CrossRef]
11. Moarref, A.E.; Sedighizadeh, M.; Esmaili, M. Multi-objective voltage and frequency regulation in autonomous microgrids using Pareto-based Big Bang-Big Crunch algorithm. *Control Eng. Pract.* **2016**, *55*, 56–68. [CrossRef]
12. Sedighizadeh, M.; Esmaili, M.; Eisapour-Moarref, A. Voltage and frequency regulation in autonomous microgrids using Hybrid Big Bang-Big Crunch algorithm. *Appl. Soft Comput.* **2017**, *52*, 176–189. [CrossRef]
13. Qazi, S.H.; Mustafa, M.W.; Sultana, U.; Mirjat, N.H.; Soomro, S.A.; Rasheed, N. Regulation of Voltage and Frequency in Solid Oxide Fuel Cell-Based Autonomous Microgrids Using the Whales Optimisation Algorithm. *Energies* **2018**, *11*, 1318. [CrossRef]
14. Jumani, T.A.; Mustafa, M.W.; Rasid, M.M.; Mirjat, N.H.; Leghari, Z.H.; Saeed, M.S. Optimal Voltage and Frequency Control of an Islanded Microgrid using Grasshopper Optimization Algorithm. *Energies* **2018**, *11*, 3191. [CrossRef]
15. Killingsworth, N.; Krstic, M. Auto-tuning of PID controllers via extremum seeking. In Proceedings of the American Control Conference, Portland, OR, USA, 8–10 June 2005; pp. 2251–2256.
16. Seborg, D.E.; Edger, T.F.; Mellichamp, D.A. *Process Dynamics and Control*, 2nd ed.; John Wiley & Sons: Chichester, UK, 2004.
17. Association, I.S. *IEEE 1547 Standard for Interconnecting Distributed Resources with Electric Power Systems*; IEEE Standards Association: Piscataway, NJ, USA, 2003.

11

Economic Dispatch of Multi-Energy System Considering Load Replaceability

Tao Zheng [1,2,3,*], Zemei Dai [1,2,3], Jiahao Yao [4], Yufeng Yang [1,2,3] and Jing Cao [1,2,3]

1. NARI Group Corporation (State Grid Electric Power Research Institute), Nanjing 211106, China
2. NARI Technology Development Co. Ltd., Nanjing 211106, China
3. State Key Laboratory of Smart Grid Protection and Control, Nanjing 211106, China
4. Key Laboratory of Measurement and Control of Complex Systems of Engineering, Ministry of Education, School of Automation, Southeast University, Nanjing 210096, China

* Correspondence: zhengtao2@sgepri.sgcc.com.cn

Abstract: By integrating gas, electricity, and cooling and heat networks, multi-energy system (MES) breaks the bondage of isolated planning and operation of independent energy systems. Appropriate scheduling of MES is critical to the operational economy, and it is essential to design scheduling strategies to achieve maximum economic benefits. In addition to the emergence of energy conversion systems, the other main novelty of MES is the multivariate of load, which offers a great optimization potential by changing load replaceability (flexibly adjusting the composition of loads). In this paper, by designing load replaceability index (LRI) of composite load in MES, its interaction mechanism with scheduling optimum is systematically analyzed. Through case studies, it is proven that the optimum can be improved by elevating load replaceability.

Keywords: multi-energy system; economic dispatch; load replaceability; multi-energy conversion

1. Introduction

Energy flexibility is crucial to the construction of more environmental-friendly and efficient power generation (consumption) patterns. Unlike conventional single-form energy-based energy system, which is usually detached from other types of energy systems from both the planning and operational perspectives, multi-energy system (MES) comprises various forms of subsystems including electricity, heat, cooling and gas networks, contributing to the interactions and corresponding energy flexibility through energy converters. Therefore, MES technologies gradually become irreplaceably key components in building the modern energy systems.

Compared with single-from energy-based system, the most significant difference of MES is the energy conversion function achieved by energy conversion equipments, which serve as in-between interfaces and assist in the formation of coupling relations. Energy converters are foundations of transformation, fusion and decomposition of energy flows among different subsystems. Because of the possibility of convertibility of different energy forms, MES outperforms systems of pure energy form by complementing the advantages of each subsystem.

An important concept in MES is the energy hub (EH), which gives an matrix model of production, conversion, consumption and storage of different energy carriers. Each of the matrix elements represent the abstract connection and conversion coefficients of internal components, which can be embodied by various types of multi-generators, among which combined heat and power (CHP) or combined heat cooling and power (CHCP) plants are the most widely investigated cogeneration or trigeneration plants.

Various research studies have recently begun to show intensive concern for the design and operation of MES, either from the perspective of microscopic (CHP and CHCP plants) or macroscopic (EHs). The main idea of optimal operation of CHP (CHCP) plants is similar to that of power systems, both starts from the optimization model building and ends with model solving by specific optimization algorithms like mathematical or evolutionary programming methods. Like economic electrical power dispatch, the first priority of multi-generation is economy. Some researchers leverage the strong computing capabilities of meta-heuristics to solve economic dispatch of CHP plants [1,2]. Apart from deterministic optimization, uncertainties stemming from volatile renewable energy are gradually introduced by stochastic programming. For example, a novel chance constrained programming model is used to formulate economic dispatch with CHP plants and wind power simultaneously [3]. Li groundbreakingly investigates how to use the temperature dynamics of heating networks to enhance the utilization of wind power, and a CHP dispatch model considering operational economy and wind power integration is built and solved by iterative methods [4]. As for economic dispatch of CHCP, a nonlinear economic optimal operation model is developed and solved by lingo solver [5]. Backward dynamic programming technique is used to obtain the optimal operational set-points of combined heat, cooling and power systems [6].

Besides designing operation schedules solely in pursuit of the economic potential elevation, optimization of (CHP) CCHP plants considering multiple operational output performance begins to capture researchers' interests. Under most circumstances, economy, energy utilization efficiency and environmental preservation are the three main optimization goals. By employing Weighting method and fuzzy optimum selection theory, the integrated performances of operation of CCHP plants considering these objectives are evaluated [7]. Dynamic object method-particle swarm optimization method is used to achieve the goal of carbon emission reduction, energy efficiency elevation and system cost minimization [8]. Additionally, some new evaluation indices evaluating the average useful output and total heat transfer area are studied, correspondingly, a multi-objective optimization-based schedule is designed [9]. Modified Bacterial Foraging Optimization method is adopted to minimize the operational cost and pollutant emission [10].

Optimal system-level operation of MES is usually implemented under EH framework by designing the synergies among generation, consumption and storage components. An optimal operation strategy of MES containing CHP plants, boiler, battery and water tank is studied [11], and particle swarm optimization (PSO) algorithm is used to obtain optimal scheduling of each device. An optimal expansion planning of EH containing CHP plants and natural gas furnaces is designed to minimize investment cost [12]. Under some circumstances MES have some uncertainties involving demand fluctuation, cost dynamics and converter efficiency change. In this situation deterministic framework no longer adapts to uncertain environments, hence robust optimization approach is used to increase the robustness towards uncertain bounded parameters [13]. In [14], uncertainties of demand, market price and renewable energy are considered by stochastic programming, and optimal operation set-points are obtained to minimize operation costs. Instead of treating EH as an individual system, studies of multi-EH systems take into consideration the mutual influence and restriction. By analyzing competitive and cooperative relation between multi-EHs, game theoretic optimal scheduling is established through quantum particle swarm optimization method [15]. Moreover, resilience of MES is researched in last several years, and coordination between different subsystems in post-disaster repair is investigated to reduce load shedding and repair duration [16].

Most of the existing MES research focuses on optimal planning or operation by considering multi-energy generation or conversion on the generation side. Recently, some research studies begin to consider the characteristics of multi-energy loads, which play essential roles in MES planning & operation as well as other applications like load prediction and demand response programs. Some researchers investigate the influence of multi-energy loads upon MES planning, and the capacity of gas turbine under different thermoelectric ratios is studied [17]. In respect to load prediction, load prediction for integrated cooling and heat system is implemented by considering the intrinsic coupling among multi-energy loads using multivariate phase space reconstruction technique [18]. By considering the operating attributes of loads, a quick cooling and heating load prediction model is constructed [19]. Deep-learning technique is adopted to predict various types of load in regional energy system integration [20]. Some researchers also study how to adjust the consumption pattern to better participate in the demand response program [21,22].

Conventionally, the load composition from each energy network is assumed to be fixed without replaceability, which is characterized by the feasibility of substitution of one form of load with another one. Load considering replaceability can be regarded as a composite load, it characterizes the multi-energy consumption on the load side and is essential to flexible and efficient energy consumption. Composite load can adjust the proportion of heterogeneous energy consumption based on the external excitation signal such as market price, thus achieving simultaneous operation cost reduction and the load and generation balance.

In this paper, an optimal scheduling approach for MES consisting of gas, electricity, cooling and heat networks considering load replaceability is designed to maximize the economic potential of operation. For the convenience of analysis of the influence of load replaceability on the optimal scheduling, load replaceability index (LRI), which measures the overall load replaceability of MES, is analyzed in respect to the optimal scheduling results. The intrinsic relations between load replacement capability and optimum under various operating conditions are investigated to improve the scheduling performance by observing specific rules of load replaceability adjustment.

The remaining of the paper is organized as follows: Section 2 addresses load replaceability and the definition of LRI in MES; with the aid of load replaceability, Section 3 presents the economic scheduling model for an integrated electricity, gas, heat and cooling networks; Section 4 studies the relation between optimal scheduling results and load replacement capability; concluding remarks are given in Section 5.

2. Load Replaceability in Multi-Energy System

In this section, load replaceability and its relation with optimal scheduling of MES are systematically analyzed. The definition and calculation method of LRI are presented.

2.1. Load Replaceability and Composite Load

Load replaceability is used to characterize the selection margin of multi-energy consumption. If one specific load can choose two or more than two types of energy flow to meet its own load demand, and then this load is called replaceable load. From the perspective of consumers, the greater the number of energy flow types is and the greater the capacity of multi-generators is, the more the selection margin of multi-energy consumption is.

Generally speaking, loads in MES can be categorized as cooling, heat, power and gas load, corresponding to their respective type of energy network. Usually, due to the natural barrier between heterogenous energy systems in the isolated operating mode, the load can only be supplied by homogenous energy flow. Nevertheless, with the coupling and integration of heterogenous systems assisted by energy conversion systems, it should be emphasized that the load can be satisfied by heterogenous energy flow as well.

In this section, heat load is taken as an example to illustrate the replaceability phenomenon. In independent heat network, heat load can only be supported by the heat flow through steam pipelines. While heat load in MES can be supplied by various forms of energy. For example, on the one hand heat energy can still be transmitted through pipelines just like in independent heat network; on the other hand electric furnaces or gas heaters can transform electric or gas into heat energy. In this case, in addition to the thermal attribute, heat load has electric attribute (electric heater) and gaseous attribute (gas furnace). Electricity and gas are just intermediate transition states. Seen from the exterior, the final state is still heat load since heat flow is the only energy form in the consumption phase.

Though the final load state belongs to heat load in either segregated independent system or integrated dependent system, the introduction of intermediate transition states makes the load property change from monotype load in independent system into composite load in dependent system. The main specialty of composite load is flexibility, especially the market response flexibility. The operator can adjust the proportion of each component in the composite load, based on the market price of each form of energy flow at different operating interval, thus reducing operation cost. The emergence of composite load enhances the optimization space in the consumption phase and further increases economic benefits. In order to vividly illustrate how load replaceability (composite load) influences the operation economy, an exemplary integrated gas power and cooling system in Figure 1 is briefly studied.

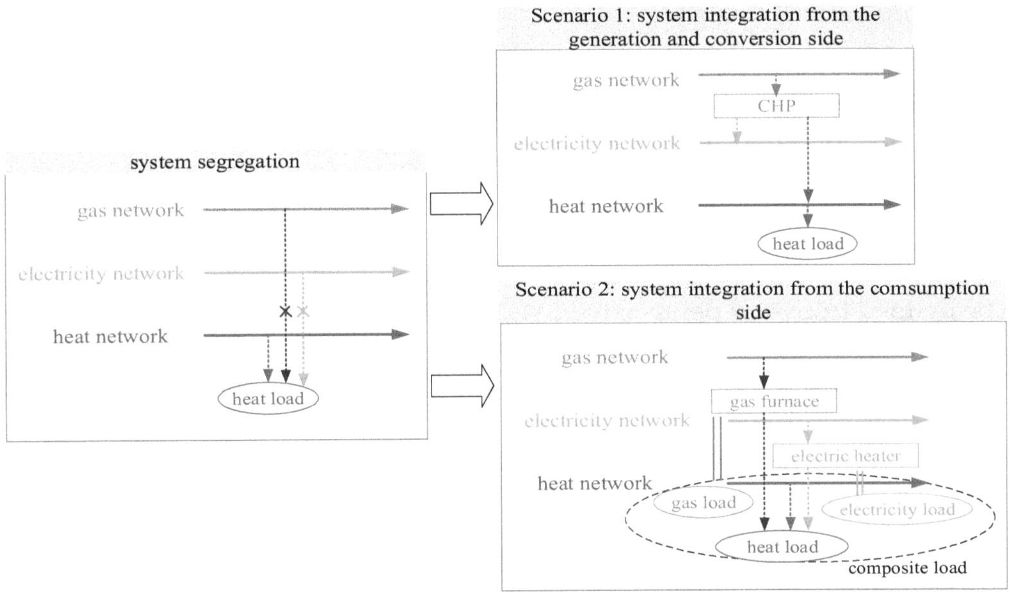

Figure 1. Schematic diagram of two-area classic system.

Suppose the current market prices of gas electricity and heat are 6 \$/kWh, 1 \$/kWh and 15 \$/kWh, respectively. And the amount of heat load demand (100 kWh) is much more than that of electricity or gas load demand, which is set by zero for simplicity. The maximal transmission capacity of each network is 50 kW. The following two scenarios are considered for operation cost calculation.

- **Scenario 1:** There exists no replaceable heat load in the system; the multi-generation plant is the CHP plant. The overall energy conversion ratio is 0.8; the energy proportion of electricity generated by the CHP plant is 0.6; the energy proportion of heat generated by the CHP plant is 0.4. And the maximal capacity of the CHP plant is supposedly to be much bigger than the maximal transmission capacity of the network.
- **Scenario 2:** There exists replaceable heat load in the system, and the load is composite load which contains electric heater and gas furnace. The maximal capacity of both devices is supposedly to be much bigger than the maximal transmission capacity of power and gas network. The energy efficiency ratio for both devices is 0.5.

In Scenario 1, the operation cost is

$$6 \times 50 \times 0.8 \times 0.4 + 15 \times (100 - 50 \times 0.8 \times 0.4) = 1356. \tag{1}$$

In Scenario 2, the operation cost is

$$1 \times 50 \times 0.5 + 6 \times 50 \times 0.5 + 15 \times (100 - 50) = 925. \tag{2}$$

As can be seen, due to the load replaceability, consumers can satisfy the demand by consuming much cheaper transformed energy. In this case it is the transformed heat flow from electric heaters and gas furnaces, thus reducing the overall operation cost. Nevertheless, the multi-generation plant (the CHP plant) can only supplement heat flow to a very limited degree. Heat flow is just a byproduct during the intermediate generation section and the multi-generation plant is not ultimately consumption-oriented, which cannot sufficiently use the price difference between different energy forms to reduce operation cost. As is shown by the results in (1) and (2), operation cost (925$) considering load replaceability is obviously smaller than that is only with multi-generation plants.

2.2. Load Replaceability Index

In Section 2.1, basic information of load replaceability and composite load is addressed. From the deduction of a simple 2-scenario integrated energy system, it can be learned that load replaceability enhances the optimization space from the consumption side. In this section, instead of focusing on substitutes of one specific type of load, we give some quantitative indices to evaluate the broad system-level load replaceability. The LRI can reflect to what extent one type of load (the final load state) can be substituted by the other (the intermediate load state).

LRI can be defined from two perspectives: (1) the potential load replaceability and (2) the actual load replaceability. The former evaluates the potential of load substitution of the system before it is ever put into operation. The latter evaluates the actual load substitution ability during the operation.

2.2.1. Load Replaceability Index Reflecting Potential Load Substitution Ability

As for an integrated gas, power, heat and cooling system, there exist four possible types of load in the system. Denote the maximal convertible load demand of the ith composite load by L_i^{max}, and denote the load component for the ith load from each network by X_i^{load}, P_i^{load} and G_i^{load}. X_i^{load} can either be C_i^{load} or H_i^{load} based on the final energy form. LRI of the ith load can be calculated by

$$\alpha_i = \frac{\left(x_i X_i^{load} + u_{i1} p_i P_i^{load} + u_{i2} g_i G_i^{load} - L_i^{\max}\right)}{L_i^{\max}}, \tag{3}$$

where x_i, p_i, and g_i represents the energy efficiency ratio of respective intermediate load. Binary variables $u_{i1}, u_{i2} \in \{0,1\}$ represent ith load is $(u_{i1}, u_{i2} = 1)$ or is not $(u_{i1}, u_{i2} = 0)$ supplied by specific intermediate loads (P_i^{load} or G_i^{load}). Suppose there exist m loads in the integrated energy system, the overall LRI can be expressed as

$$\alpha = \frac{\sum\limits_{i=1}^{m} \alpha_i L_i^{\max}}{\sum\limits_{i=1}^{m} L_i^{\max}}. \tag{4}$$

LRI in (4) actually uses the proportion of remaining alternative load reserve in the maximal load demand to quantify the substitution potential.

2.2.2. Load Replaceability Index Reflecting Actual Load Substitution Ability

As can be seen from (4), all the parameters are actually the upper bounds of corresponding replaceable loads. That is to say, (4) reflects the remaining load conversion reserve at critical states (the maximum values). Nevertheless, replaceable loads in the actual scheduling would not certainly be equal to the maximum; values of replaceable loads should depend on the scheduling, which will be presented in detail in the next section. Therefore, a modified online scheduling-based LRI of the ith load is expressed by

$$\beta_i = \frac{u_{i1} p_i P_i^{load} + u_{i2} g_i G_i^{load}}{L_i}, \tag{5}$$

where P_i^{load} and G_i^{load} represent the actual intermediate load amount in the scheduling; L_i represents the actual composite load demand. Then the system-level LRI can be expressed by

$$\beta = \frac{\sum_{i=1}^{m} \beta_i L_i}{\sum_{i=1}^{m} L_i}. \tag{6}$$

Notice that the replaceable load term $x_i X_i^{load}$ disappears in (5); it is because that $x_i X_i^{load} + u_{i1} p_i P_i^{load} + u_{i2} g_i G_i^{load} = L_i$ in the scheduling. If (3) is still used, LRI is always zero. In this case only the proportion of energy converted by intermediate loads in the composite load is used to quantify the degree of flexibility.

3. Optimal Economic Scheduling of Multi-Energy System Considering Load Replaceability

It can be learned from Section 2 that load replaceability offers redundant energy flow path and the optimization space of multi-energy complementary coordination. In response to the market price of different energy flow, the consumption side can participate in economic dispatch by adaptively adjusting the proportion of the intermediate load of various energy form in the composite load. Unlike conventional demand response program in power systems, because of the load replaceability capability through conversion systems, this multi-energy demand response in MES does not cause the change in the ultimate total energy consumption. The consumers only choose the intermediate energy source from which they buy to obtain the ultimate energy rather than change the consumption pattern like they do in electric utility demand response programs. That is to say, operation economy along with customer satisfaction on energy supply is guaranteed.

In this section, by considering load replaceability the economic dispatch model is established for MES in Figure 2 consisting of gas, power, and cooling and heat networks.

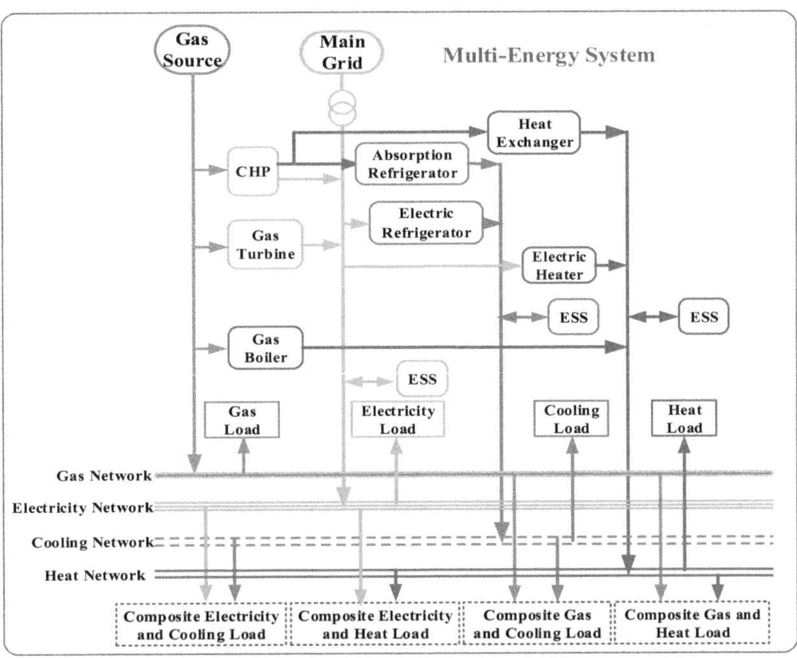

Figure 2. Schematic diagram of a multi-energy system (MES) consisting of gas, power, and cooling and heat networks. CHP = combined heat and power; ESS = energy storage system.

3.1. Objective Function of Optimal Economic Scheduling of MES

From the perspective of economy, operation cost mainly comes from three aspects including the cost of purchasing gas from external gas networks, the cost of purchasing electricity from external electric grids and depreciation expense of the multi-energy storage systems.

The cost of gas purchase O_G is written by the accumulative product of gas price and purchase amount (which is procured from the external gas source shown in Figure 2) during the whole dispatching intervals

$$O_G = \sum_{t=1}^{T} r_{gas}(t) \frac{G_{ext}(t)}{R_{gas}} t, \tag{7}$$

where t represents the dispatch interval; T represents the total number of dispatch intervals; $r_{gas}(t)$ represents gas price at time t; $G_{ext}(t)$ represents gas purchase amount at time t; $R_{gas} = 9.77\ (kw \cdot h/m^3)$ represents the calorific value.

The supplementary electricity is procured from the external main grid as shown in Figure 2. Similarly, we can compute the cost of electricity purchase O_P as

$$O_P = \sum_{t=1}^{T} r_{ele}(t) P_{ext}(t)\, t, \tag{8}$$

where t and T have the same meaning as in (7); $r_{ele}(t)$ represents electricity price at time t; and $P_{ext}(t)$ represents electricity purchase amount at time t.

The depreciation cost steams from ageing of storage system during the repeated energy charge/discharge processes. In this paper, it is assumed that gas is either directly supplied to gas load or converted to other energy flow, and there is no gas storage system. For simplicity, it is expressed as the linear combination of storage amount of different energy flow:

$$O_S = \sum_{t=1}^{T} \left[r_{PS} P_{PS}^{S}(t) + r_{HS} H_{HS}^{S}(t) + r_{CS} C_{CS}^{S}(t) \right], \tag{9}$$

where r_{PS}, r_{HS}, and r_{CS} represent the depreciation rate per unit quantity of power, and heat and cooling energy; $P_{PS}^{S}(t)$, $H_{HS}^{S}(t)$ and $C_{CS}^{S}(t)$ represent the electricity, and heat and cooling energy stored at time t.

3.2. Constraints of Optimal Economic Scheduling of MES

3.2.1. Energy Supply and Demand Balance

Similar to power balance equation in electric grids, the energy supply and demand balance should be guaranteed at the first place. Nevertheless, due to the coupling enabled by conversion systems, the energy generation side might contain transformed energy flow (the output of conversion systems), the energy consumption side might contain energy flow to be transformed (the input of conversion systems), which is exactly the intermediate load previously addressed in Section 2. Therefore, under the circumstances of MES, the load for each network can be categorized by convertible load and inconvertible load, which are equivalent to intermediate and ultimate load in Section 2.

The energy supply and demand balance equation for the cooling system can be written by

$$C_{AR}(t) + C_{ER}(t) + C_{CS}^{R}(t) - C_{CS}^{S}(t) = C_{load}^{D}(t) + C_{load}^{PC}(t) + C_{load}^{GC}(t), \tag{10}$$

where t represents the dispatch interval; $C_{AR}(t)$ represents cooling energy generated by the absorption refrigerator; $C_{ER}(t)$ represents cooling energy generated by the electric refrigerator; $C_{CS}^{R}(t)$ represents cooling energy released by cooling energy storage system; $C_{CS}^{S}(t)$ represents cooling energy stored by cooling energy storage system; $C_{load}^{D}(t)$ represents inconvertible cooling load; $C_{load}^{PC}(t)$ represents convertible electricity-based cooling load; and $C_{load}^{GC}(t)$ represents convertible gas-based cooling load.

Similarly, the energy supply and demand balance equation for the heat system can be written by

$$H_{HE}(t) + H_{GB}(t) + H_{EH}(t) + H_{HS}^{R}(t) - H_{HS}^{S}(t) = H_{load}^{D}(t) + H_{load}^{PH}(t) + H_{load}^{GH}(t), \qquad (11)$$

where $H_{HE}(t)$ represents heat energy generated by heat exchanger; $H_{GB}(t)$ represents heat energy generated by gas boiler; $H_{EH}(t)$ represents heat energy generated by electric heater; $H_{HS}^{R}(t)$ represents heat energy released by heat energy storage system; $H_{HS}^{S}(t)$ represents heat energy stored by heat energy storage system; $H_{load}^{D}(t)$ represents inconvertible heat load; $H_{load}^{PH}(t)$ represents convertible power-based heat load; and $H_{load}^{GH}(t)$ represents convertible gas-based heat load.

The energy supply and demand balance equation for the gas system can be written by

$$G_{ext}(t) - G_{CHP}(t) - G_{GT}(t) - G_{GB}(t) = G_{load}^{D}(t) + G_{load}^{GC}(t) + G_{load}^{GH}(t), \qquad (12)$$

where $G_{CHP}(t)$ represents gas energy consumed by the CHP plant; $G_{GT}(t)$ represents gas energy consumed by the gas turbine; $G_{GB}(t)$ represents gas energy consumed by the gas boiler; $G_{load}^{D}(t)$ represents inconvertible gas load; $G_{load}^{GC}(t)$ represents convertible cooling-oriented gas load; and $G_{load}^{GH}(t)$ represents convertible heat-oriented gas load.

The energy supply and demand balance equation for the power system can be written by

$$P_{ext}(t) + P_{CHP}(t) + P_{GT}(t) + P_{PS}^{R}(t) - P_{PS}^{S}(t) - P_{EH}(t) - P_{ER}(t) = P_{load}^{D}(t) + P_{load}^{PC}(t) + P_{load}^{PH}(t), \quad (13)$$

where $P_{CHP}(t)$ represents electricity energy generated by the CHP plant; $P_{GT}(t)$ represents electricity energy generated by the gas turbine; $P_{PS}^{R}(t)$ represents electricity energy released by the electricity energy storage system; $P_{PS}^{S}(t)$ represents electricity energy stored by the electricity energy storage system; $P_{EH}(t)$ represents electricity energy consumed by the electric heater; $P_{ER}(t)$ represents electricity energy consumed by the electric refrigerator; $P_{load}^{D}(t)$ represents inconvertible power load; $P_{load}^{PC}(t)$ represents convertible cooling-oriented power load; and $P_{load}^{PH}(t)$ represents convertible heat-oriented power load.

As can be seen from (10)–(13), there exist inconvertible load $C_{load}^{D}(t)$, $H_{load}^{D}(t)$, $G_{load}^{D}(t)$, and $P_{load}^{D}(t)$; meanwhile, the convertible load during the intermediate transition state are $C_{load}^{PC}(t)$, $C_{load}^{GC}(t)$, $H_{load}^{PH}(t)$, $H_{load}^{GH}(t)$, $G_{load}^{GC}(t)$, $G_{load}^{GH}(t)$, $P_{load}^{PC}(t)$, and $P_{load}^{PH}(t)$. Moreover, the first four $C_{load}^{PC}(t)$, $C_{load}^{GC}(t)$, $H_{load}^{PH}(t)$, and $H_{load}^{GH}(t)$ are actually the final state of the convertible load, while the last four $G_{load}^{GC}(t)$, $G_{load}^{GH}(t)$, $P_{load}^{PC}(t)$, and $P_{load}^{PH}(t)$ are actually the initial state of the convertible load.

The total amount of energy of composite load in the final energy form (cooling or heat) can thus be expressed by

$$\begin{cases} p_{load}^{PC}P_{load}^{PC}(t) + C_{load}^{PC}(t) = L_{load}^{PC}(t) \\ g_{load}^{GC}G_{load}^{GC}(t) + C_{load}^{GC}(t) = L_{load}^{GC}(t) \\ p_{load}^{PH}P_{load}^{PH}(t) + H_{load}^{PH}(t) = L_{load}^{PH}(t) \\ g_{load}^{GH}G_{load}^{GH}(t) + H_{load}^{GH}(t) = L_{load}^{GH}(t) \end{cases}, \qquad (14)$$

where p_{load}^{PC} represents the energy efficiency ratio of convertible cooling-oriented power load; $L_{load}^{PC}(t)$ represents the composite power and cooling load; g_{load}^{GC} represents the energy efficiency ratio of convertible cooling-oriented gas load; $L_{load}^{GC}(t)$ represents the composite power and gas load; p_{load}^{PH} represents the energy efficiency ratio of convertible heat-oriented power load; $L_{load}^{PH}(t)$ represents the composite power and heat load; g_{load}^{GH} represents the energy efficiency ratio of convertible heat-oriented gas load; and $L_{load}^{GH}(t)$ represents the composite gas and heat load.

3.2.2. Energy Balance of Conversion Systems

As can be seen from Figure 2, the conversion systems comprise the CHP plant, gas turbine, gas boiler, electric heater and electric furnace. In this section, the input and output energy balance of respective conversion system is presented.

The CHP plant is the use of natural gas to generate electricity and useful heat simultaneously. The relation between input gas power and output electricity or heat power of the back pressure CHP plant is

$$P_{CHP}(t) = p_{CHP} G_{CHP}(t) \tag{15}$$

$$Q_{CHP}(t) = q_{QP} P_{CHP}(t), \tag{16}$$

where p_{CHP} represents the power generation efficiency of the CHP plant; q_{QP} represents the ratio of electricity-to-heat; G_{CHP} represents the input gas power of the CHP plant; and Q_{CHP} represents the surplus heat, which is transformed to cooling and heat energy through the absorption refrigerator and heat exchanger, respectively.

$$C_{AR}(t) = \eta_{AR}(1 - \gamma_{HC}) Q_{CHP}(t), \tag{17}$$

$$H_{HE}(t) = \eta_{HE} \gamma_{HC} Q_{CHP}(t). \tag{18}$$

where η_{AR} and η_{HE} represent the energy efficiency ratio of the absorption refrigerator and heat exchanger; γ_{HC} represents the proportion ratio.

The gas turbine is used to transform gas energy into electricity energy

$$P_{GT}(t) = \eta_{GT} G_{GT}(t), \tag{19}$$

where η_{GT} represents the energy efficiency ratio of the gas turbine.

The gas boiler is used to transform gas energy into heat energy

$$H_{GB}(t) = \eta_{GB} G_{GB}(t), \tag{20}$$

where η_{GB} represents the energy efficiency ratio of the gas boiler.

The electric refrigerator is used to transform electricity energy into cooling energy

$$C_{ER}(t) = \eta_{ER} P_{ER}(t), \tag{21}$$

where η_{ER} represents the energy efficiency ratio of the the electric refrigerator.

The electric heater is used to transform electricity energy into heat energy

$$H_{EH}(t) = \eta_{EH} P_{EH}(t), \tag{22}$$

where η_{EH} represents the energy efficiency ratio of the electric heater.

3.2.3. Energy Storage Systems

Operating constraints of the energy storage system mainly comprise the storage state and some output constraints no matter it is the electricity, heat or cooling storage system. Take the electricity energy storage system as an example, the typical operating model can be written as

$$\begin{cases} S_{PS}(t) = (1-\eta_{PS}) S_{PS}(t-1) + \eta^{S}_{PS} P^{S}_{PS}(t) - P^{R}_{PS}(t)/\eta^{R}_{PS} \\ S^{\min}_{PS} \le S_{PS}(t) \le S^{\max}_{PS} \\ S_{PS}(t_1) = S_{PS}(t_{end}) \\ U^{S}_{PS}(t) P^{\min}_{PS} \le P^{S}_{PS}(t) \le U^{S}_{PS}(t) P^{\max}_{PS} \\ U^{R}_{PS}(t) P^{\min}_{PS} \le P^{R}_{PS}(t) \le U^{R}_{PS}(t) P^{\max}_{PS} \\ U^{R}_{PS}(t) + U^{S}_{PS}(t) \le 1 \\ U^{R}_{PS}(t), U^{S}_{PS}(t) \in (0,1) \end{cases}, \quad (23)$$

where $S_{PS}(t)$ represents the remaining power in the electricity energy storage system; η_{PS} represents the attrition rate during the charge/discharge process; η^{S}_{PS} represents the charge efficiency; η^{R}_{PS} represents the discharge efficiency; $U^{S}_{PS}(t)$ represents the on/off charge state; $U^{R}_{PS}(t)$ represents the on/off discharge state; $S^{\min}_{PS}$ and $S^{\max}_{PS}$ represents the capacity limit; $S_{PS}(t_1)$ represents the initial power in the beginning of one cycle of operation; and $S_{PS}(t_{end})$ represents the final power in the end of one cycle of operation. When considering the periodicity of the operation of ESS (energy storage system), the final power of the previous cycle should be equal to the initial power of the next cycle; hence, $S_{PS}(t_{end}) = S_{PS}(t_1)$ should be satisfied.

3.2.4. Other Constraints

In addition to constraints characterizing energy supply and demand balance, input and output energy relation of conversion systems, other mentionable operating constraints are those that describe the output limit of specific components in MES.

Firstly, the convertible load in each energy network has its lower/upper limit:

$$\begin{cases} \underline{P}^{\mathrm{PC}}_{\mathrm{load}} \le P^{\mathrm{PC}}_{\mathrm{load}}(t) \le \overline{P}^{\mathrm{PC}}_{\mathrm{load}} \\ \underline{C}^{\mathrm{PC}}_{\mathrm{load}} \le C^{\mathrm{PC}}_{\mathrm{load}}(t) \le \overline{C}^{\mathrm{PC}}_{\mathrm{load}} \\ \underline{P}^{\mathrm{PH}}_{\mathrm{load}} \le P^{\mathrm{PH}}_{\mathrm{load}}(t) \le \overline{P}^{\mathrm{PH}}_{\mathrm{load}} \\ \underline{H}^{\mathrm{PH}}_{\mathrm{load}} \le H^{\mathrm{PH}}_{\mathrm{load}}(t) \le \overline{H}^{\mathrm{PH}}_{\mathrm{load}} \\ \underline{G}^{\mathrm{GC}}_{\mathrm{load}} \le G^{\mathrm{GC}}_{\mathrm{load}} \le \overline{G}^{\mathrm{GC}}_{\mathrm{load}} \\ \underline{C}^{\mathrm{GC}}_{\mathrm{load}} \le C^{\mathrm{GC}}_{\mathrm{load}}(t) \le \overline{C}^{\mathrm{GC}}_{\mathrm{load}} \\ \underline{G}^{\mathrm{GH}}_{\mathrm{load}} \le G^{\mathrm{GH}}_{\mathrm{load}} \le \overline{G}^{\mathrm{PH}}_{\mathrm{load}} \\ \underline{H}^{\mathrm{GH}}_{\mathrm{load}} \le H^{\mathrm{GH}}_{\mathrm{load}}(t) \le \overline{H}^{\mathrm{GH}}_{\mathrm{load}} \end{cases}. \quad (24)$$

Secondly, the output of conversion system has its lower/upper limit:

$$\begin{cases} \underline{P}_{\mathrm{CHP}} \le P_{\mathrm{CHP}}(t) \le \overline{P}_{\mathrm{CHP}} \\ \underline{P}_{\mathrm{GT}} \le P_{\mathrm{GT}}(t) \le \overline{P}_{\mathrm{GT}} \\ \underline{H}_{\mathrm{GB}} \le H_{\mathrm{GB}}(t) \le \overline{H}_{\mathrm{GB}} \\ \underline{C}_{\mathrm{ER}} \le C_{\mathrm{ER}}(t) \le \overline{C}_{\mathrm{ER}} \\ \underline{H}_{\mathrm{EH}} \le H_{\mathrm{EH}}(t) \le \overline{H}_{\mathrm{EH}} \end{cases}. \quad (25)$$

Thirdly, the purchase amount of electricity $P_{ext}(t)$ and gas $G_{ext}(t)$ cannot be unlimited; it is assumed that they cannot surpass the allowable maximum.

$$\begin{cases} 0 \le P_{ext}(t) \le P^{\max}_{ext} \\ 0 \le G_{ext}(t) \le G^{\max}_{ext} \end{cases}. \quad (26)$$

Based on (26) it can be learned that the minimal purchase amount from external electricity or gas network is nonnegative. Particularly with respect to gas purchase, since gas is the only crude source which is used as the primitive input energy of all conversion systems and multi-generation plants, it cannot be less than zero. Otherwise, it means MES can autonomously inject gas toward external network out of nothing, which is against the law of conservation of energy. On the other hand, the electricity purchased takes the role of auxiliary, which can act as supplementary energy provider in due time. Because of the existence of gas load, electricity purchased cannot single-handedly solve the energy supply problem. Nevertheless, the energy supplier for heat and cooling load could be electricity purchased.

The model (7)–(26) belongs to mixed integer linear programming (MILP) and is solved by commercial optimization software package CPLEX under general algebraic modelling system (GAMS).

4. Case Study

In this section, MES in Figure 2 is used for numerical analyses. Specifically, the intrinsic relation between load replaceability (as is quantified by (4) and (6)) and optimal scheduling results is studied, and thus giving a priori knowledge of improving economic scheduling performance by reformulating the composite load.

4.1. Simulation Settings

Parameters of main system components are given in Table 1.

Table 1. Parameters and corresponding values of MES.

Parameter	Value
maximum allowable electricity procurement P_{ext}^{max}	12,000 kW
maximum allowable gas procurement G_{ext}^{max}	60,000 kW
maximum output of electric heater $\overline{H}_{EH}$	2000 kW
maximum output of electric refrigerator $\overline{C}_{ER}$	2000 kW
maximum output electricity of CHP plant $\overline{P}_{CHP}$	8000 kW
maximum output of heat exchanger $\overline{H}_{HE}$	4000 kW
maximum output of absorption refrigerator $\overline{C}_{AR}$	6000 kW
maximum output of gas boiler $\overline{H}_{GB}$	2000 kW
maximum output of gas turbine $\overline{P}_{GT}$	8000 kW
power generation efficiency of CHP plant p_{CHP}	0.3
electricity-to-heat ratio of CHP plant q_{QP}	2/3
energy efficiency ratio of heat exchanger η_{HE}	0.9
energy efficiency ratio of absorption refrigerator η_{AR}	0.9
energy efficiency ratio of electric heater η_{EH}	0.95
energy generation proportion ratio γ_{HC}	0.6
energy efficiency ratio of electric refrigerator η_{ER}	3.5
energy efficiency ratio of gas boiler η_{GB}	0.9
maximum charge/discharge energy of electricity ESS P_{PS}^{max}	2000 kW
maximum charge/discharge energy of heat ESS H_{HS}^{max}	1000 kW
maximum charge/discharge energy of cooling ESS P_{PS}^{max}	2000 kW
maximum remaining energy of electricity ESS S_{PS}^{max}	8000 kWh
maximum remaining energy of heat ESS S_{HS}^{max}	4000 kWh
maximum remaining energy of cooling ESS S_{CS}^{max}	8000 kWh
attrition rate of electricity ESS η_{PS}	0.2%
attrition rate of heat ESS η_{HS}	0.3%
attrition rate of cooling ESS η_{CS}	0.2%
operating cost of electricity ESS r_{PS}	0.32 $/kWh
operating cost of heat ESS r_{HS}	0.25 $/kWh
operating cost of cooling ESS r_{CS}	0.2 $/kWh

The load status described by loading rate at every scheduling interval is shown in Figure 3.

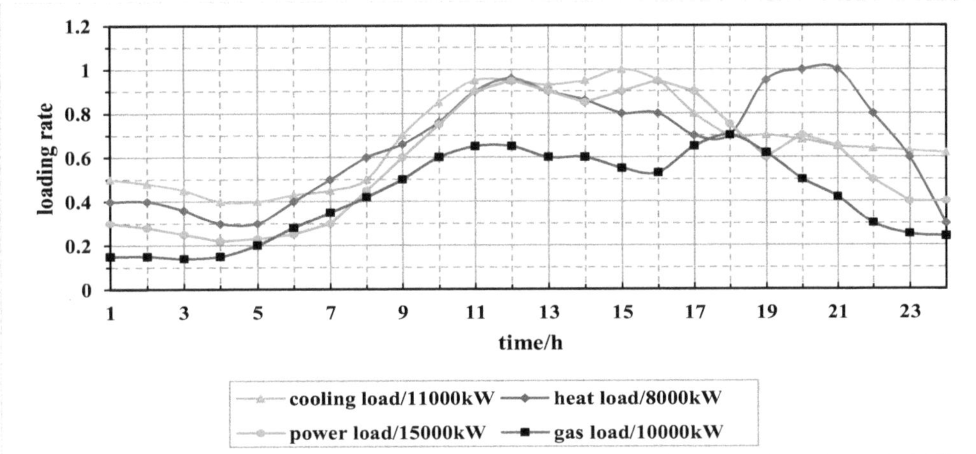

Figure 3. Hourly load status of gas, power, and heat and cooling load of MES.

The rated power of cooling, heat, and power and gas load is 11,000 kW, 8000 kW, 15,000 kW, and 10,000 kW, respectively. The day-ahead price information from the electricity and gas market is given in Table 2.

Table 2. Time-of-Use price of gas and electricity in a day.

Periods	r_{gas}	Periods	r_{ele}
00:00–07:00	2.7	00:00–08:00	0.35
07:00–12:00	3.3	08:00–12:00	1.65
12:00–16:00	3.0	12:00–17:00	0.95
16:00–20:00	3.3	17:00–21:00	1.65
20:00–24:00	3.0	21:00–24:00	0.95

4.2. Simulation Results and Discussions

After presenting the basic simulation settings, optimal scheduling and its relation with load replaceability is studied by calculating the scheduling model in (7)–(9) in Section 3 and LRI in (3)–(6) in Section 2.2. As previously addressed in Section 2.2, there exist two ways of describing the load replacement capability of MES; therefore, by special setting of composite load in (14), we change the values of LRI both in (4) and (6), and analyze the resulting influence upon optimal scheduling results through a sensitivity analysis tests. Specifically, the following 2 scenarios are considered.

4.2.1. Scenario 1

It is supposed that the composite load is of electricity-cooling and electricity-heat types. As for composite electricity-cooling load, the rated capacity of alternative electricity load is fixed to 2000 kW, while the capacity of alternative cooling load is increased from 600 kW to 6000 kW at the step size of 600 kW. Similarly, the rated capacity of alternative electricity load is fixed to 3750 kW in composite electricity-heat load, while the the capacity of alternative heat load is increased from 300 kW to 3000 kW at the step size of 300 kW. Consequently, 10 LRIs can be calculated based on (4) and (6), and the results are shown in Figure 4.

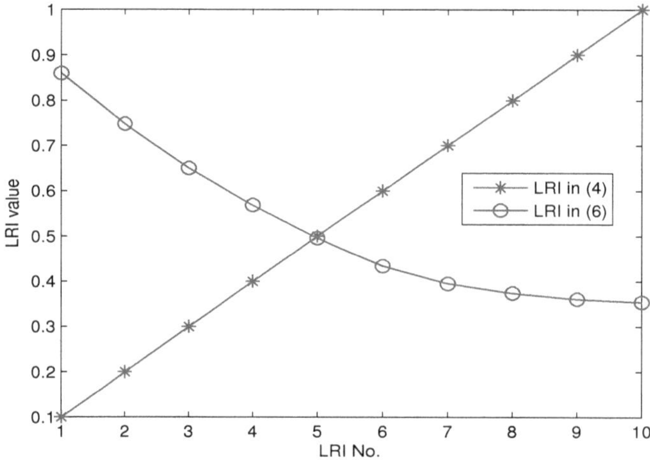

Figure 4. load replaceability index (LRI) with and without optimal scheduling in Scenario 1.

As can be seen from the blue curve in Figure 4, with the increase of alternative cooling and heat load, the remaining alternative load reserve is elevated from 900 kW to 9000 kW, thus enhancing LRI from 0.1 to 1. Meanwhile, the red curve shows that the proportion of energy converted by intermediate loads in the actual scheduling drops as the rated capacity of alternative heat (cooling) load increases, which means the operator chooses to obtain heat or cooling energy through redundant conversion paths. Take the electricity-cooling load as an example, the redundant conversion path is shown in dashed black line in Figure 5.

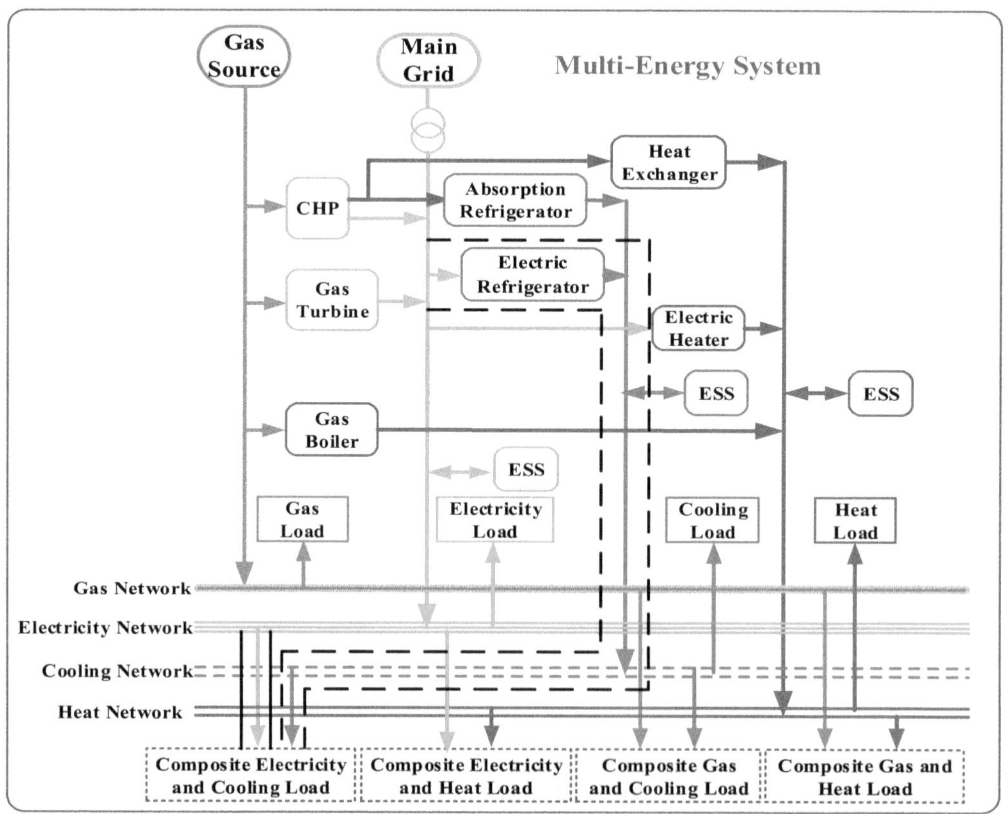

Figure 5. Redundant conversion path of electricity-cooling load in Scenario 1.

Due to the fact that energy efficiency ratio of electric refrigerator is much higher than that of alternative electricity load in the composite electricity and cooling load. The operator chooses redundant path such that more cooling energy can be harvested with equal amount of electricity energy. Similar conclusions can be made in respect to composite electricity and heat load. Therefore, the proportion of energy converted by intermediate electricity load in the composite load is decreased with the increase of transmission capacity of redundant paths, which is equivalent to the drop of LRI (from 0.86 to 0.35). Furthermore, the relation between LRI and optimal scheduling results in Scenario 1 is shown in Figures 6 and 7.

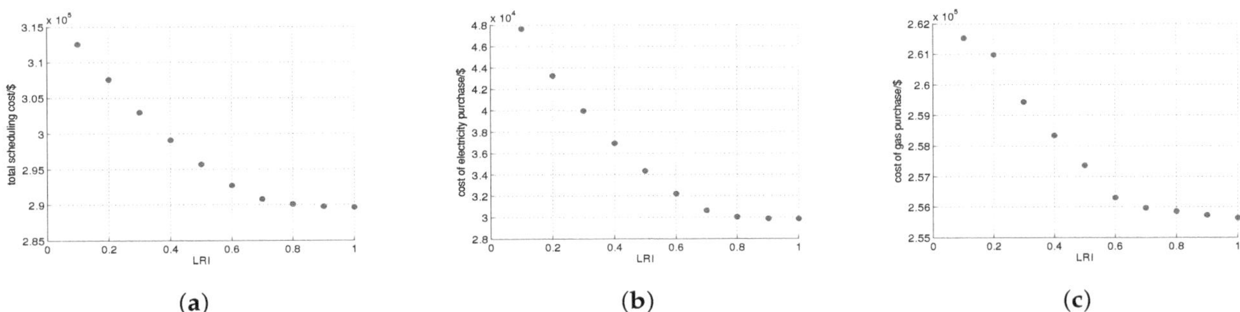

(a) (b) (c)

Figure 6. Relation between offline potential-based LRI and optimal scheduling results in Scenario 1. (**a**) The relation between LRI and optimal energy cost. (**b**) The relation between LRI and optimal cost of electricity purchase. (**c**) The relation between LRI and optimal cost of gas purchase.

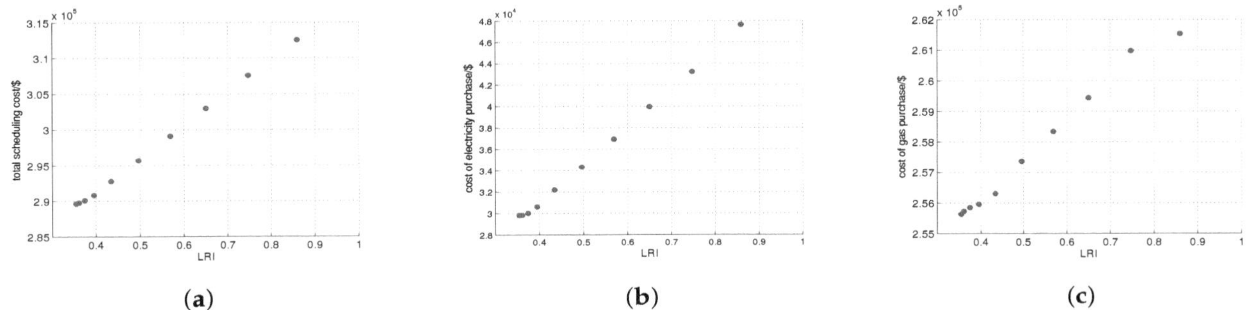

(a) (b) (c)

Figure 7. Relation between online scheduling-based LRI and optimal scheduling results in Scenario 1. (**a**) The relation between LRI and optimal energy cost. (**b**) The relation between LRI and optimal cost of electricity purchase. (**c**) The relation between LRI and optimal cost of gas purchase.

As can be seen from Figure 6, the cost including cost of gas and electricity purchase decreases with the increase of offline potential-based LRI. Contrarily, there exists a positive correlation between energy cost and online scheduling-based LRI. It verifies that the optimization space can be enhanced by elevating the capacity of alternative load, thus enabling the decrease of minimal scheduling costs.

4.2.2. Scenario 2

It is supposed that the composite load is of gas-cooling and gas-heat types. As for composite gas-cooling load, the rated capacity of alternative gas load is fixed to 5000 kW, while the capacity of alternative cooling load is increased from 600 kW to 6000 kW at the step size of 600 kW. Similarly, the rated capacity of alternative gas load is fixed to 6000 kW in composite gas-heat load, while the the capacity of alternative heat load is increased from 300 kW to 3000 kW at the step size of 300 kW. Consequently, 10 LRIs can be calculated based on (4) and (6), and the results are shown in Figure 8.

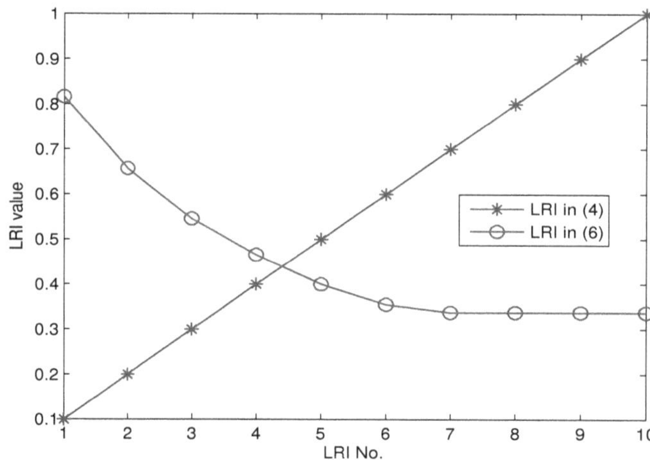

Figure 8. LRI with and without optimal scheduling in Scenario 2.

The relation between LRI and optimal scheduling results in Scenario 2 is shown in Figures 9 and 10. Similar to Scenario 1, it can be learned that the optimal energy cost can be decreased by elevating the offline potential based LRI, which quantifies the maximum alternative load capacity of the system. In addition, it can be found that the optimal cost does not decrease when LRI reaches the critical state. It means that there is no unrestricted expansion of optimization space by single-handedly elevating alternative load capacity, which could improve the expectancy of optimum under the joint force of other factors such as system configuration and market price signals.

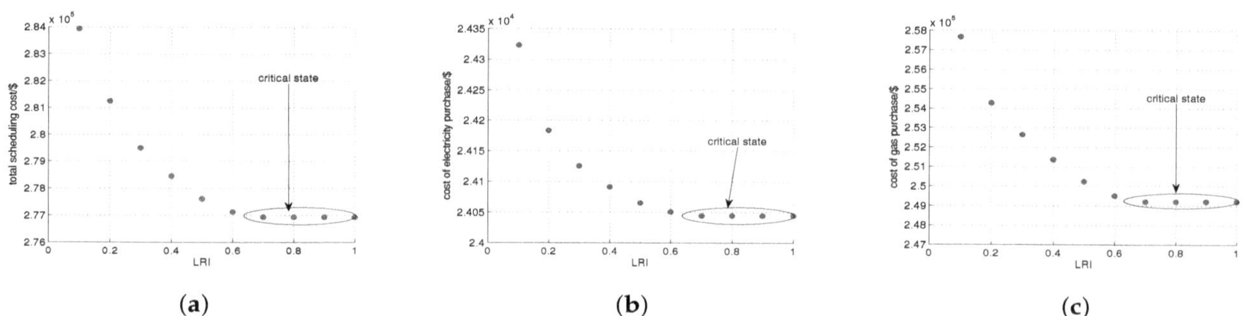

(a) (b) (c)

Figure 9. Relation between offline potential-based LRI and optimal scheduling results in Scenario 2. (**a**) The relation between LRI and optimal energy cost. (**b**) The relation between LRI and optimal cost of electricity purchase. (**c**) The relation between LRI and optimal cost of gas purchase.

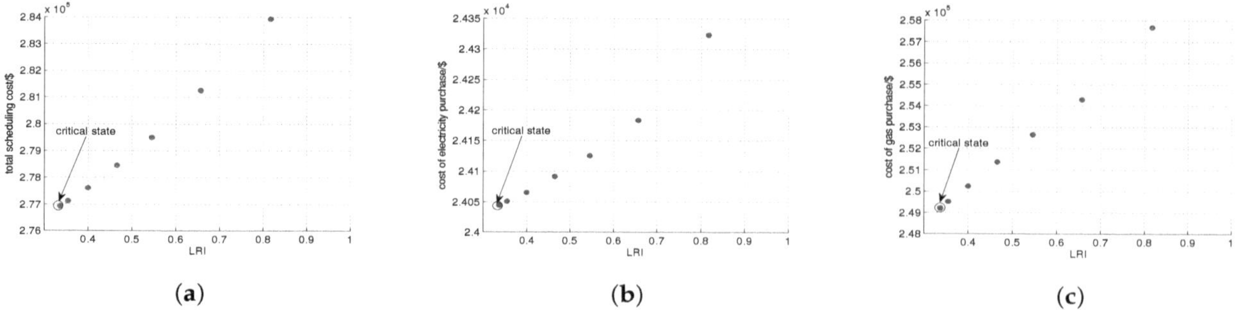

(a) (b) (c)

Figure 10. Relation between online scheduling-based LRI and optimal scheduling results in Scenario 2. (**a**) The relation between LRI and optimal energy cost. (**b**) The relation between LRI and optimal cost of electricity purchase. (**c**) The relation between LRI and optimal cost of gas purchase.

5. Conclusions

In this paper, optimal economic scheduling of multi-energy systems considering load replaceability is presented. Two types of LRI reflecting the offline potential and online scheduling performance are designed, and the relation between LRI and optimal scheduling results is analyzed. Through numerical analysis, it can be found that optimal scheduling performance can be improved by increasing the alternative load capacity (offline potential-based LRI). And this improvement is not unrestricted and could only maintain by changing load replaceability in a limited range, after which the optimum reaches the saturation phase.

Author Contributions: T.Z. and Z.D. conceived and designed the experiments; J.Y. performed the experiment; T.Z. wrote the paper; Y.Y. and J.C. reviewed the paper.

References

1. Vasebi, A.; Fesanghary, M.; Bathaee, S. Combined heat and power economic dispatch by harmony search algorithm. *Int. J. Electr. Power Energy Syst.* **2007**, *29*, 713–719. [CrossRef]
2. Vögelin, P.; Koch, B.; Georges, G.; Boulouchos, K. Heuristic approach for the economic optimisation of combined heat and power (CHP) plants: Operating strategy, heat storage and power. *Energy* **2017**, *121*, 66–77. [CrossRef]
3. Wu, C.; Jiang, P.; Sun, Y.; Zhang, C.; Gu, W. Economic dispatch with chp and wind power using probabilistic sequence theory and hybrid heuristic algorithm. *J. Renew. Sustain. Energy* **2017**, *9*, 013303. [CrossRef]
4. Li, Z.; Wu, W.; Shahidehpour, M.; Wang, J.; Zhang, B. Combined heat and power dispatch considering pipeline energy storage of district heating network. *IEEE Trans. Sustain. Energy* **2015**, *7*, 12–22. [CrossRef]
5. Hashemi, R. A developed offline model for optimal operation of combined heating and cooling and power systems. *IEEE Trans. Energy Convers.* **2009**, *24*, 222–229. [CrossRef]
6. Facci, A.L.; Andreassi, L.; Ubertini, S. Optimization of CHCP (combined heat power and cooling) systems operation strategy using dynamic programming. *Energy* **2014**, *66*, 387–400. [CrossRef]
7. Li, M.; Mu, H.; Li, N. Optimal design and operation strategy for integrated evaluation of CCHP (combined cooling heating and power) system. *Energy* **2016**, *99*, 202–220. [CrossRef]
8. Tan, Z.; Guo, H.; Lin, H.; Tan, Q.; Yang, S.; Gejirifu, D.; Ju, L.; Song, X. Robust scheduling optimization model for multi-energy interdependent system based on energy storage technology and ground-source heat pump. *Processes* **2019**, *7*, 27. [CrossRef]
9. Wang, M.; Wang, J.; Zhao, P.; Dai, Y. Multi-objective optimization of a combined cooling, heating and power system driven by solar energy. *Energy Convers. Manag.* **2015**, *89*, 289–297. [CrossRef]
10. Motevasel, M.; Seifi, A.R.; Niknam, T. Multi-objective energy management of CHP (combined heat and power)-based micro-grid. *Energy* **2013**, *51*, 123–136. [CrossRef]
11. Li, K.; Yan, H.; He, G.; Zhu, C.; Liu, K.; Liu, Y. Seasonal Operation Strategy Optimization for Integrated Energy Systems with Considering System Cooling Loads Independently. *Processes* **2018**, *6*, 202. [CrossRef]
12. Zhang, X.; Shahidehpour, M.; Alabdulwahab, A.; Abusorrah, A. Optimal expansion planning of energy hub with multiple energy infrastructures. *IEEE Trans. Smart Grid* **2015**, *6*, 2302–2311. [CrossRef]
13. Parisio, A.; Del Vecchio, C.; Vaccaro, A. A robust optimization approach to energy hub management. *Int. J. Electr. Power Energy Syst.* **2012**, *42*, 98–104. [CrossRef]
14. Vahid-Pakdel, M.; Nojavan, S.; Mohammadi-Ivatloo, B.; Zare, K. Stochastic optimization of energy hub operation with consideration of thermal energy market and demand response. *Energy Convers. Manag.* **2017**, *145*, 117–128. [CrossRef]
15. Huang, Y.; Zhang, W.; Yang, K.; Hou, W.; Huang, Y. An optimal scheduling method for multi-energy hub systems using game theory. *Energies* **2019**, *12*, 2270. [CrossRef]
16. Lin, Y.; Chen, B.; Wang, J.; Bie, Z. A combined repair crew dispatch problem for resilient electric and natural gas system considering reconfiguration and DG islanding. *IEEE Trans. Power Syst.* **2019**, *34*, 2755–2767. [CrossRef]

17. Liu, Z.; Yang, P.; Xu, Z. Capacity allocation of integrated energy system considering typical day economic operation. *Electr. Power Constr.* **2017**, *38*, 51–59.
18. Zhao, F.; Sun, B.; Zhang, C. Cooling, heating and electrical load forecasting method for CCHP system based on multivariate phase space reconstruction and Kalman filter. *Proc. CSEE* **2016**, *36*, 399–406.
19. Chen, F.; Xu, J.; Wang, C. Research on building cooling and heating load prediction model on user's side in energy internet system. *Proc. CSEE* **2015**, *35*, 3678–3684.
20. Shi, J.; Tan, T.; Guo, J. Multi-task learning based on deep architecture for various types of load forecasting in regional energy system integration. *Power Syst. Technol.* **2018**, *42*, 698–706.
21. Sheikhi, A.; Rayati, M.; Bahrami, S. Integrated demand side management game in smart energy hubs. *IEEE Trans. Smart Grid* **2015**, *6*, 675–683. [CrossRef]
22. Bahrami, S.; Sheikhi, A. From demand response in smart grid toward integrated demand response in smart energy hub. *IEEE Trans. Smart Grid* **2016**, *7*, 650–658. [CrossRef]

12

Simulation-Based Design and Economic Evaluation of a Novel Internally Circulating Fluidized Bed Reactor for Power Production with Integrated CO_2 Capture

Jan Hendrik Cloete [1,*], Mohammed N. Khan [2], Schalk Cloete [1] and Shahriar Amini [1,2,*]

1 Flow Technology Research Group, SINTEF Industry, 7465 Trondheim, Norway; schalk.cloete@sintef.no
2 Department of Energy and Process Engineering, Norwegian University of Science and Technology, 7491 Trondheim, Norway; mkhan@ntnu.no
* Correspondence: henri.cloete@sintef.no (J.H.C.); shahriar.amini@sintef.no (S.A.)

Abstract: Limiting global temperature rise to well below 2 °C according to the Paris climate accord will require accelerated development, scale-up, and commercialization of innovative and environmentally friendly reactor concepts. Simulation-based design can play a central role in achieving this goal by decreasing the number of costly and time-consuming experimental scale-up steps. To illustrate this approach, a multiscale computational fluid dynamics (CFD) approach was utilized in this study to simulate a novel internally circulating fluidized bed reactor (ICR) for power production with integrated CO_2 capture on an industrial scale. These simulations were made computationally feasible by using closures in a filtered two-fluid model (fTFM) to model the effects of important subgrid multiphase structures. The CFD simulations provided valuable insight regarding ICR behavior, predicting that CO_2 capture efficiencies and purities above 95% can be achieved, and proposing a reasonable reactor size. The results from the reactor simulations were then used as input for an economic evaluation of an ICR-based natural gas combined cycle power plant. The economic performance results showed that the ICR plant can achieve a CO_2 avoidance cost as low as $58/ton. Future work will investigate additional firing after the ICR to reach the high inlet temperatures of modern gas turbines.

Keywords: chemical looping combustion; power production; carbon capture; internally circulating reactor; reactor design; fluidization; techno-economics; computational fluid dynamics; filtered two-fluid model; coarse-grid simulations

---

1. Introduction

Several high-profile studies have shown that carbon capture and storage must play a central role in the future energy mix to reach the goal of limiting the global temperature increase to well below 2 C above preindustrial limits at a reasonable cost [1–3]. A low-cost pathway to limiting global CO_2 emissions will be essential to prevent the negative consequences of climate change, while allowing for continued development in developing nations where billions of people still live in poverty.

Many different technologies have been proposed to capture CO_2 from fossil-fuel power plants, after which the CO_2 can either be stored or utilized in other industrial processes. However, a major challenge of such processes is the energy penalty associated with CO_2 capture. An increased energy penalty requires more fuel to be used to achieve the same power output, increasing operating and capital costs, but also increasing the amount of CO_2 that must be dealt with.

A promising group of technologies for capturing CO_2 are those based on chemical looping combustion (CLC) [4], as they can essentially eliminate the energy penalty of CO_2 and potentially even offer efficiency improvements in comparison to unabated plants [5]. Traditionally, the CLC process is performed in a dual circulating fluidized bed (CFB) configuration. In the oxidation reactor, a metallic oxygen carrier is oxidized, providing large amounts of heat. The thermal energy in the gas phase is used for power production, whereas the hot particles are transported to the fuel reactor where the oxidized particles are reduced by a fuel, producing CO_2 and steam. The CLC process therefore keeps the CO_2 stream separate from the nitrogen-containing air stream, allowing an almost pure CO_2 stream for storage to be obtained simply by knocking out the water.

A drawback of the dual circulating fluidized bed CLC approach is that efficient power production with CO_2 storage requires high pressure operation. However, progress on the scale-up of pressurized CLC systems has been limited [6] due to the complexity of pressurizing the two reactors, loop seals, and cyclones, and in maintaining the required solids circulation between the reactors. Consequently, several alternative CLC configurations have been proposed to overcome the challenges of the pressurized dual CFB CLC system. These include gas switching technologies [7,8], rotating bed reactors [9,10], packed bed chemical looping [11,12], and internally circulating reactors (ICRs) [13–15], which will be the focus of the present study.

The internally circulating reactor concept replaces the loop seals and cyclones that separate reactors in the dual CFB with simple ports connecting two sections of a reactor vessel, allowing the oxygen carrier to circulate between the reducing and oxidizing sections. This allows the CLC process to take place within a single unit, significantly simplifying pressurization and scale-up. The disadvantage is that gas will leak through the ports along with the circulating solids, reducing the CO_2 capture efficiency and the purity of the captured CO_2. However, it has been shown that the detrimental effect of gas leakage can be limited by controlling the fluidization velocity ratio of the two sections and the bed loading [13], achieving CO_2 capture efficiencies greater than 95% and purities greater than 92%.

Academia has been prolific in proposing novel processes and reactors for CO_2 capture. Unfortunately, implementation of new technologies in the process and energy industry has traditionally been slow, requiring several decades from process conception to commercial reality. The urgency of climate change will require very rapid scale-up and industrialization of these novel CO_2 capture technologies, starting once governments start imposing strong policies to reduce carbon emissions (the IEA Sustainable Development Scenario assumes CO_2 prices of $63/ton and $140/ton in 2025 and 2040, respectively [16]). This is also valid for other industries—rapid innovation and implementation of new process technologies will be necessary in a world that is increasingly environmentally and resource-constrained.

Simulation-based engineering will be an essential tool in enabling such rapid innovation by decreasing the number of costly and time-consuming experimental scale-up steps, and computational fluid dynamics (CFD) is the most suitable tool for investigating the chemical reactors common in the process and energy industries. However, although CFD has proven extremely useful in better understanding flow processes on the lab-scale, a common challenge to industrial simulation is the fact that important phenomena may occur on time- and length-scales that are several orders of magnitude smaller than those associated with the industrial processes [17]. This is especially relevant in multiphase processes and for the fluidized beds used in the ICR reactor studied here, where gas bubbles and particle clusters of length-scales in the order of ten particle diameters play an important role on the overall fluidized bed behavior. Using small enough grid cells and time steps to resolve these small-scale phenomena remains impossible for parametric studies of industrial-scale devices, even with large, modern computational clusters.

Multiscale methods are necessary to overcome this challenge—allowing the use of coarse computational grids to achieve reasonable computational times by using closures for unresolved subgrid effects to maintain acceptable accuracy. The filtered two-fluid model (fTFM) [18] is a common approach for multiscale modeling of fluidized beds. In the fTFM, the governing equations of the two-fluid model (TFM) closed by the kinetic theory of granular flow, where the solids phase is assumed to behave as a continuum and closures capture the effects of random particle collisions and translation, is spatially averaged, revealing subgrid terms that require closure. Several groups have strived to develop such closures. Most of the work has focused on the subgrid correction to the drag [19–26], which substantially reduces the drag compared to the drag law evaluated at the resolved quantities, although several other closures are necessary for fluidized bed hydrodynamics [27]. Research on closures for reactive flow has been limited. Most studies have investigated the influence of subgrid effects on the effective reaction rate of first-order solids-catalyzed reactions [28–30], where mass-transfer limitations imposed by the bubbles and clusters drastically reduce the effective reaction rates. Closures are also required for the dispersion of scalars (such as species and enthalpy) due to subgrid velocity fluctuations and for the effective interphase heat transfer rate [31,32].

The present study aimed to demonstrate how multiscale CFD simulations can be used to assist the evaluation of novel reactor concepts on an industrial scale, focusing on an internally circulating fluidized bed reactor for power production with CO_2 capture. Firstly, some improvements were proposed for existing fTFM closures, improving the accuracy and simplicity of existing hydrodynamics closures [27,33] and, most importantly, proposing a generalized reactive fTFM closure. The latter is important, since existing closures [28–30] are only valid for simple first-order solids-catalyzed reaction equations. Next, an fTFM accounting for all important subgrid effects in reactive flows was used to evaluate the effect of several design and operating parameters on the ICR behavior. It can be noted that, to the best of the authors' knowledge, this is the most complete implementation—in terms of the number of subgrid effects accounted for—of a reactive fTFM to date. Then, results from the reactor simulations were combined with previously published power plant simulations by the same authors [34] to conduct an economic assessment of the ICR concept for low carbon power production from natural gas. Finally, the results are used to discuss the future of virtual prototyping of novel reactors using multiscale CFD simulations, as well as the potential of the ICR to combat climate change.

2. Materials and Methods

The present study utilizes both multiscale CFD reactor modeling and process modeling to inform the economic evaluation of power production with CO_2 capture using the ICR concept. Figure 1 shows how information flows between these three parts of the study, and the subsequent sections describe each part in detail.

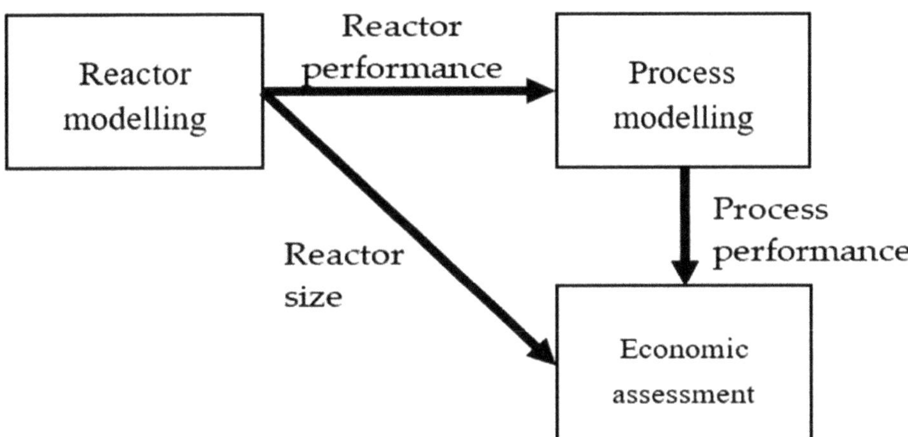

Figure 1. Information flow between different parts of the present study.

2.1. Reactor Modeling

2.1.1. The Filtered Two-Fluid Model (fTFM)

The fTFM solves the spatially-averaged (or filtered) form of the governing equations for the two-fluid model closed by the kinetic theory of granular flow [35,36]. This section briefly presents the filtered governing equations, as well as the closures that are used for the subgrid terms. The interested reader may find a more complete discussion of the derivation of the filtered equations in earlier studies [33,37].

The filtered continuity equations are given below. S_R is a source term due to the mass transfer during the reduction and oxidation of the oxygen carrier. Closures for the filtered reaction rates, which can be used to calculate the source term, are discussed in Section 2.1.2.

$$\frac{\partial}{\partial t}\left(\overline{\alpha_g}\rho_g\right) + \nabla \cdot \left(\overline{\alpha_g}\rho_g \widetilde{\vec{v}_g}\right) = S_R, \tag{1}$$

$$\frac{\partial}{\partial t}\left(\overline{\alpha_s}\rho_s\right) + \nabla \cdot \left(\overline{\alpha_s}\rho_s \widetilde{\vec{v}_s}\right) = -S_R. \tag{2}$$

Next, the filtered momentum equations are shown in Equations (3) and (4):

$$\frac{\partial}{\partial t}\left(\rho_g\overline{\alpha_g}\widetilde{\vec{v}_g}\right) + \nabla \cdot \left(\rho_g\overline{\alpha_g}\widetilde{\vec{v}_g}\widetilde{\vec{v}_g}\right) = -\overline{\alpha_g}\nabla p - \nabla \cdot \left(\overline{\rho_g\alpha_g\vec{v}_g{''}\vec{v}_g{''}}\right) + \nabla \cdot \overline{\overline{\tau}}_g + \overline{\alpha_g}\rho_g\vec{g} + \overline{K_{sg}\left(\vec{v}_s - \vec{v}_g\right)} - \overline{\alpha_g{'}\nabla p{'}}, \tag{3}$$

$$\frac{\partial}{\partial t}\left(\rho_s\overline{\alpha_s}\widetilde{\vec{v}_s}\right) + \nabla \cdot \left(\rho_s\overline{\alpha_s}\widetilde{\vec{v}_s}\widetilde{\vec{v}_s}\right) = -\overline{\alpha_s}\nabla p - \nabla \cdot \left(\overline{\rho_s\alpha_s\vec{v}_s{''}\vec{v}_s{''}}\right) - \nabla\overline{p_s} + \nabla \cdot \overline{\overline{\overline{\tau}}}_s + \overline{\alpha_s}\rho_s\vec{g} + \overline{K_{gs}\left(\vec{v}_g - \vec{v}_s\right)} - \overline{\alpha_s{'}\nabla p{'}}. \tag{4}$$

In both filtered momentum equations, the second term on the right-hand side represents the stresses due to subgrid velocity fluctuations (arising from gas bubbles and solids clusters), which add to the diffusive momentum transport. The gas-phase subgrid stress is usually relatively small compared to the solids stresses due to the large difference in the phase densities [38] and can safely be neglected. However, the solids subgrid stresses are accounted for by means of an anisotropic stress closure [39], which has been shown to offer significant improvements compared to Boussinesq approximation-based closures using isotropic independent variables [39,40]. In the filtered solids momentum equation, the filtered kinetic theory stresses (third and fourth terms on the right-hand side) are small at the grid sizes that are relevant for industrial-scale fluidized beds [39], which was used in the present study. Therefore, the filtered kinetic theory stresses were estimated on the basis of the unfiltered granular temperature equation, as it was previously shown to be sufficient [41].

The second-to-last term on the right-hand side of both momentum equations represents the filtered drag force, where subgrid effects generally reduce the drag compared to that in a homogenous suspension. This is due to the tendency of fluidized particles to form solids clusters and gas bubbles, which are not resolved on a coarse grid. These meso-scale structures vary in size and shape due to local flow conditions. Gas will tend to pass through dilute regions, reducing the effective drag on the solids clusters, the effect of which must be accounted for in a closure. A modified version of a 3-marker anisotropic closure published previously [27] was used to close the filtered drag force. It was found that the 3-marker closure could be simplified significantly, while maintaining similar accuracy, by eliminating the filtered slip velocity as a marker. More information about the development and verification of this new closure can be found in the Supplementary Material.

Finally, the last term on the right-hand side of both momentum equations is due to subgrid pressure gradient fluctuations and is referred to here as the meso-scale interphase force. This contribution arises from the redistribution of the pressure gradient over subgrid gas bubbles and solids clusters [42] and tends to add to the effective drag force [33]. For the present study, an older anisotropic closure [33] was improved on by drawing an analogy to the closure for the meso-scale solids stresses [39], where it was found that a filtered co-variance term can be accurately closed as a function of the relevant gradients. The Supplementary Material also details the development and verification of this new closure.

Next, Equation (5) gives the filtered species transport equation for reactant A, which is consumed in an nth order reaction.

$$\frac{\partial}{\partial t}\left(\rho_g\overline{\alpha_g}\widetilde{X_A}\right) + \nabla \cdot \left(\rho_g\overline{\alpha_g}\widetilde{X_A}\widetilde{\vec{v}_g}\right) = \nabla \cdot \left(\rho_g\overline{D\alpha_g\nabla X_A}\right) - \nabla \cdot \left(\overline{\rho_g\alpha_g X_A{''}\vec{v}_g{''}}\right) - \overline{k_A\alpha_s C_A^n M_A}. \tag{5}$$

The species dispersion due to the filtered microscopic diffusion (the first term on the right-hand side) was expected to be small relative to meso-scale dispersion, as well as convective transport. Therefore, in line with previous work regarding scalar dispersion in fTFMs [31], it was simply evaluated as $\nabla \cdot \left(\rho_g D \overline{\alpha_g} \nabla \widetilde{X}_A\right)$ in the present study. The subgrid species dispersion rate (the second term on the right-hand side) tends to disperse the species due to sub-grid velocities arising from unresolved gas bubbles and solids clusters. This effect was accounted for using the closures of Agrawal et al. [31] but has been shown to only have a minor effect on the overall reaction rate [28]. The filtered reaction rate (third term on the right-hand side) is typically substantially reduced by subgrid bubbles and clusters and is essential to model [28]. This is because, for gas-solid reactions, the reactant will be consumed faster inside dense regions, creating a mass transfer limitation due to the finite rate at which reactants are transported to these dense regions. A limited number of studies have investigated reactive fTFM closures [28,30,41], but they have all focused on reactions that are solids catalyzed and first-order with respect to the gaseous reactant. The next section of the present study therefore proposes a novel, simplified approach for accounting for different reaction orders and for reactions where the solids phase participate in the reaction. Finally, it can be noted that the filtered solids species equations are similar to those of the gas-phase species and are thus treated in a similar way. Consequently, they are not discussed separately.

Finally, Equation (6) shows the filtered enthalpy transport equation for the gas-phase.

$$\frac{\partial}{\partial t}\left(\rho_g \overline{\alpha_g} \widetilde{h_g}\right) + \nabla \cdot \left(\rho_s \overline{\alpha_g} \widetilde{h_g} \widetilde{\vec{v}_g}\right) = \nabla \cdot \left(\overline{\kappa_g \alpha_g \nabla T_g}\right) - \nabla \cdot \left(\overline{\rho_s \alpha_s h_g'' \vec{v}_g''}\right) + \overline{\gamma\left(T_s - T_g\right)} + \overline{k_A \alpha_s C_A^n M_A \Delta H_{r,A}}. \quad (6)$$

Here, as with the species diffusion, the filtered microscopic conductivity (first term on the right-hand side) is small compared to the enthalpy dispersion from sub-grid velocity fluctuations and is simply approximated as $\overline{\kappa_g \alpha_g} \nabla \widetilde{T}_g$. The subgrid enthalpy dispersion rate (second term on the right-hand side) and the filtered heat transfer rate (third term on the right-hand side) were modeled using the closures of Agrawal et al. [31]. The physical behavior of these contributions is analogous to that of the species dispersion rate and filtered reaction rate, due to the similarity of mass and heat transfer. The enthalpy source term due to reaction (fourth term on the right-hand side) was evaluated at the filtered reaction rate modeled in Equation (5), assuming that the heat of reaction is uniform for each cell in the coarse-grid simulations. This is a reasonable assumption based on the good mixing and fast heat transfer in fluidized beds. Finally, it can be noted that the solids-phase filtered enthalpy equation was treated in a similar way, and it is therefore not discussed here separately.

2.1.2. Reaction Modeling

In the present study, Ni/NiO (supported on Al_2O_3) was used as oxygen carrier due to its high reactivity [43] and its ability to tolerate high operating temperatures [44]. In the fuel section, the oxygen carrier was reduced by the fuel according to:

$$4\text{NiO(s)} + CH_4(g) \rightarrow 4Ni(s) + 2H_2O(g) + CO_2(g). \quad (7)$$

In the air side of the ICR, the oxygen carrier re-oxidizes as:

$$2\text{Ni(s)} + O_2(g) \rightarrow 2NiO(s). \quad (8)$$

The reactions are implemented in the filtered species conservation equations, as follows, for species i taking part in reaction k, where v_i is the stoichiometric constant:

$$\frac{\partial}{\partial t}\left(\rho_g \overline{\alpha_g} \widetilde{X}_i\right) + \nabla \cdot \left(\rho_g \overline{\alpha_g} \widetilde{X}_i \widetilde{\vec{v}_g}\right) = \nabla \cdot \left(\rho_g D \overline{\alpha_g} \nabla \widetilde{X}_i\right) - \nabla \cdot \left(\overline{\rho_g \alpha_g X_i'' \vec{v}_g''}\right) + \sum\nolimits_{k=0}^{n} v_i R_k^H M_i. \quad (9)$$

The effective reaction rate, R_k^H, in units of $mol/\left(m^3 s\right)$ was calculated as shown in Equation (10), where gas species A reacts with solids species B (here, B represents NiO in reduction and Ni in oxidation).

$$R_{\mathrm{k}}^{H} = \eta \overline{\alpha_s} \rho_s a_0 \left(\widetilde{X_{Ni}} + \widetilde{X_{NiO}}\right) k_s \widetilde{C_A}^{n} \left(1 - \frac{\widetilde{X_B}}{\widetilde{X_{Ni}} + \widetilde{X_{NiO}}}\right)^{\frac{2}{3}}. \tag{10}$$

The solid particles are porous, and the reaction can be considered to be kinetically controlled following the shrinking core model applied to microscopic grains inside the porous particle [45]. Application of the shrinking core model with reaction rate control [46] is evident in the final factor of Equation (10). The reaction rate constant, k_s, is expressed as follows, where the detrimental effect of increasing pressure is accounted for in the pre-exponential factor:

$$k_s = \frac{k_0}{\overline{p}^q} e^{-\frac{E_a}{RT}}. \tag{11}$$

The kinetic parameters for the reduction and oxidation reactions, as well as the oxygen carrier properties, were obtained from the experimental work of Abad et al. [45]. It can be noted that the aforementioned study found no intraparticle mass transfer limitations, as may be expected for such small, porous particles.

In the fTFM, the subgrid bubbles and clusters impose an additional mass transfer limitation on the reactions, since the gaseous reactants have to be transported into the dense solid clusters for the reactions to occur. This effect is modeled in Equation (10) by means of an effectiveness factor, η. In the present study, η was first modeled for a reference first-order solids catalyzed reaction with a fixed reaction rate constant, as in previous studies [28,41]. It was then found that the reference closure can be effectively scaled to different reaction rate constants and reaction orders by drawing an analogy with packed bed theory and defining a cluster-scale Thiele modulus, as follows:

$$\phi = \sqrt{\frac{n+1}{2} \frac{k'\left(d_p \mathcal{L}\right)^2}{D}}. \tag{12}$$

Here, $\mathcal{L}$ is the average ratio of the cluster diameter to the particle diameter, which requires closure, and ϕ is the Thiele modulus [47]. The effective reaction rate constant, k', was obtained by re-writing the reaction equation as first-order with respect to the gaseous reactant and the solids volume fraction. This approach has previously been shown to be useful to extend effectiveness factors from intraparticle mass transfer theory to various reaction orders [48]. For the example of Equation (10), the effective reaction rate constant becomes:

$$k' = \rho_s a_0 \left(\widetilde{X_{Ni}} + \widetilde{X_{NiO}}\right) k_s \widetilde{C_A}^{n-1} \left(1 - \frac{\widetilde{X_B}}{\widetilde{X_{Ni}} + \widetilde{X_{NiO}}}\right)^{\frac{2}{3}}. \tag{13}$$

The effectiveness factor for a spherical particle can then be written as follows [49]:

$$\eta = \frac{1}{\phi}\left(\frac{1}{\tanh(3\phi)} - \frac{1}{3\phi}\right). \tag{14}$$

This relation is exact for a first-order reaction in a porous particle with no convective transport. Relatively small discrepancies arise for reactions of different order, but the largest uncertainty in this application is the constant deformation of the clusters in the fluidized bed, as well as the convective species transport taking place inside the cluster.

The basic premise of the approach proposed in the present study is that the effectiveness factor in Equation (14) is analogous to the effectiveness factor of a particle cluster at the largest achievable mass transfer resistance (smallest effectiveness factor). This will typically occur at intermediate filtered solids volume fractions when maximum phase segregation is achieved and clusters are relatively large. As the filtered solids volume fraction tends to the limits of zero or maximum packing, clustering disappears, and the mass transfer resistance tends to zero ($\eta = 1$).

A hypothesis can then be formulated that, for different cluster-scale Thiele moduli, the minimum effectiveness factor (η_{min}) can be scaled by using Equation (15) when a filtered effectiveness factor closure (η_{ref}) is derived from resolved simulation data for a reference Thiele modulus (ϕ_{ref}).

$$\eta_{new,min} = \eta_{ref,min} \frac{\frac{1}{\phi_{new}}\left(\frac{1}{\tanh(3\phi_{new})} - \frac{1}{3\phi_{new}}\right)}{\frac{1}{\phi_{ref}}\left(\frac{1}{\tanh(3\phi_{ref})} - \frac{1}{3\phi_{ref}}\right)}. \quad (15)$$

Then, the new effectiveness factor (η_{new}) can be calculated as follows, assuming that the tendency towards $\eta = 1$ will be proportional to the tendency of η_{min} to unity (no subgrid correction):

$$\eta_{new} = 1 - \left(1 - \eta_{ref}\right)\frac{\left(1 - \eta_{new,min}\right)}{\left(1 - \eta_{ref,min}\right)}. \quad (16)$$

It was found that the suggested hypothesis holds well and that this approach is essential to accurately model reactions that are not simple first-order solids catalyzed reactions in the fTFM. Consequently, the proposed approach was used to model the reactions in the present study. The complete development and verification of the generalized reactive fTFM closure is presented in the Appendix A.

Finally, it can be noted that the effectiveness factor closure presented here does not account for the Stefan flow (one mole of methane produces three moles of gas products) occurring in the fuel section of the ICR and investigation of this topic is recommended for future work. However, considering that the reduction reactions are extremely fast (see Section 3.1.1) and occur only near the inlet, it is not expected to have a large impact on the overall reactor behavior.

2.1.3. Simulation Geometry and Mesh

Figure 2 shows the reactor geometry that was considered for the ICR. In the base case, the reactor consists of a cylinder with a height of 6.92 m and a diameter of 3.46 m. These sizes were selected to yield a fluidization velocity of roughly 1 m/s in the freeboard, which is a typical value for vigorous bubbling fluidization. An aspect ratio of 2, typical of fluidized beds, was chosen. A thin wall separates the reactor into the reduction and oxidation sections, consisting of a 2 m high vertical section at the center of the bed and a section sloping at an angle of 30° with the vertical axis to the reactor wall to ensure that solids will not deposit on this surface. Two ports allow the oxygen carrier to circulate between the sections. Reduced oxygen carrier travels through the bottom port (height of 0.6 m) to the air section, whereas oxidized oxygen carrier is carried through the top port (terminating 0.7 m above the bottom of the reactor) to the reduction section. In the base geometry, the width of the square ports (see Figure 2b) is 20 cm. The gas outlet from each section was sized to yield a velocity of roughly 50 m/s, accounting for the much larger flow rate in the air section, which is a typical value for gas transport.

A cut-cell mesh was used to mesh the complex ICR geometry. Long simulations, in the order of 2500 simulated seconds, were necessary to achieve steady reactor behavior; therefore, the average grid size was chosen to yield a coarse mesh of approximately 50,000 cells. A minimum of five cells across the gaps in the ports and the outlets were specified to resolve the most important flow gradients.

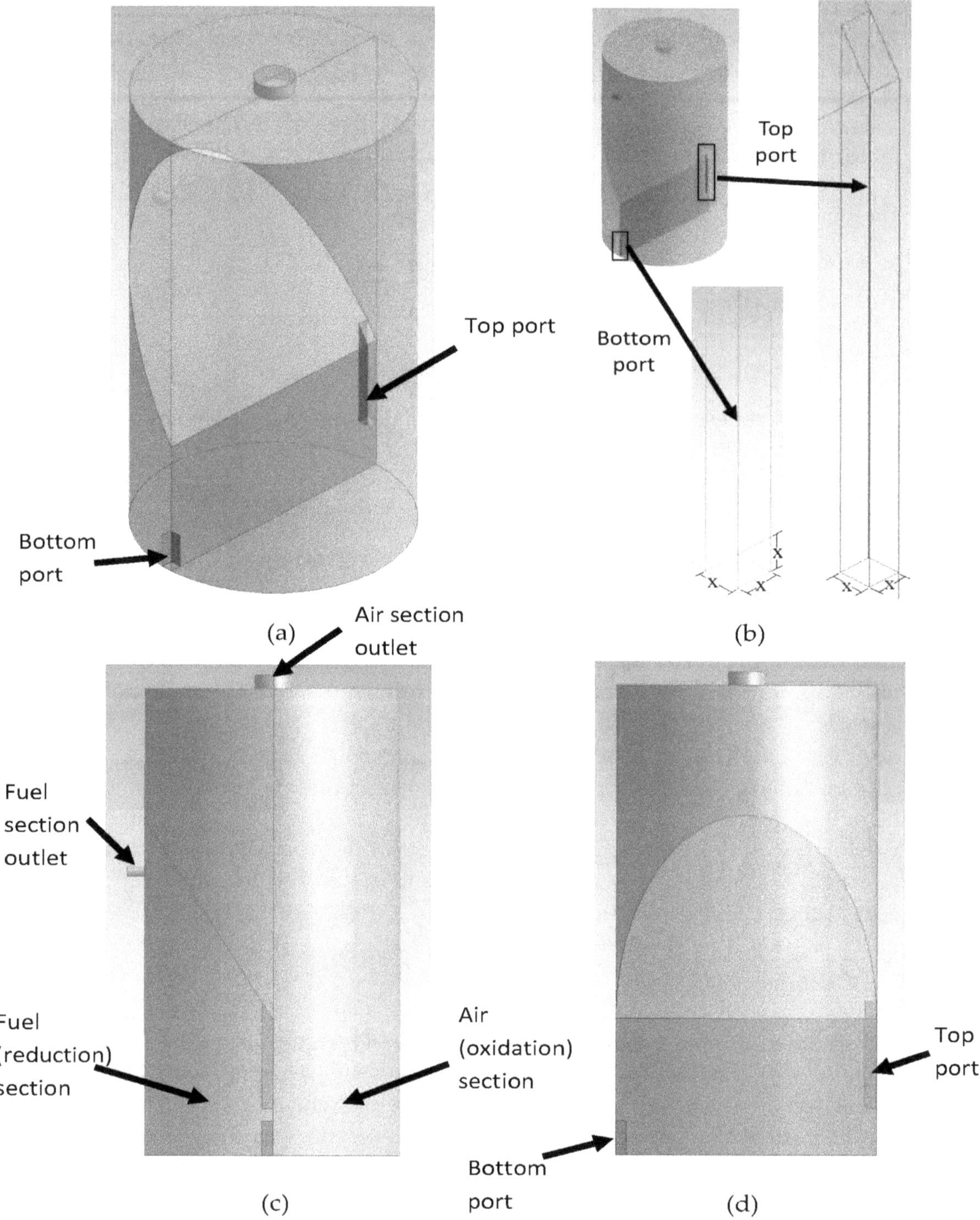

Figure 2. Simulated geometry for the internally circulating reactor (ICR): (**a**) isometric view; (**b**) details of the ports connecting the reactor sections, where x is the port width; (**c**) front view; (**d**) side view.

2.1.4. Reactor Operating Conditions

NiO particles supported on Al_2O_3 were used as oxygen carrier. The oxygen carrier particles were considered to have a diameter of 150 μm, a typical value for bubbling fluidization, a density of 3446 kg/m^3, and an active mass fraction of 0.4 [45]. It can be noted that the reactor model assumes monodisperse particles, due to the complexity of accounting for particle size distributions in fTFMs and due to the limited state of development of subgrid closures accounting for polydispersity [50]. Additionally, the simulations assume the particle density to be constant during the reactor operation, since the density changes will be small (mostly less than 5%) due to the high inert content of the oxygen carrier. The loading of the bed corresponds to an initial bed height of 1 m at a solids volume fraction of 0.6.

Fluidization gas was added uniformly at the bottom of the reactor to the two reactor sections, assuming a perfect gas distributor. Additionally, the effect of the inlet conditions on the subgrid behavior were not accounted for in the fTFM closures. These simplifications are necessary, since none of the state-of-the-art fTFMs have thus far included these effects in their closures. However, considering the large dimensions of the reactors considered, inlet effects are expected to have a relatively small influence on the overall reactor behavior, thereby minimizing the error associated with these simplifications. The other inlet boundary conditions are listed in Table 1 (note that the natural gas used in the process simulations was replaced with an equivalent amount of methane in the reactor simulations).

Table 1. Summary of the conditions for the inlets of the two reactor sections.

Inlet	Oxidation	Reduction
Mass flow rate (kg/s)	41.15	0.698
Temperature (°C)	422	434
Composition	Air	Methane

Uniform pressure outlet boundary conditions were considered for the fuel and air section outlets. For the air section outlet, a pressure of 18 bar (absolute) was considered, which results from the air compressor pressure ratio of 18 [51] employed in the process simulations, which is a typical value for standard, large-scale, F-class gas turbines [52]. For the fuel section outlet, a relatively small overpressure relative to the air section outlet was employed to achieve a target flow rate. This is discussed in more detail in Section 3.1.2.

A no-slip boundary condition was specified for the gas at the walls, whereas partial slip boundary conditions with a specularity coefficient of 0.1 was employed at the walls, based on the model of Johnson and Jackson [53]. It can be noted that, technically, a subgrid closure is required for the particle–wall interaction. However, such closures have not yet been developed in the fTFM community and were therefore neglected in the present study. The effect of the particle–wall boundary condition is expected to be small for the large reactor dimensions considered in the present study; therefore, neglecting the sub-grid effects is a reasonable assumption.

2.1.5. Solver

The reactor simulations were performed in the commercial CFD solver, ANSYS FLUENT 19.2, using user defined functions to implement the subgrid closures of the fTFM. The phase-coupled SIMPLE algorithm [54] was used for pressure-velocity coupling, and all other equations were discretized based on the QUICK scheme [55].

2.2. Process Modeling

This study conducted an economic assessment of the ICR integrated into a natural gas combined cycle (NGCC) power plant, as recently evaluated for CO_2 capture using CLC [34]. The interested reader is referred to that study for details about the process modeling methodology. One important change from this previous work was the inclusion of the gas leakage between reactor sections in the ICR. The mixing between the outlet streams of the oxidation and reduction reactors was adjusted to yield 95% CO_2 capture and purity (molar percentage and dry basis), based on the reactor simulations (see Section 3.1.3) for the conditions considered.

The layout of the simulated plant is shown in Figure 3, where 20 parallel ICR reactors were needed to accommodate the required air throughput. Natural gas is pre-heated and fed to the fuel section of the ICR reactors where it is converted to CO_2 and H_2O, which is expanded to generate some power.

After H_2O is condensed out, the remaining CO_2 is compressed and pumped to 110 bar for transport and storage. The air section of the ICR replaces the combustor for the main gas turbine. Air from the main compressor reacts with the reduced oxygen carrier in a highly exothermic reaction and is heated to 1150 °C in the base case. This temperature was selected based on material limitations, and a sensitivity analysis of this value was performed in the economic evaluation in Section 3.2. The hot depleted air stream is then expanded in the main gas turbine before being sent to a heat recovery steam generator for extra power production using a steam cycle. The results of this plant were compared to the reference NGCC plant detailed in Khan et al. [34].

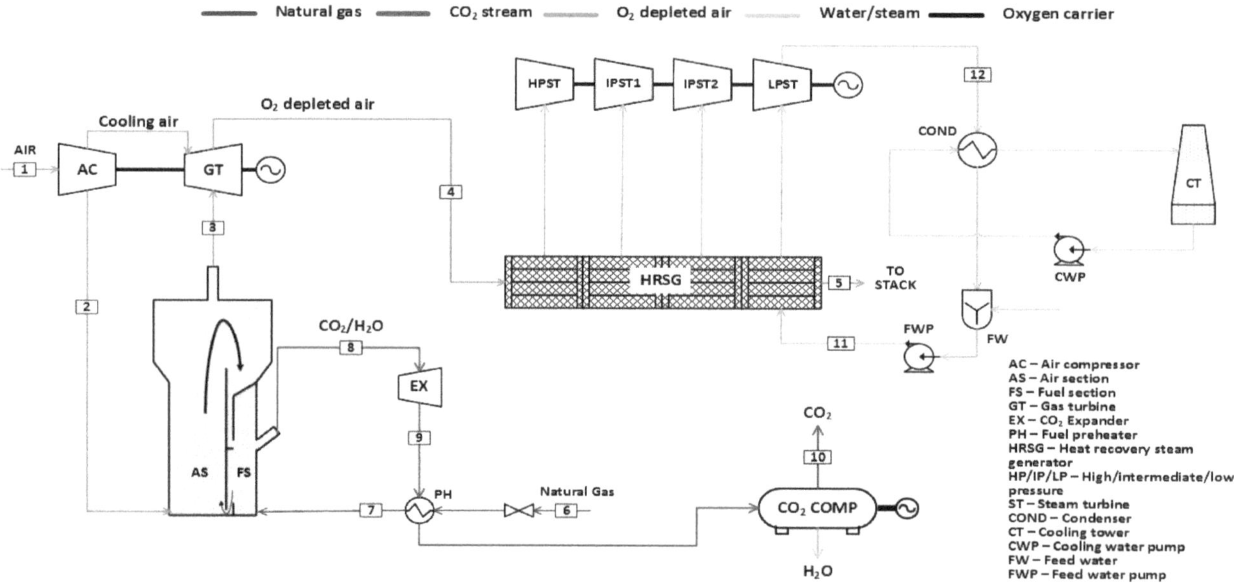

Figure 3. Process flowsheet of the ICR integrated into a combined cycle. It can be noted that the reactor in the flowsheet represents a cluster of ICR reactors and that the outputs from these reactors were combined in stream 3.

2.3. Economic Assessment

Capital costs: The total cost of the combined power cycle was obtained directly from the PEACE component in Thermoflex. This includes direct component costs and several additional cost components accounting for construction, engineering, contingencies, and other cost components.

Costs related to the CO_2 compressors and intercoolers were estimated using installed cost data from Aspen Plus. This cost was increased by approximately 74% to account for engineering, contingencies, and owner's costs, based on the methodology of Gerdes et al. [56].

The ICR capital costs were estimated based on cost correlations for process vessels from Turton et al. [57]. Each ICR was composed of two process vessels: (1) an inner vessel to carry the temperature, attrition, and corrosive loads constructed from an expensive Ni-alloy, and (2) a thick pressure shell carrying the pressure load constructed from carbon steel. An insulation layer of 0.4 m thickness was inserted between these two vessels. To account for the relatively complex ICR geometry, the cost of the inner vessel was increased by a factor of three. This was a somewhat arbitrary adjustment, and a sensitivity of total plant economics to ICR cost is therefore presented later. Costs for auxiliaries and contingencies were subsequently added according to Turton et al. [57] to yield the total reactor cost. A breakdown of the different components of the cost of the 20 ICR units required for the base case is shown in Figure 4. It can be noted that the capital cost associated with the oxygen carrier was only for the initial loading (replacement of the oxygen carrier was considered under operating and maintenance costs). Further, the number of ICR reactors selected to deliver the required process throughput at the reactor operating conditions is specified in Section 2.1.4.

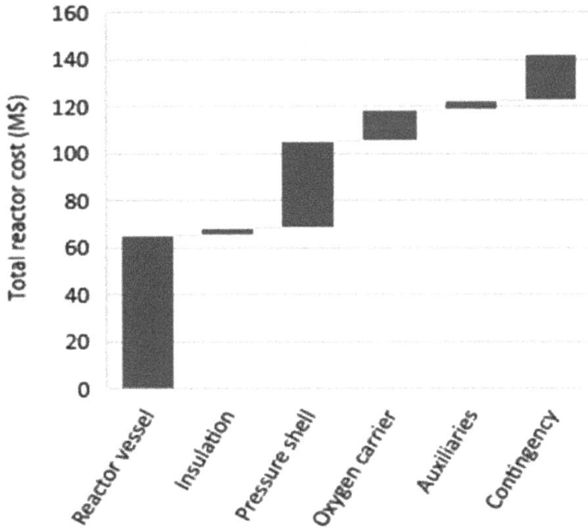

Figure 4. Breakdown of the cost of 20 ICR units.

The total costs of these three main components were then added together to yield the total plant cost. All costs are reported in 2019 US dollars using the chemical engineering plant cost index (CEPCI).

Operating and maintenance (O&M) costs: Fixed O&M labor costs were calculated assuming 11 personnel per shift for the NGCC reference plant and 13 personnel per shift for the ICR plant using the methodology of Peters and Timmerhaus [58]. A $45/hour rate was used with appropriate increases for benefits, maintenance labor, and overheads. In addition, 1.5% of total plant costs per year was added for insurance and property taxes. The key assumptions for variable O&M costs are summarized in Table 2. Costs for the oxygen carrier [59] and water [60] were taken from the literature, whereas the oxygen carrier lifetime was specified based on discussions with catalyst suppliers. CO_2 transport and storage costs can vary widely based on the transport distance and the type of transport and storage, but a reasonable average value was selected based on costs provided in two IEA reports [61,62]. Natural gas prices are known to vary widely, and a value representative of Europe was assumed for this study.

Table 2. Variable O&M cost assumptions.

Natural Gas	8 $/GJ
Oxygen carrier (OC)	15 $/kg
OC replacement period	2 years
Process water	2 $/m^3
Cooling water	0.35 $/m^3
CO_2 transport and storage costs	12 $/ton CO_2

Capital and O&M costs were then used to calculate the levelized cost of electricity (Equation (17)) and the CO_2 avoidance costs (Equation (18)) using a discount rate of 8%, a plant economic lifetime of 30 years, and a construction period of 2 years for NGCC and 3 years for the ICR plants (investment is assumed to be linear over the construction period).

$$\text{LCOE } (\$/\text{MWh}) = \frac{\sum_{t=1}^{m} \frac{I_t+M_t+F_t}{(1+r)^t}}{\sum_{t=1}^{m} \frac{E_t}{(1+r)^t}}, \tag{17}$$

$$\text{CAC } (\$/\text{ton}) = \frac{LCOE_{CCS} - LCOE_{ref}}{e_{ref} - e_{CCS}}. \tag{18}$$

Here, the summations are done for each year during construction and operation (t) up to the end of the plant economic lifetime (m). I is the investment expenditures, M is the O&M expenditures, F is the fuel expenditure, E is the electricity generation, r is the discount rate, and e is the plant-specific emissions (ton/MWh).

3. Results

This section outlines how the fTFM described in Section 2.1 was first used to optimize and size an industrial-scale ICR reactor for power production with integrated CO_2 capture. Subsequently, the ICR process was then evaluated economically using the reactor size and performance suggested by the simulations.

3.1. Reactor Optimization

3.1.1. Characteristics of ICR Operation

In this study, plots and animations from the reactor simulations were used to introduce important characteristics of ICR operation for a typical case. Firstly, Figure 5 (as well as the associated Video S1 in the Supplementary Material) demonstrates the circulation of the oxygen carrier between the two reactor sections. In the reduction section, the relatively low fluidization velocity from the fuel feed results in a dense bubbling bed. Due to the very fast reaction of the oxygen carrier with the methane, most of the conversion takes place near the inlet, leading to a highly reduced oxygen carrier. However, the mixing is very fast in the fluidized bed and a relatively uniform distribution of the oxygen carrier is rapidly attained in the rest of the bed on the fuel side. Oxygen carrier particles, reduced by the fuel, pass through the bottom port to the air (oxidation) section, where they are rapidly oxidized and mixed into the rest of the particles. Owing to the much larger molar flow rate on the air side, a more vigorous fluidization occurs, lifting the particles to the freeboard, including a diameter expansion (which helps to reduce particle elutriation), and allowing them to pass back to the fuel section through the top port. Again, the oxidized particles mix rapidly into the reduction side bed, where they are reduced by the fuel.

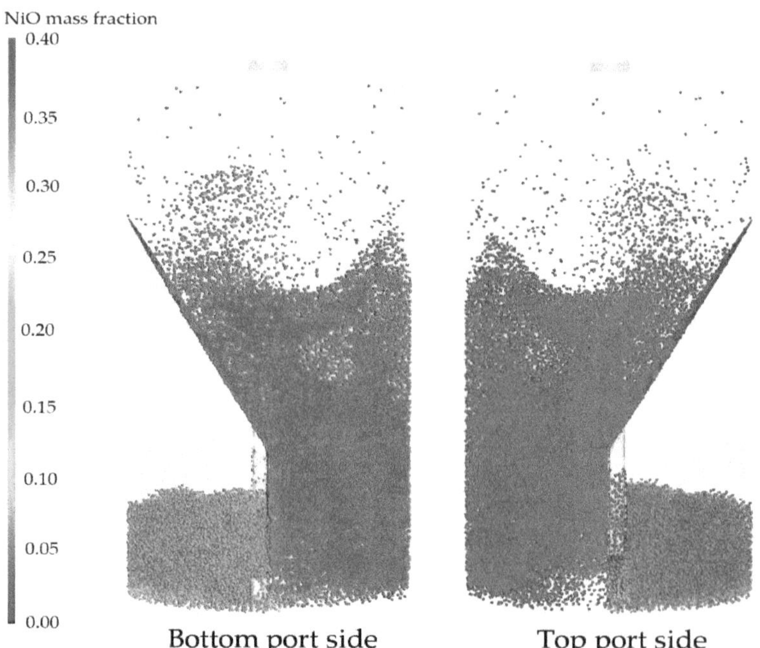

Figure 5. Particle plot of the instantaneous NiO mass fraction. A corresponding animation is provided in Video S1. Note that the particles pictured are tracers following the continuous solids flow for visualization purposes and do not influence the simulation solution.

Some further comments can be made about the nature of the solids flow through the ports. In the bottom port, the flow is quite dense, with a time-average solids volume fraction of about 0.5. The flow through the top port is more dilute, with a time-averaged solids volume fraction between 0.3 and 0.4. The animations show some transient fluctuations of solids in both ports; therefore, the flow in not completely steady and there is a risk of backflow, which might reduce the reactor performance. It may be noted that no problems with blockage of the ports have been experienced during extensive experimental evaluations of the ICR concept [14,15].

Figure 6 (and Video S2) shows that the solids circulation between the reactor sections is associated with undesired gas leakage—CO_2 leaks from the fuel section to the air section, reducing the CO_2 capture efficiency of the reactor, and N_2 leaks from the air section to the fuel section, reducing the purity of the CO_2. One of the most important criteria for designing and operating the ICR is therefore to minimize the amount of gas leakage between the reactor sections, while maintaining sufficient oxygen carrier circulation to ensure that the fuel is completely converted in the reduction sector.

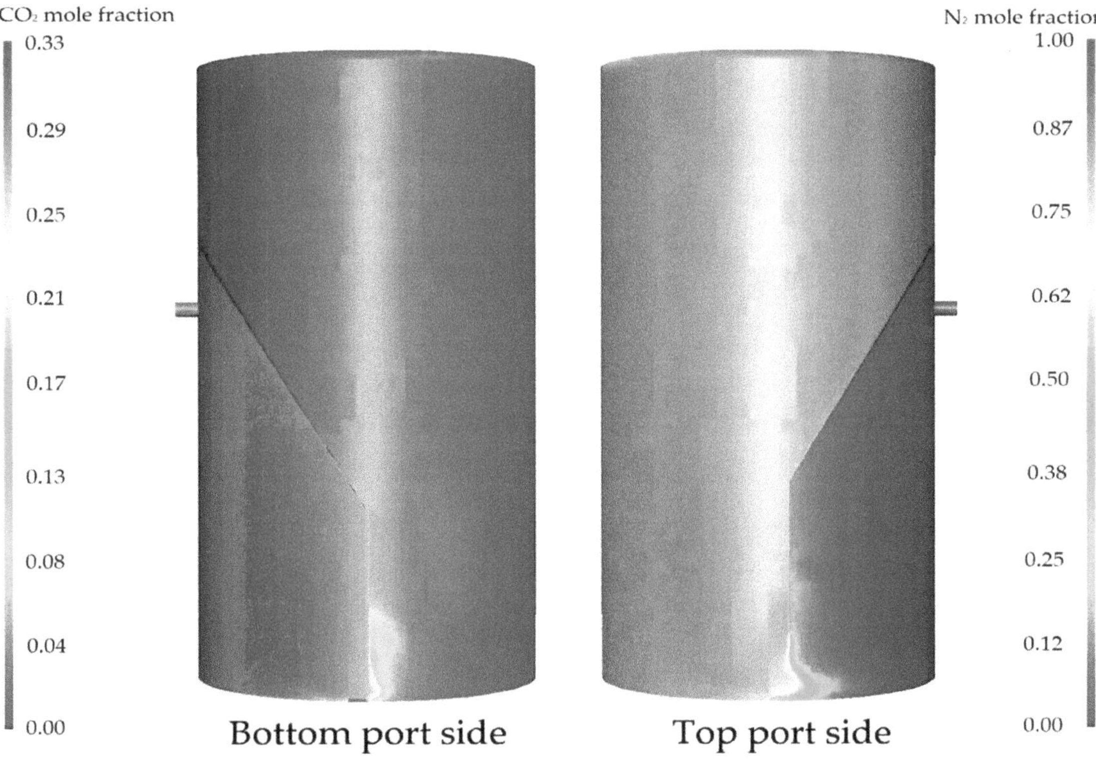

Figure 6. Contour plot of the instantaneous CO_2 and N_2 mole fractions at the outer wall of the ICR showing the undesired gas leakage between the two sections of the reactor. A corresponding animation is provided in Video S2.

Many design and operating parameters can influence the ICR performance. These include, but are not limited to, the solids loading, the particle size and distribution, the gas flow rates to the reactor sections, the operating pressure and temperature, and several dimensions of the reactor and internals. Due to the complexity of simultaneously optimizing these parameters, the scope of the present study was limited to three important factors that will be investigated in the subsequent sections:

- The pressure difference between the two reactor outlets, which can be used to control the solids distribution between the reactor sections, as well as the solids circulation.
- The size of the ports connecting the reactor sections, which can be sized to allow sufficient solids circulation while limiting undesired gas leakage.

- The overall reactor size, which primarily determines the amount of solids elutriation from the reactor.

3.1.2. Reduction Section Overpressure

During ICR operation, the most practical way to control the reactor behavior would be to tune the pressure difference between the reduction and oxidation section outlets. For example, applying an overpressure at the reduction side outlet will lead to more gas exiting the reactor on the oxidation side. To satisfy the mass balance of the reactor, this means that relatively more gas must pass through the bottom port to reach the oxidation section. As is shown in this section, this gas flow influences the solids circulation between the sections, as well as the distribution of solids in the two sections. Both these factors have a critical effect on the reactor performance.

To better understand this behavior, ICR simulations were performed at different overpressures applied at the reduction section outlet. Specifically, reduction outlet flow ratios (ROFR) from 0.96 to 1 were investigated. The ROFR is defined as the ratio of the reduction outlet molar flow rate to the ideal outlet molar flow rate that would occur in case of no gas leakage between the reactor sections and complete fuel conversion. A lower ROFR implies a higher overpressure in the fuel section, with the ROFR = 0.96 case corresponding to an overpressure of 0.34 bar. It can also be noted that the figures and animations presented in the previous section are for the case of ROFR = 0.98.

Figure 7 (and Video S3) shows the effect of the reduction section outlet overpressure on the ICR behavior. Firstly, it is interesting to note from the solids volume fraction values that, despite the large grid sizes employed, a substantial amount of phase segregation is still resolved in the more vigorously fluidized air section. In the slowly fluidized fuel section, the resolved solids distribution is nearly homogenous, and the effects of particle clusters and bubbles are therefore nearly completely accounted for in the subgrid closures of the fTFM.

Increasing the overpressure (corresponding to lower ROFR values), more gas will pass through the bottom port to the air section, which also increases the solids flow rate through the bottom port. This causes the bed loading on the oxidation side to increase, which creates the hydrostatic pressure buildup required to achieve solids flow through the top port against the overpressure imposed in the fuel section.

Therefore, cases with a lower ROFR will reach a pseudo-steady state (where the time-averaged solids flow rates through the top and bottom ports are equal) with a larger fraction of the oxygen carrier on the oxidation side, as shown in Figure 8b. The solids elutriation rate (Figure 8a), which occurs almost entirely from the oxidation side, is therefore greatest at lower ROFR values.

Furthermore, Figure 8a shows that a maximum solids circulation rate occurs at an ROFR of 0.98. At first, when decreasing the ROFR from 1, the solids circulation rate increases due to more gas flow through the bottom port, thereby entraining more solids, as well as more solids flowing through the top port as a result of the higher bed loading on the air side. However, as the ROFR further decreases, the bed on the fuel side becomes so low that the solids barely reach the top of the bottom port, thus limiting the achievable circulation rate through the bottom port. Regular backflow through the bottom port also starts to happen during these cases, since the low bed height on the fuel side does not provide sufficient hydrostatic pressure to maintain a steady flow through the bottom port. If the solids circulation rate becomes too low, significant fuel slip, that is, incomplete fuel conversion, starts to occur (Figure 8c), since not enough oxygen is available in the fuel section to convert all of the fuel and the low bed height reduces the gas residence time in the bed.

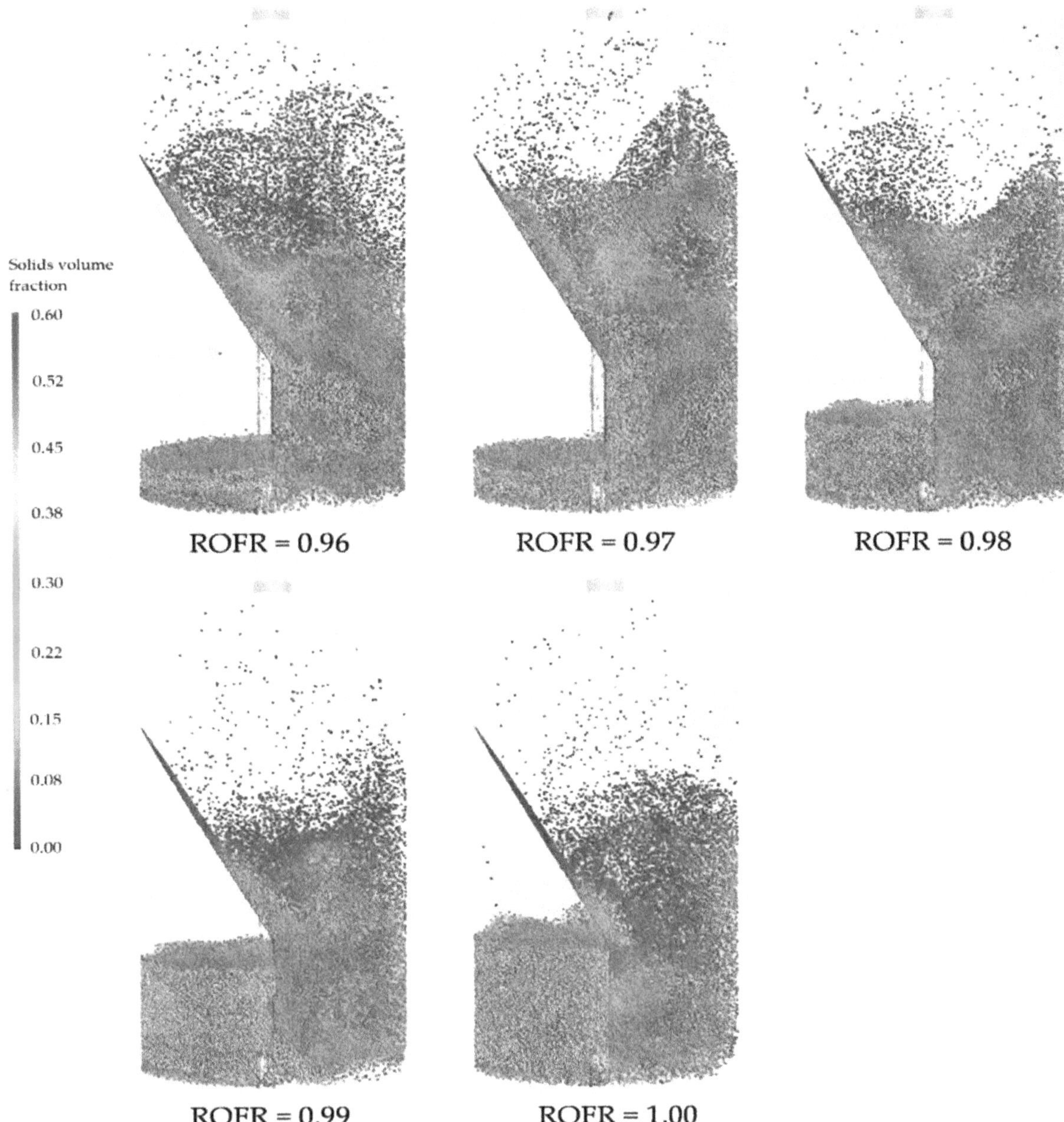

Figure 7. Instantaneous particle plot of the cases with varying reduction outlet flow ratios, where the particles are colored by the particle volume fraction. A corresponding animation is provided in Video S3. Note that the particles pictured are tracers following the continuous solids flow for visualization purposes and do not influence the simulation solution.

Figure 8c also shows that the CO_2 capture percentage decreases as the ROFR decreases, due to the increased gas flow through the bottom port, allowing more CO_2 to exit with the depleted air at the oxidation section outlet. The CO_2 purity behavior is more complex—it remains relatively constant when lowering the ROFR from 1 to 0.98, despite the increasing solids flow rate through the top port. This indicates that the gas-to-solids leakage ratio through the top port increases at high ROFR values, resulting in more gas leakage per unit of solids circulation. The purity increases in the ROFR = 0.97 case due to the lowering solids flow rate but decreases in the ROFR = 0.96 case due to backflow through the bottom port.

Based on the results in this section, the ROFR = 0.98 case (corresponding to a reduction outlet overpressure of 0.26 bar) was chosen for further investigation, primarily due to the high solids circulation rate that was obtained. Furthermore, this case showed a good compromise between decreasing solids elutriation and decreasing CO_2 capture with decreasing ROFR.

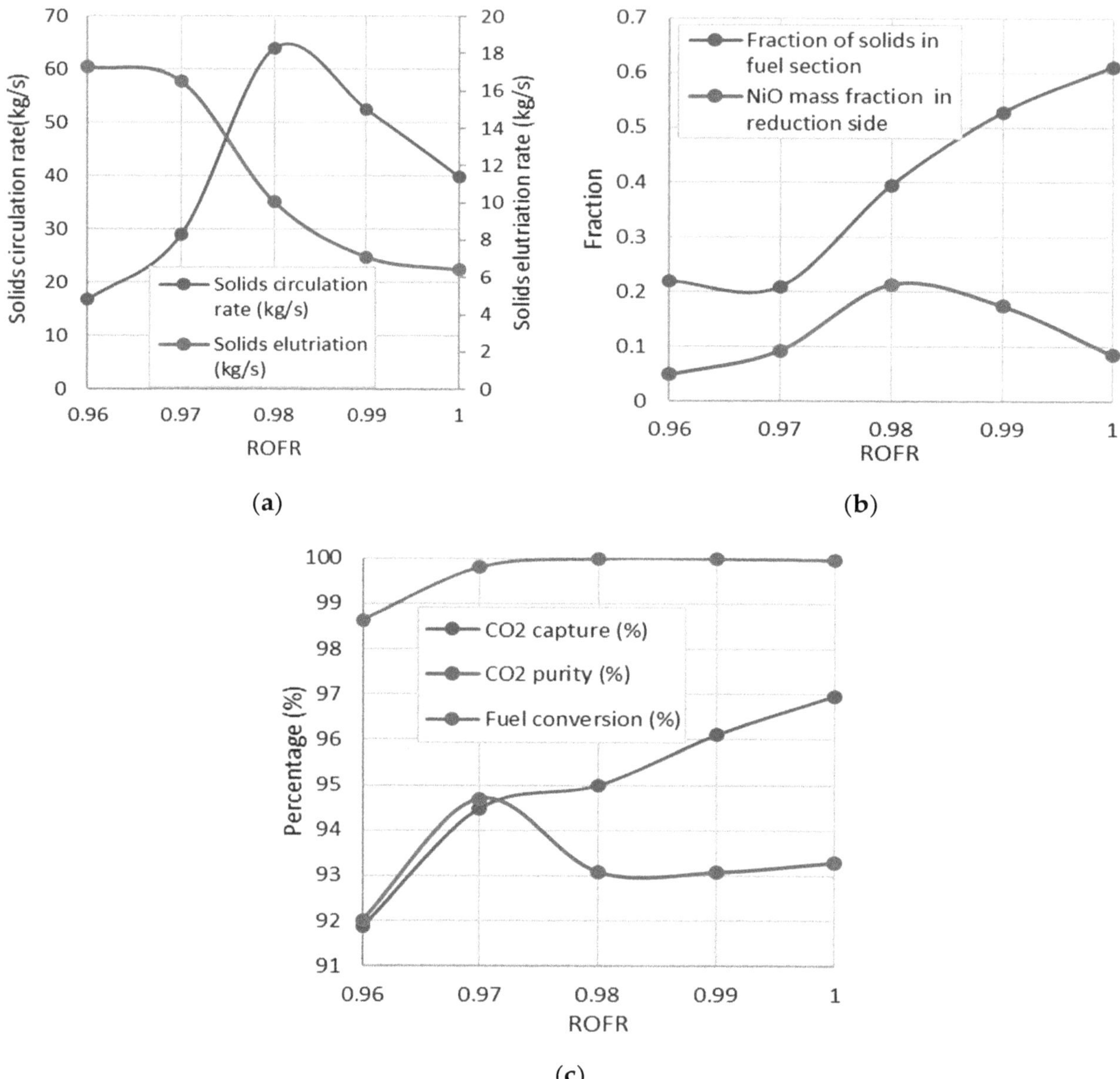

Figure 8. Summary of important time-averaged properties of the ICR as a function of the reduction outlet flow ratio: (**a**) solids circulation and elutriation rates; (**b**) fraction of solids and NiO mass fraction in the reduction side; (**c**) CO_2 capture, CO_2 purity (molar percentage on a dry basis), and fuel conversion.

3.1.3. Port Size

The previous section revealed that the primary criterion for achieving complete fuel conversion is a sufficient solids circulation rate to transport enough oxygen to the fuel reactor to oxidize the methane. However, since the ROFR = 0.98 case from the previous section had a much higher than necessary solids circulation rate, there is the potential to further increase the reactor performance by decreasing the size of the ports (dimension x in Figure 2). This will decrease the solids circulation rate between the reactor sections, but also decrease the associated undesired gas leakage.

Simulations were therefore performed at an ROFR of 0.98 while decreasing the port size dimensions, as shown in Figure 9. As expected, the results showed a decreasing solids circulation rate with decreasing port size. Consequently, the average NiO mass fraction in the fuel section was also reduced. In the case with a port size of 16 cm, not enough NiO was present in the reduction section to fully oxidize the fuel, leading to a large fuel slip, which would be unacceptable in practice.

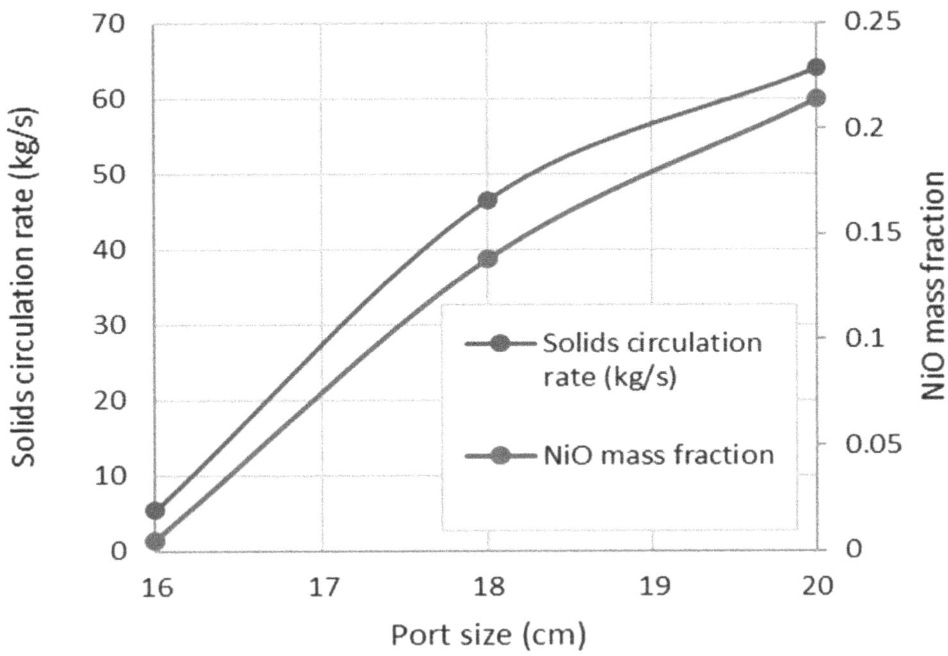

Figure 9. Plot of the solids circulation rate and the NiO mass fraction in the fuel section with changing port size.

The case with ROFR = 0.98 and a port size of 18 cm was consequently chosen as the best ICR case. Compared to the case with ROFR = 0.98 and a port size of 20 cm, the CO_2 capture efficiency increased from 95.0% to 95.7%, and the CO_2 purity increased from 93.1% to 95.2% while still preserving complete fuel conversion. Consequently, CO_2 capture efficiencies and purities of 95% are assumed as reasonable for the economic assessment in Section 3.2.

3.1.4. Reactor Size

One of the main criteria for reactor sizing is usually to ensure complete reactant conversion, with the increasing reactor diameter and height both serving to increase the gas residence time (the latter by decreasing the superficial velocity of a fixed mass flow rate of gas) for complete conversion. However, as previously noted, the oxidation reaction of methane with NiO is very fast for the temperatures considered for ICR operation. Additionally, the height of the bed on the fuel side is limited to a minimum due to the height of the bottom port. Consequently, the solids circulation rate between the reactor sections is primarily responsible for ensuring complete fuel conversion, and the overall reactor size is of lesser importance.

Nonetheless, the overall reactor size has an important effect on the particle elutriation. In the best case investigated so far (ROFR = 0.98 and a port size of 18 cm), the predicted solids elutriation rate was 11 kg/s, corresponding to 2.0% of the total solids loading per hour. Consequently, a rather large amount of elutriated solids will have to be separated using cyclones and filters downstream of the reactor. The amount of solids elutriation can be reduced by increasing the reactor height (more space for solids to fall back down) and diameter (lower superficial gas velocities).

To investigate the effect of increasing the reactor size, all the reactor dimensions were multiplied by a scaling factor (SF) from 1 to 1.2 (compared to the base case geometry discussed in Section 2.1.3. with a port size of 18 cm). Figure 10 (and Video S4) visualize the change in reactor behavior with increasing scaling factor, showing clearly that a larger reactor leads to a slower fluidization and less solids elutriation on the air side.

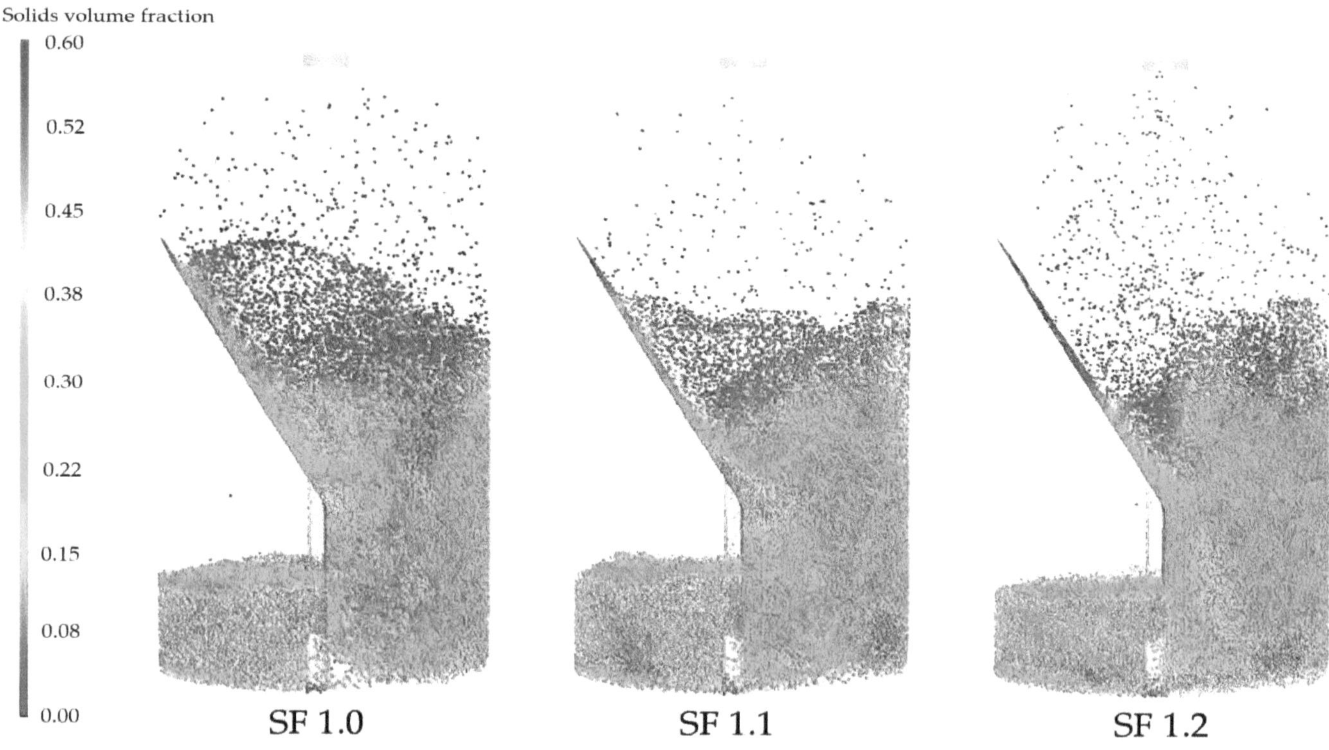

Figure 10. Instantaneous particle plot of the cases with reactor scaling factors ranging from 1 to 1.2, where the particles are colored by the particle volume fraction. A corresponding animation is provided in Video S4. Note that the particles pictured are tracers following the continuous solids flow for visualization purposes and do not influence the simulation solution.

The decreasing elutriation rate is quantified in Figure 11a, with the solids elutriation (in percentage of solids loading per hour) decreasing from 2.0% to 0.49% at a scaling factor of 1.1, and 0.20% at a scaling factor of 1.2. However, it was also found that the CO_2 capture and purity decreases with increasing scaling factor (Figure 11c). This is because the port sizes were also scaled along with the other reactor dimensions, leading to an increase in the solids circulation rate (Figure 11a) and the associated gas leakage. However, the results also showed an increase in the average NiO fraction in the fuel section (Figure 11b), indicating higher than necessary solids circulation rate. As a result, it is expected that CO_2 capture efficiencies and purities of more than 95% can be obtained by again decreasing the port size to achieve a solids circulation rate close to the minimum required for complete fuel conversion, as was done in Section 3.1.3.

In summary, the preferred ICR reactor size will in practice be determined by the trade-off in increasing capital cost of the reactor and the decreasing cost of handling elutriated solids with increasing reactor size. The cost of the former is quantified in the subsequent economic assessment, whereas quantification of the latter is left for a future study.

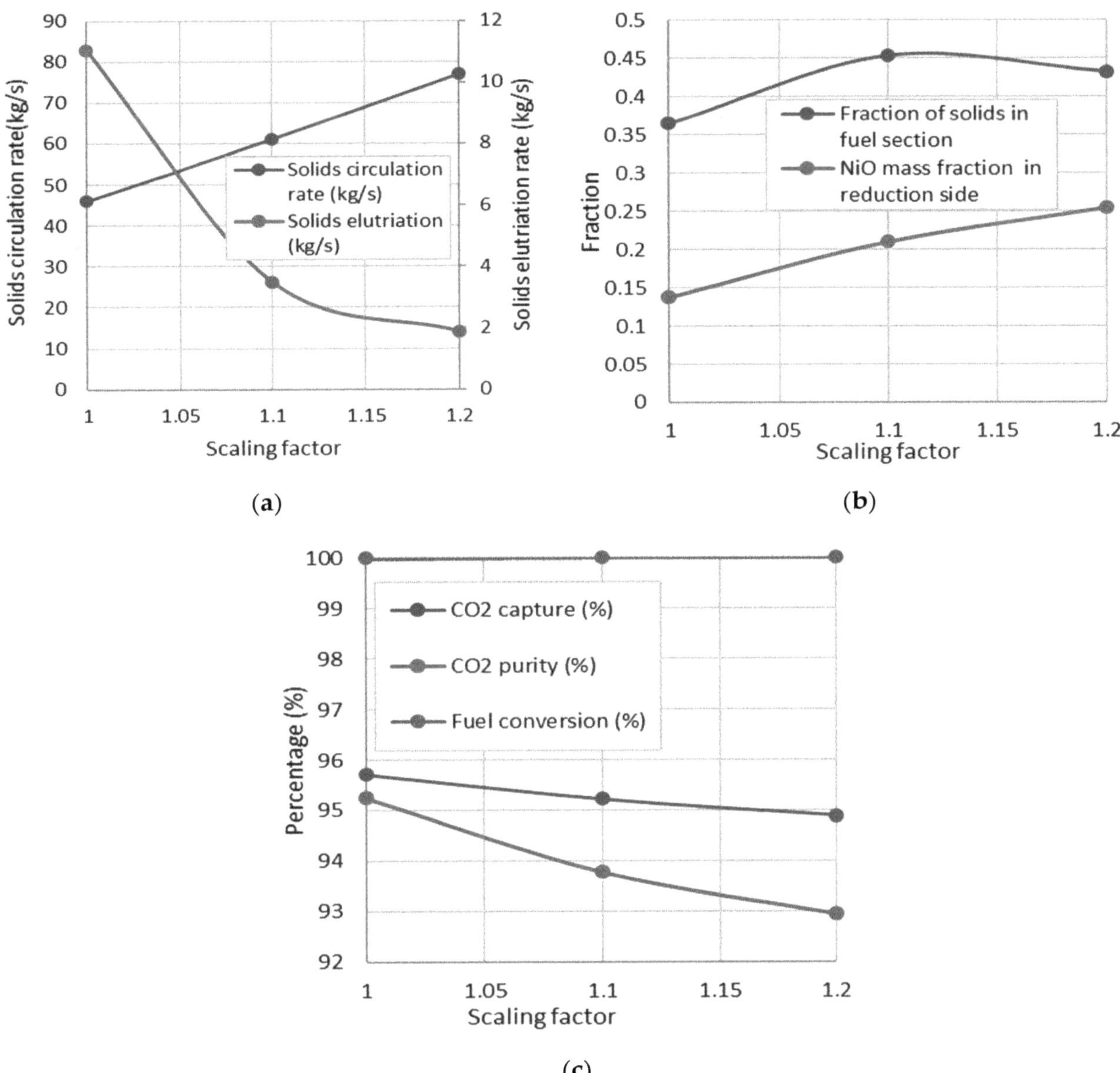

Figure 11. Summary of important time-averaged properties of the ICR as a function of the reactor size scaling factor: (**a**) solids circulation and elutriation rates; (**b**) fraction of solids and NiO mass fraction in the reduction side; (**c**) CO_2 capture, CO_2 purity, and fuel conversion.

3.2. Economic Evaluation

In this section, an ICR-based NGCC plant is evaluated economically based on the CO_2 capture and purity estimated in the reactor simulations, as well as the size of reactor that was found to be adequate (SF = 1). In addition, an unabated NGCC plant is assessed as a reference.

Three ICR cases were investigated with different assumptions about the maximum reactor temperature. In combined cycles, it is crucial to maximize the turbine inlet temperature to maximize net electric efficiency and minimize the required air flow rate through the system. However, the reactor, oxygen carrier, and downstream filters will impose constraints on the maximum allowable temperature. The oxygen carrier material may be the most important constraint. To date, the highest successfully demonstrated operating temperature has been a NiO-based oxygen carrier successfully operated at 1185 °C [44]. However, modern gas turbines can operate with combustor outlet temperatures well over 1600 °C, implying that this is a very important constraint. The economic impact of this constraint will be illustrated in this work.

To this end, the cases listed in Table 3 were evaluated in this study. It can be noted that the reactor simulations were performed only for the ICR-1150 case, since it was expected that the gas leakage would not vary much between the cases at different temperatures.

It is immediately evident from the results in Table 3 that the achieved plant efficiency is strongly dependent on the ICR outlet temperature. When the temperature can reach up to 1300 °C, an attractively low energy penalty of 3.6% points is attained, but this increases to an unacceptable 12.3% points in the case with an ICR outlet temperature of 1000 °C. Since fuel costs are typically the major cost component in a natural gas fired plant, this is expected to have a major impact on the economic performance.

Table 3. Performance of the different plants evaluated in this study.

Plant	NGCC	ICR-1000	ICR-1150	ICR-1300
Combustor/reactor outlet temperature (°C)	1416	1000	1150	1300
Thermal input (MW)	765.0	697.5	697.5	697.5
Gas turbine (MW)	292.5	220.9	223.0	235.5
Steam turbine (MW)	161.6	84.3	106.9	121.5
CO_2 expander (MW)		36.3	41.4	46.4
CO_2 compressors (MW)		−15.9	−15.7	−15.5
Auxiliaries (MW)	−8.5	−4.6	−6.0	−6.6
Net power (MW)	445.6	321.0	349.7	381.3
Net electric efficiency (%)	58.3	46.0	50.1	54.7
CO_2 intensity (kg/MWh)	352.2	22.3	20.5	18.8

The levelized cost of electricity and CO_2 avoidance cost for the different cases are shown in Figure 12. Clearly, the ICR outlet temperature has a large impact on the economic performance of the plant. Interestingly, the relative increase in capital costs from the ICR-1300 case to the ICR-1000 case (50%) is greater than the relative increase in fuel costs (25%). This is due to the large increase in air flowrate required by the cases with lower ICR outlet temperatures. The larger air flowrate increases the costs of ICR reactors, gas turbines, and heat recovery steam generators, adding to the increase in specific capital costs caused by a reduction in net electric efficiency.

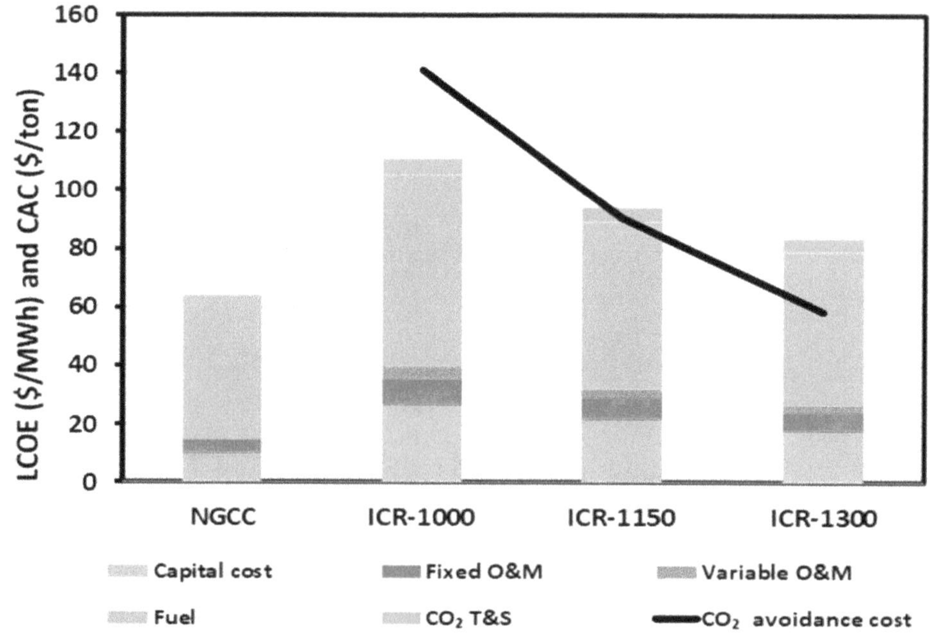

Figure 12. Levelized cost of electricity (LCOE) and CO_2 avoidance costs (CAC).

CO_2 avoidance costs range from $58/ton to $141/ton. For comparison, a thorough review by Rubin et al. [63] found the representative CO_2 avoidance cost for NGCC plants with post-combustion capture to be $107/ton (after adjusting to 2019 currency and adding CO_2 transport and storage costs). The ICR-1300 case, if it is technically possible, therefore appears attractive relative to this benchmark, whereas the $91/ton CO_2 avoidance cost of the more realistic ICR-1150 case offers marginal benefits.

As outlined earlier, all the reactor sizes investigated performed well in terms of fuel conversion, and the primary difference was the amount of particle elutriation. Preventing particle elutriation will require the addition of a cyclone (possibly an internal cyclone placed in the ICR freeboard), including smaller reactor sizes needing cyclones with a higher separation efficiency to keep escaped fines constant. Inclusion of such a cyclone was not studied here, but the effect of the change in reactor size was investigated.

Figure 13 shows that higher reactor costs have a relatively small impact on overall economic performance of the plant. An increase in reactor costs from 108 to 233 M$ only increased the levelized cost of electricity by $5.5/MWh and the CO_2 avoidance cost by $16.5/ton. The uncertainty in the reactor cost estimation was also highlighted in the methodology description, particularly the factor 3 that was used to account for the cost increase of including the ICR internal structure in a Ni-alloy process vessel. For perspective, the range of reactor costs shown in Figure 13 is equivalent to a wide range in this tuning factor of 1.7–6.4, indicating that the overall plant performance is not overly sensitive to uncertainties in the ICR cost assessment.

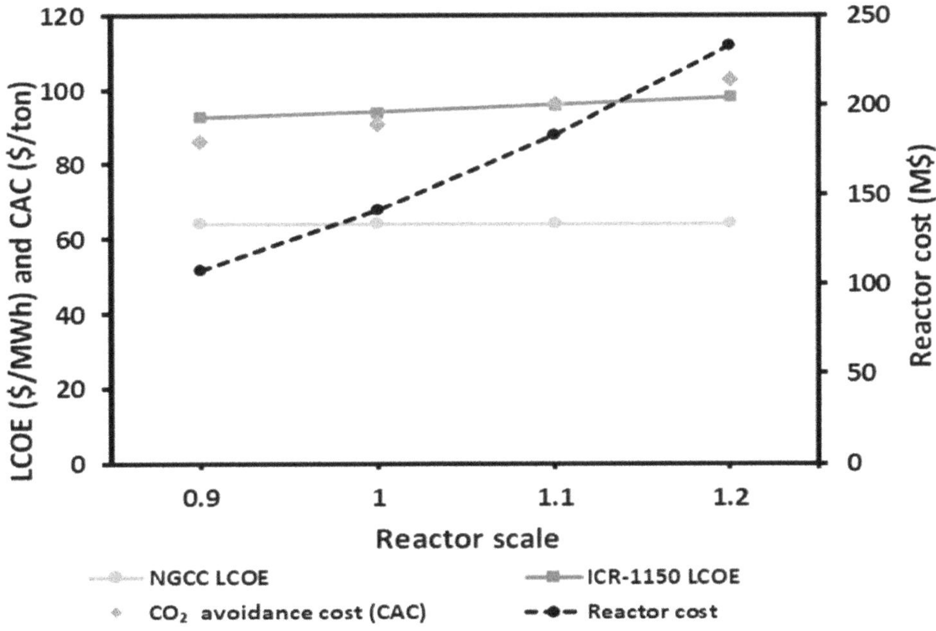

Figure 13. The effect of reactor scale factor (reactor diameter and height relative to the base case) on economic performance indicators.

4. Discussion

The energy and process industry is in a state of flux due to climate change concerns, increasingly stringent environmental regulations, and constraints on raw material availability. However, a swift response to this changing market environment is challenging due to the multidecade scale-up and demonstration timeframes typical in the industry. This study applied a simulation-based approach that could alleviate this challenge to the design and economic evaluation of a novel reactor concept for clean power production from natural gas.

Specifically, filtered two-fluid model (fTFM) simulations were used to design and further investigate a relatively complex internally circulating reactor (ICR) at an industrial scale. Naturally, there are many uncertainties with modeling such a complex multiphase reactor at industrial scales and operating conditions, but the simulation results served to increase confidence in the ICR technology and offered several insights on the practical operation of a large-scale ICR. Such insights would be very expensive and time-consuming to gain via gradual experimental scale-up and demonstration.

The ICR has many different design and operating parameters that can be adjusted to optimize performance. Simulations can allow for cost-effective completion of the very large number of design iterations that will be required to successfully optimize within this large parameter space. Model accuracy is the main uncertainty in this strategy, making validation against large-scale experimental data a high priority. Unfortunately, such data is scarce, often not publicly available, and generally not detailed enough for proper model validation. Given the large fundamental benefits of such a simulation-based design approach, efforts to collect such data from large-scale reactors for further fTFM development and validation is highly recommended. Furthermore, there is substantial scope for further development of fTFM closures for improved accuracy and generality, which would reduce the uncertainty of simulation-based investigations of industrial-scale fluidized beds. Accounting for polydispersity in fTFMs is an especially rich area for research, considering the importance of polydispersity to many industrial applications, the complexity of its subgrid modeling, and the limited amount of research on the topic to date.

Sizing of the ICR via simulations allowed for an economic assessment of the ICR integrated into a natural gas combined cycle (NGCC) power plant for clean power production using the chemical looping combustion (CLC) principle. The primary challenge of using CLC in combined cycles is the maximum temperature achievable in the reactor, which limits the turbine inlet temperature and, hence, the power cycle efficiency. Results from the economic assessment illustrated this with a decrease in CO_2 avoidance costs from $141/ton to $58/ton if the ICR operating temperature could be increased from 1000 to 1300 °C. These results emphasize that one of the main focusses of oxygen carrier development effort should be the high temperature durability of the materials.

Since it is doubtful that ICR operating temperatures can increase far beyond 1200 °C and gas turbine technology keeps progressing beyond 1600 °C to allow for higher efficiencies, the maximum reactor temperature is a serious limitation. One potential solution is to combust additional fuel after the ICR to further increase the gas temperature, as outlined in Khan et al. [34]. The use of such a combustor can eliminate most of the energy penalty, although this is at the expense of increased CO_2 emissions if natural gas is used for the extra firing, or at the expense of added costs if hydrogen is used. The economic implications of this trade-off will be explored in future work involving detailed modeling of a modern gas turbine with a combustor outlet temperature exceeding 1600 °C.

Author Contributions: Conceptualization, J.H.C. and S.C.; data curation, J.H.C. and M.N.K.; formal analysis, J.H.C., M.N.K., and S.C.; funding acquisition, S.A.; investigation, J.H.C. and M.N.K.; methodology, J.H.C., M.N.K., and S.C.; project administration, S.A.; software, J.H.C.; supervision, S.C. and S.A.; validation, J.H.C. and M.N.K.; visualization, J.H.C. and M.N.K.; writing—original draft, J.H.C., M.N.K., and S.C.; writing—review and editing, J.H.C., S.C., and S.A.

Acknowledgments: The resolved simulations used for closure development and verification in this study were performed on computing resources provided by UNINETT Sigma2—the National Infrastructure for High Performance Computing and Data Storage in Norway.

Nomenclature

Acronym Definitions

CCS	Carbon capture and storage	IEA	International Energy Agency
CEPCI	Chemical engineering plant cost index	LCOE	Levelized cost of electricity
CFB	Circulating fluidized bed	NGCC	Natural gas combined cycle
CFD	Computation fluid dynamics	OC	Oxygen carrier
CLC	Chemical looping combustion	O&M	Operating and maintenance
CAC	CO_2 avoidance cost	ROFR	Reduction outlet flow ratio
fTFM	Filtered two-fluid model	T&S	Transport and storage
GPM	Gradient product marker	TFM	Two-fluid model
ICR	Internally circulating reactor		

Main Symbol Definitions

a_0	Specific surface area (m^2/kg)	q	Pressure exponent
C	Molar concentration (mol/m^3)	R	Universal gas constant (J/(mol. K))
D	Diffusion coefficient (m^2/s)	R^H	Heterogenous reaction rate (mol/(m^3.s))
d_p	Particle diameter (m)	S_R	Mass transfer source term (kg/(m^3.s))
E_a	Activation energy (J/mol)	r	Discount rate
e	Plant specific emissions (ton/MWh)	T	Temperature (K)
F	Fuel expenditure ($/year)	t	Time (s)
$\vec{g}$	Gravitational acceleration (m/s^2)	v	Stoichiometric constant
h	Specific enthalpy (J/kg)	X	Species mass fraction
I	Investment expenditures ($/year)	α	Volume fraction
K_{gs}	Interphase momentum exchange coefficient (kg/(m.s))	γ	Heat transfer coefficient (W/(m^3.K))
k_0	Pre-exponential factor	ΔH_r	Heat of reaction (J/kg)
k	Reaction rate constant (m^{3n-3}/mol^{n-1}s)	Δ_f	Filter size (m)
k_s	Reaction rate constant (m^{3n-2}/mol^{n-1}s)	η	Effectiveness factor
k'	Effective reaction rate constant (1/s)	κ	Thermal conductivity (W/(m.K))
M	Molecular weight (kg/mol)	ρ	Density (kg/m^3)
M	O&M expenditures ($/year)	$\overline{\overline{\tau}}$	Stress tensor (Pa)
m	Plant economic lifetime (years)	v	Velocity (m/s)
n	Reaction order	ϕ	Thiele modulus
p	Pressure (Pa)	$\mathcal{L}$	Dimensionless cluster length-scale

Sub- and Superscript Definitions

f	Filter	min	Minimum
g	Gas	p	Particle
i	Species index	r	Reaction
k	Reaction index	ref	Reference
max	Maximum	s	Solids

Sub- and Superscript Definitions

$\overline{x}$	Algebraic volume average	$\widetilde{x}$	Phase-weighted volume average
x'	Fluctuation from mean (algebraic)	x^*	Scaled value
x''	Fluctuation from mean (phase-weighted)	$\vec{x}$	Vector quantity
$\hat{x}$	Dimensionless value		

Appendix A. Reactive Closure Development

Appendix A.1. Resolved Two-Fluid Model (TFM) Simulations

The subgrid reactive closure was developed from resolved TFM simulations using a similar setup, material properties, and methodology to what has been done in previous work [41]; therefore, the detailed description of the simulation setup and statistical analysis of the data is not repeated here.

However, new resolved simulations were performed with some changes in place compared to the aforementioned study in order to collect better data for reaction modeling. Firstly, the height of the periodic simulation domain was doubled to 1.28 m to increase the region for collecting data. Secondly, the species boundary conditions for the reactants were set to a fixed value at the bottom boundary. This ensured that the reactants were never depleted in the periodic simulations, allowing data to be collected continuously. However, imposing this fixed boundary condition influenced the effectiveness factors predicted near the boundary. Therefore, data was only collected from heights between 0.4 and 0.88 m, where the effectiveness factor was found to be independent of the domain height. Four simulations were performed with domain-averaged solids volume fractions of 0.05, 0.1, 0.2, and 0.4 to provide continuous data over the entire range of filtered solids volume fractions.

In the present study, isothermal conditions were assumed, and five solids-catalyzed reactions were considered. In each of the reactions, a single product gas phase species was converted to another product species. All reactions were independent from each other and each species only took part in one reaction. Three first-order reactions were considered with different rate constants, as well as a 0.5th order and a 2nd order reaction. The different reactions are summarized in Table A1. Consequently, the species transport equation can be written as follows for the reactants and products, respectively.

$$\frac{\partial}{\partial t}\left(\alpha_g \rho_g X_i\right) + \nabla \cdot \left(\alpha_g \rho_g \vec{v}_g X_i\right) = \nabla \cdot \left(\alpha_g \rho_g D_i \nabla X_i\right) - k_i \alpha_s C_i^{n_i} M_i, \tag{A1}$$

$$\frac{\partial}{\partial t}\left(\alpha_g \rho_g X_i\right) + \nabla \cdot \left(\alpha_g \rho_g \vec{v}_g X_i\right) = \nabla \cdot \left(\alpha_g \rho_g D_i \nabla X_i\right) + k_i \alpha_s C_i^{n_i} M_i. \tag{A2}$$

Table A1. Summary of the reactions that were considered.

Abbreviation	i—Reactant	i—Product	k_i ($m^{3n-3}/mol^{n-1}s$)	n
1 slow	A	B	15.8	1
1 mid	C	D	63.0	1
1 fast	E	F	252	1
0.5 mid	G	H	63.0	0.5
2 mid	I	J	63.0	2

Appendix A.2. Closure Development

It was previously established that a simple effectiveness factor closure is sufficient for use in reactive fTFMs [41]. Therefore, a simple one-marker closure, similar to the one described previously [41], was developed for the reference reaction, 1 mid. It was found that the following expression can accurately predict the effectiveness factor for the reference reaction as a function of the filter size and the filtered solids volume fraction:

$$-\log(\eta) = \left(\frac{2}{\pi}\right)^3 \operatorname{atan}\left(x_1 \Delta_f^{*\,x_2} \overline{\alpha}_s\right) \operatorname{atan}\left(x_3 \Delta_f^{*\,x_2} \max(\alpha_{s,\max} - \overline{\alpha}_s, 0)\right) \operatorname{atan}\left(x_4 \Delta_f^{*}\right) x_5. \tag{A3}$$

An excellent fit ($R^2 = 0.991$) was obtained against the binned (conditionally-averaged) data from the resolved TFM simulations, using the following model coefficient values: $x_1 = 2.162$, $x_2 = 0.3204$, $x_3 = 5.504$, $x_4 = 1.163$, $x_5 = 1.281$, and $\alpha_{s,\max} = 0.5621$.

Next, it was found that the length factor, $\mathcal{L}$, in Equation (12) can be closed as a function of the filter size. This is understandable, as the average size of the subgrid clusters will increase as the grid size increases in the coarse-grid simulations. Using the methodology described in Section 2.1.2, the best fit to the data from all five reactions was obtained with the following length factor closure:

$$\mathcal{L} = \left(\frac{2}{\pi}\right) \operatorname{atan}\left(x_1 \Delta_f^{*}\right) x_2. \tag{A4}$$

A reasonably good fit of $R^2 = 0.864$ was obtained over all the reaction data with the coefficients $x_1 = 0.08768$ and $x_2 = 71.24$, and when evaluating the minimum effectiveness factor of the reference closure (used in Equations (15) and (16)) at a filtered solids volume fraction of 0.3403.

The good fit against the binned data for such a large variety of reactions implies that applying an analogy to intraparticle mass transfer works sufficiently well for generalizing the reactive fTFM closure. This is further shown in Figure A1, which compares the binned reactions rates from the resolved simulations to the model predictions as a function of the reactant concentration. Both axes are shown on a base-10 log scale to show the entire range of data collected. It is clearly seen from Figure A1a that the importance of the effectiveness factor due to sub-grid effects increases as the reaction rate increases. Furthermore, the generalized effectiveness factor closure performs well in capturing this effect.

Next, Figure A1b shows the importance of accounting for the reaction rate order in the effectiveness factor closure, especially for the 0.5 order reaction. Although some deviations exist in the 0.5 order case, the intraparticle mass transfer analogy generally captures the effect of changing reaction order well. Furthermore, the generalized closure clearly presents a substantial improvement over simply using the closure for a first order reaction, noting that all previous fTFM closures for the reaction effectiveness factor only considered first order reactions.

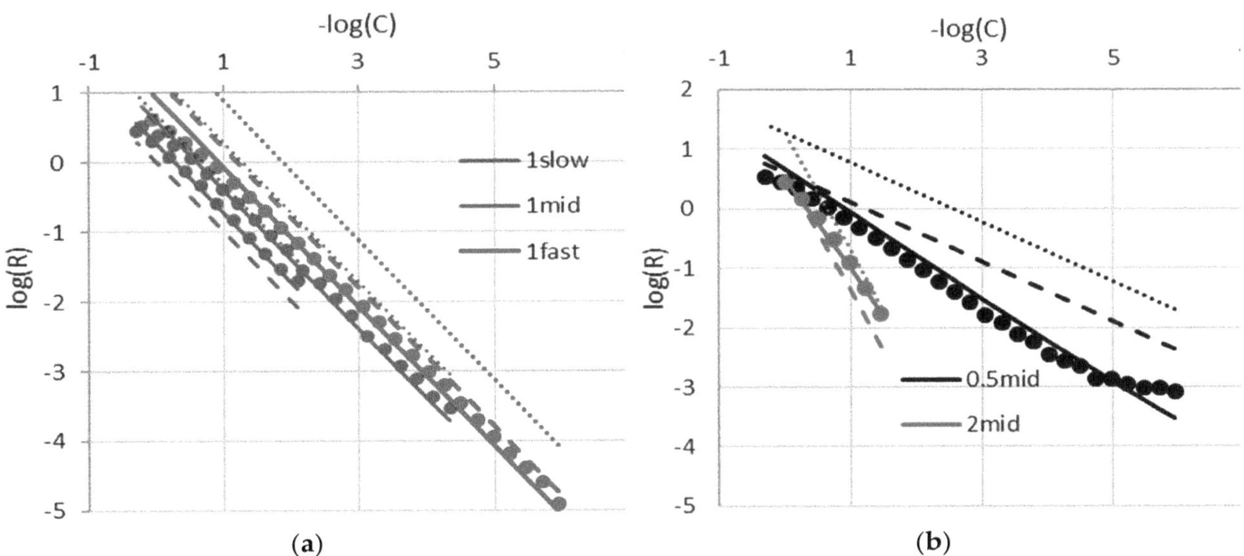

Figure A1. The reaction rate plotted against the concentration, comparing the binned data from resolved simulations (symbols) to model predictions (solid lines) for (**a**) first order and (**b**) other order reactions. The dashed lines show the predictions when using the effectiveness factor from the reference reaction (1 mid) and the dotted line the predictions without an effectiveness factor. Data is shown for the largest filter size considered ($\hat{\Delta}_f = 10.4$) and for an intermediate filtered solids volume fraction ($\overline{\alpha_s} = 0.30$).

Appendix A.3. Closure Verification

The generalized reactive effectiveness factor closure was verified by comparing coarse-grid simulation predictions to that of resolved simulations for the fast bubbling case discussed in the Supplementary Material. Figure A2a shows the importance of subgrid modeling for the reactions, showcasing large overpredictions of the scaled conversion occurring when a closure is neglected, especially for large grid sizes and for the case with large subgrid corrections (fast reaction rate and/or low reaction order). Figure A2b shows that even though both the reference and generalized effectiveness factor closures offer a significant improvement over the case with no closure, the generalized closure consistently outperforms the reference closure for all reactions (except for the reference reaction, C, where the closures are identical and the performance is the same). It is interesting to note that an excellent prediction of the conversion could be obtained despite larger inaccuracies in the hydrodynamics prediction (Figure S4b in the Supplementary Material), suggesting that the effectiveness factor closure is robust.

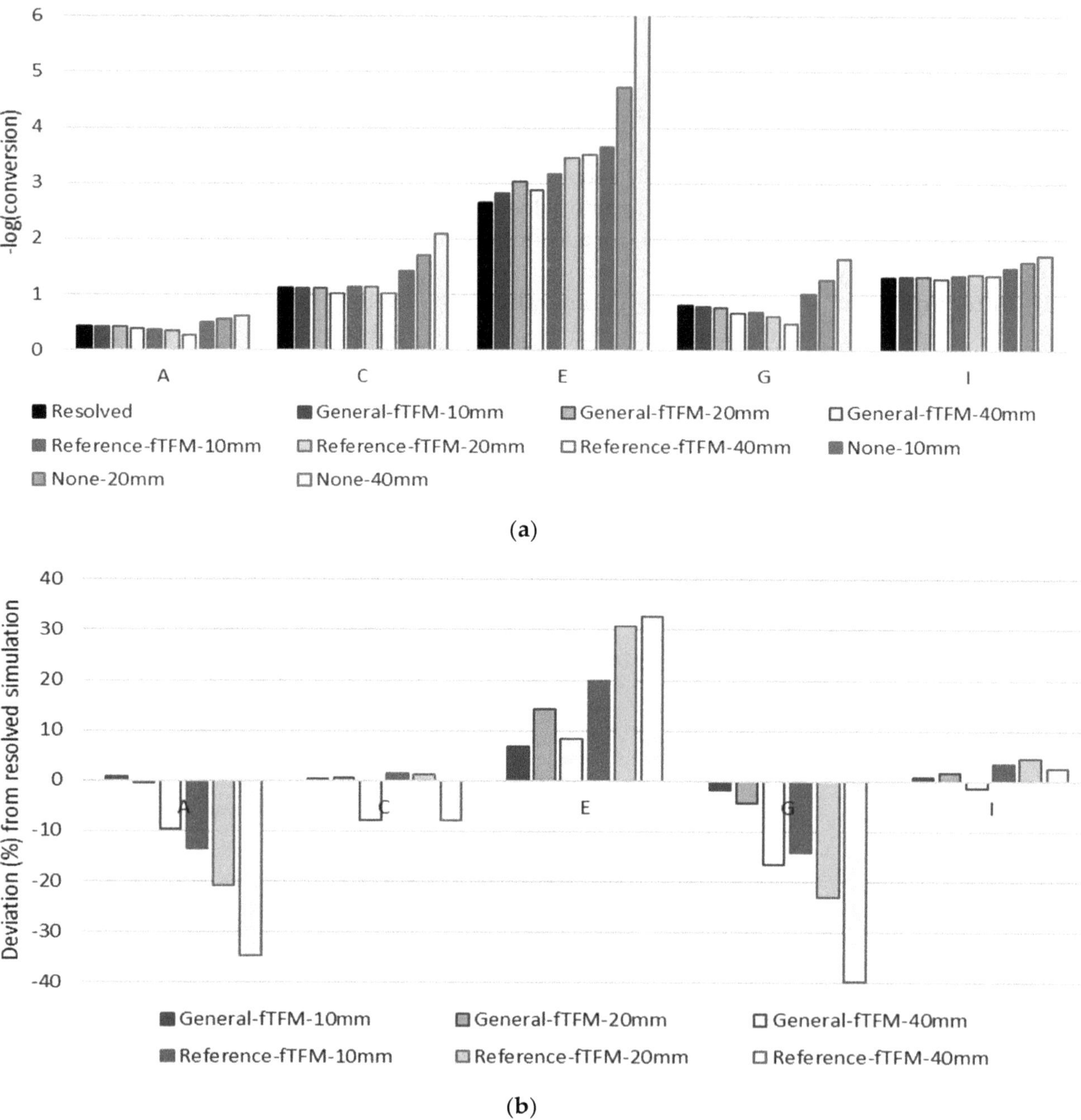

Figure A2. Evaluation of different approaches to modeling the filtered reaction rate in coarse-grid simulations of fast bubbling flow: neglecting subgrid effects on the filtered reaction rate (None), using an effectiveness factor for a reference first-order reaction (Reference), and using the generalized effectiveness factor closure described in this paper (General). Three different grid sizes were considered in the coarse-grid simulations. (**a**) Comparison of the scaled conversion for the different cases; (**b**) comparison of the percentage error of the generalized and reference closure predictions when compared to the benchmark resolved simulations.

The generalized closure is reasonably accurate in predicting the scaled conversion for all reactions, although some discrepancies occur, especially at large grid sizes. However, it is likely that these discrepancies are mainly due to the grid size being too coarse to resolve important macro-scale hydrodynamic flow structures [27], and may therefore not be primarily due to shortcomings of the reactive closure. Future studies will focus on a more detailed verification of reactive fTFM closures.

The results of this section emphasize the benefit of the generalized effectiveness factor closure; therefore, such a closure is highly recommended for future studies of industrial-scale fluidized bed reactors using reactive fTFM simulations.

References

1. IPCC. *Fifth Assessment Report: Mitigation of Climate Change*; Intergovernmental Panel on Climate Change: Cambridge University Press: Cambridge, UK; New York, NY, USA, 2014.
2. Bauer, N.; Calvin, K.; Emmerling, J.; Fricko, O.; Fujimori, S.; Hilaire, J.; Eom, J.; Krey, V.; Kriegler, E.; Mouratiadou, I.; et al. Shared Socio-Economic Pathways of the Energy Sector–Quantifying the Narratives. *Glob. Environ. Chang.* **2017**, *42*, 316–330. [CrossRef]
3. IEAGHG. *CCS in Energy and Climate Scenarios*; IEA Greenhouse Gas R & D Programme: Cheltenham, UK, 2019.
4. Ishida, M.; Zheng, D.; Akehata, T. Evaluation of a chemical-looping-combustion power-generation system by graphic exergy analysis. *Energy* **1987**, *12*, 147–154. [CrossRef]
5. Arnaiz del Pozo, C.; Cloete, S.; Cloete, J.H.; Jiménez Álvaro, Á.; Amini, S. The potential of chemical looping combustion using the gas switching concept to eliminate the energy penalty of CO_2 capture. *Int. J. Greenh. Gas Control* **2019**, *83*, 265–281. [CrossRef]
6. Mattisson, T.; Keller, M.; Linderholm, C.; Moldenhauer, P.; Rydén, M.; Leion, H.; Lyngfelt, A. Chemical-looping technologies using circulating fluidized bed systems: Status of development. *Fuel Process. Technol.* **2018**, *172*, 1–12. [CrossRef]
7. Zaabout, A.; Cloete, S.; Johansen, S.T.; van Sint Annaland, M.; Gallucci, F.; Amini, S. Experimental Demonstration of a Novel Gas Switching Combustion Reactor for Power Production with Integrated CO_2 Capture. *Ind. Eng. Chem. Res.* **2013**, *52*, 14241–14250. [CrossRef]
8. Cloete, S.; Romano, M.C.; Chiesa, P.; Lozza, G.; Amini, S. Integration of a Gas Switching Combustion (GSC) system in integrated gasification combined cycles. *Int. J. Greenh. Gas Control* **2015**, *42*, 340–356. [CrossRef]
9. Håkonsen, S.F.; Blom, R. Chemical Looping Combustion in a Rotating Bed Reactor–Finding Optimal Process Conditions for Prototype Reactor. *Environ. Sci. Technol.* **2011**, *45*, 9619–9626. [CrossRef]
10. Håkonsen, S.F.; Grande, C.A.; Blom, R. Rotating bed reactor for CLC: Bed characteristics dependencies on internal gas mixing. *Appl. Energy* **2014**, *113*, 1952–1957. [CrossRef]
11. Noorman, S.; Gallucci, F.; van Sint Annaland, M.; Kuipers, J.A.M. Experimental Investigation of Chemical-Looping Combustion in Packed Beds: A Parametric Study. *Ind. Eng. Chem. Res.* **2011**, *50*, 1968–1980. [CrossRef]
12. Spallina, V.; Gallucci, F.; Romano, M.C.; Chiesa, P.; Lozza, G.; van Sint Annaland, M. Investigation of heat management for CLC of syngas in packed bed reactors. *Chem. Eng. J.* **2013**, *225*, 174–191. [CrossRef]
13. Zaabout, A.; Cloete, S.; Amini, S. Innovative Internally Circulating Reactor Concept for Chemical Looping-Based CO_2 Capture Processes: Hydrodynamic Investigation. *Chem. Eng. Technol.* **2016**, *39*, 1413–1424. [CrossRef]
14. Osman, M.; Zaabout, A.; Cloete, S.; Amini, S. Internally circulating fluidized-bed reactor for syngas production using chemical looping reforming. *Chem. Eng. J.* **2018**. [CrossRef]
15. Osman, M.; Zaabout, A.; Cloete, S.; Amini, S. Mapping the operating performance of a novel internally circulating fluidized bed reactor applied to chemical looping combustion. *Fuel Process. Technol.* **2020**, *197*, 106183. [CrossRef]
16. IEA. *World Energy Outlook*; International Energy Agency: Paris, France, 2018.
17. Ge, W.; Chang, Q.; Li, C.; Wang, J. Multiscale structures in particle–fluid systems: Characterization, modeling, and simulation. *Chem. Eng. Sci.* **2019**, *198*, 198–223. [CrossRef]
18. Igci, Y.; Andrews, A.T.; Sundaresan, S.; Pannala, S.; O'Brien, T. Filtered two-fluid models for fluidized gas-particle suspensions. *AIChE J.* **2008**, *54*, 1431–1448. [CrossRef]
19. Igci, Y.; Sundaresan, S. Constitutive Models for Filtered Two-Fluid Models of Fluidized Gas–Particle Flows. *Ind. Eng. Chem. Res.* **2011**, *50*, 13190–13201. [CrossRef]
20. Sarkar, A.; Milioli, F.E.; Ozarkar, S.; Li, T.; Sun, X.; Sundaresan, S. Filtered sub-grid constitutive models for fluidized gas-particle flows constructed from 3-D simulations. *Chem. Eng. Sci.* **2016**, *152*, 443–456. [CrossRef]

21. Ozel, A.; Fede, P.; Simonin, O. Development of filtered Euler–Euler two-phase model for circulating fluidised bed: High resolution simulation, formulation and a priori analyses. *Int. J. Multiph. Flow* **2013**, *55*, 43–63. [CrossRef]
22. Ozel, A.; Gu, Y.; Milioli, C.C.; Kolehmainen, J.; Sundaresan, S. Towards filtered drag force model for non-cohesive and cohesive particle-gas flows. *Phys. Fluids* **2017**, *29*, 103308. [CrossRef]
23. Jiang, Y.; Kolehmainen, J.; Gu, Y.; Kevrekidis, Y.G.; Ozel, A.; Sundaresan, S. Neural-network-based filtered drag model for gas-particle flows. *Powder Technol.* **2018**. [CrossRef]
24. Schneiderbauer, S. A spatially-averaged two-fluid model for dense large-scale gas-solid flows. *AIChE J.* **2017**, *63*, 3544–3562. [CrossRef]
25. Schneiderbauer, S.; Saeedipour, M. Approximate deconvolution model for the simulation of turbulent gas-solid flows: An a priori analysis. *Phys. Fluids* **2018**, *30*, 023301. [CrossRef]
26. Gao, X.; Li, T.; Sarkar, A.; Lu, L.; Rogers, W.A. Development and validation of an enhanced filtered drag model for simulating gas-solid fluidization of Geldart A particles in all flow regimes. *Chem. Eng. Sci.* **2018**, *184*, 33–51. [CrossRef]
27. Cloete, J.H.; Cloete, S.; Radl, S.; Amini, S. On the choice of closure complexity in anisotropic drag closures for filtered Two Fluid Models. *Chem. Eng. Sci.* **2019**, *207*, 379–396. [CrossRef]
28. Cloete, J.H.; Cloete, S.; Radl, S.; Amini, S. Verification of filtered Two Fluid Models for reactive gas-solid flows. In Proceedings of the CFD 2017, Trondheim, Norway, 30 May–1 June 2017.
29. Cloete, J.H.; Cloete, S.; Radl, S.; Amini, S. Verification study of anisotropic filtered Two Fluid Model Closures. In Proceedings of the AIChE Annual Meeting, Minneapolis, MN, USA, 29 October–3 November 2017.
30. Holloway, W.; Sundaresan, S. Filtered models for reacting gas–particle flows. *Chem. Eng. Sci.* **2012**, *82*, 132–143. [CrossRef]
31. Agrawal, K.; Holloway, W.; Milioli, C.C.; Milioli, F.E.; Sundaresan, S. Filtered models for scalar transport in gas–particle flows. *Chem. Eng. Sci.* **2013**, *95*, 291–300. [CrossRef]
32. Huang, Z.; Zhang, C.; Jiang, M.; Zhou, Q. Development of a Filtered Interphase Heat Transfer Model Based on Fine-Grid Simulations of Gas-Solid Flows. *AIChE J.* **2019**. [CrossRef]
33. Cloete, J.H.; Cloete, S.; Municchi, F.; Radl, S.; Amini, S. Development and verification of anisotropic drag closures for filtered Two Fluid Models. *Chem. Eng. Sci.* **2018**, *192*, 930–954. [CrossRef]
34. Khan, M.N.; Cloete, S.; Amini, S. Efficiency Improvement of Chemical Looping Combustion Combined Cycle Power Plants. *Energy Technol.* **2019**. in Press. [CrossRef]
35. Gidaspow, D.; Bezburuah, R.; Ding, J. Hydrodynamics of Circulating Fluidized Beds, Kinetic Theory Approach. In Proceedings of the 7th Engineering Foundation Conference on Fluidization, Brisbane, Australia, 3–8 May 1992; pp. 75–82.
36. Lun, C.K.K.; Savage, S.B.; Jeffrey, D.J.; Chepurniy, N. Kinetic Theories for Granular Flow: Inelastic Particles in Couette Flow and Slightly Inelastic Particles in a General Flow Field. *J. Fluid Mech.* **1984**, *140*, 223–256. [CrossRef]
37. Cloete, J.H.; Cloete, S.; Municchi, F.; Radl, S.; Amini, S. The sensitivity of filtered Two Fluid Model to the underlying resolved simulation setup. *Powder Technol.* **2017**, *316*, 265–277. [CrossRef]
38. Milioli, C.C.; Milioli, F.E.; Holloway, W.; Agrawal, K.; Sundaresan, S. Filtered two-fluid models of fluidized gas-particle flows: New constitutive relations. *AIChE J.* **2013**, *59*, 3265–3275. [CrossRef]
39. Cloete, J.H.; Cloete, S.; Radl, S.; Amini, S. Development and verification of anisotropic solids stress closures for filtered Two Fluid Models. *Chem. Eng. Sci.* **2018**, *192*, 906–929. [CrossRef]
40. Cloete, S.; Cloete, J.H.; Amini, S. Hydrodynamic validation study of filtered Two Fluid Models. *Chem. Eng. Sci.* **2018**, *182*, 93–107. [CrossRef]
41. Cloete, J.H. Development of Anisotropic Filtered Two Fluid Model Closures. Ph.D. Thesis, Norwegian University of Science and Technology, Trondheim, Norway, 2018.
42. Zhang, D.Z.; VanderHeyden, W.B. The effects of mesoscale structures on the macroscopic momentum equations for two-phase flows. *Int. J. Multiph. Flow* **2002**, *28*, 805–822. [CrossRef]
43. Zaabout, A.; Cloete, S.; Amini, S. Autothermal operation of a pressurized Gas Switching Combustion with ilmenite ore. *Int. J. Greenh. Gas Control* **2017**, *63*, 175–183. [CrossRef]
44. Kuusik, R.; Trikkel, A.; Lyngfelt, A.; Mattisson, T. High temperature behavior of NiO-based oxygen carriers for Chemical Looping Combustion. *Energy Procedia* **2009**, *1*, 3885–3892. [CrossRef]

45. Abad, A.; Adánez, J.; García-Labiano, F.; de Diego, L.F.; Gayán, P.; Celaya, J. Mapping of the range of operational conditions for Cu-, Fe-, and Ni-based oxygen carriers in chemical-looping combustion. *Chem. Eng. Sci.* **2007**, *62*, 533–549. [CrossRef]
46. Levenspiel, O. *Chemical Reaction Engineering*, 3rd ed.; John Wiley & Sons: Hoboken, NJ, USA, 1999.
47. Thiele, E.W. Relation between Catalytic Activity and Size of Particle. *Ind. Eng. Chem.* **1939**, *31*, 916–920. [CrossRef]
48. Yang, W.; Cloete, S.; Morud, J.; Amini, S. An Effective Reaction Rate Model for Gas-Solid Reactions with High Intra-Particle Diffusion Resistance. *Int. J. Chem. React. Eng.* **2016**, *14*, 331. [CrossRef]
49. Rawlings, J.B.; Ekerdt, J.G. Chapter 7. In *Chemical Reactor Analysis and Design Fundamentals*; Nob Hill Publishing: Madison, WI, USA, 2002.
50. Chevrier, S. Development of Subgrid Models for a Periodic Circulating Fluidized Bed of Binary Mixture of Particles. Ph.D. Thesis, Université de Toulouse, Toulouse, France, 2017.
51. Naqvi, R.; Wolf, J.; Bolland, O. Part-load analysis of a chemical looping combustion (CLC) combined cycle with CO_2 capture. *Energy* **2007**, *32*, 360–370. [CrossRef]
52. EBTF. *European Best Practice Guide for Assessment of CO_2 Capture Technologies*; European Benchmark Task Force, European Commission: Brussels, Belgium, 2011.
53. Johnson, P.C.; Jackson, R. Frictional-Collisional Constitutive Relations for Granular Materials, with Application to Plane Shearing. *J. Fluid Mech.* **1987**, *176*, 67–93. [CrossRef]
54. Patankar, S. *Numerical Heat Transfer and Fluid Flow*; Hemisphere Publishing Corporation: New York, NY, USA, 1980.
55. Leonard, B.P.; Mokhtari, S. *ULTRA-SHARP Nonoscillatory Convection Schemes for High-Speed Steady Multidimensional Flow*; NASA Lewis Research Center: Cleveland, OH, USA, 1990.
56. Gerdes, K.; Summers, W.M.; Wimer. *Quality Guidelines for Energy System Studies: Cost Estimation Methodology for NETL Assessments of Power Plant Performance*; DOE/NETL-2011/1455 United States 10.2172/1513278 NETL-IR English; NETL: Pittsburgh, PA, USA, 2011; p. Medium: ED.
57. Turton, R.; Bailie, R.C.; Whiting, W.B.; Shaeiwitz, J.A.; Bhattacharyya, D. Appendix A. In *Analysis, Synthesis and Design of Chemical Processes*; Pearson Education: Upper Saddle River, NJ, USA, 2008.
58. Peters, M.S.; Timmerhaus, K.D. *Plant Design and Economics for Chemical Engineers*; McGraw-Hill: New York, NY, USA, 1991. [CrossRef]
59. Adanez, J.; Abad, A.; Garcia-Labiano, F.; Gayan, P.; de Diego, L.F. Progress in Chemical-Looping Combustion and Reforming technologies. *Prog. Energy Combust. Sci.* **2012**, *38*, 215–282. [CrossRef]
60. Spallina, V.; Pandolfo, D.; Battistella, A.; Romano, M.C.; Van Sint Annaland, M.; Gallucci, F. Techno-economic assessment of membrane assisted fluidized bed reactors for pure H2 production with CO_2 capture. *Energy Convers. Manag.* **2016**, *120*, 257–273. [CrossRef]
61. IEAGHG. *The Costs of CO_2 Transport: Post-Demonstration CCS in the EU*; European Technology Platform for Zero Emission Fossil Fuel Power Plants: Brussels, Belgium, 2011.
62. IEAGHG. *The Costs of CO_2 Storage: Post-Demonstration CCS in the EU*; European Technology Platform for Zero Emission Fossil Fuel Power Plants: Brussels, Belgium, 2011.
63. Rubin, E.S.; Davison, J.E.; Herzog, H.J. The cost of CO_2 capture and storage. *Int. J. Greenh. Gas Control* **2015**, *40*, 378–400. [CrossRef]

Permissions

All chapters in this book were first published in MDPI; hereby published with permission under the Creative Commons Attribution License or equivalent. Every chapter published in this book has been scrutinized by our experts. Their significance has been extensively debated. The topics covered herein carry significant findings which will fuel the growth of the discipline. They may even be implemented as practical applications or may be referred to as a beginning point for another development.

The contributors of this book come from diverse backgrounds, making this book a truly international effort. This book will bring forth new frontiers with its revolutionizing research information and detailed analysis of the nascent developments around the world.

We would like to thank all the contributing authors for lending their expertise to make the book truly unique. They have played a crucial role in the development of this book. Without their invaluable contributions this book wouldn't have been possible. They have made vital efforts to compile up to date information on the varied aspects of this subject to make this book a valuable addition to the collection of many professionals and students.

This book was conceptualized with the vision of imparting up-to-date information and advanced data in this field. To ensure the same, a matchless editorial board was set up. Every individual on the board went through rigorous rounds of assessment to prove their worth. After which they invested a large part of their time researching and compiling the most relevant data for our readers.

The editorial board has been involved in producing this book since its inception. They have spent rigorous hours researching and exploring the diverse topics which have resulted in the successful publishing of this book. They have passed on their knowledge of decades through this book. To expedite this challenging task, the publisher supported the team at every step. A small team of assistant editors was also appointed to further simplify the editing procedure and attain best results for the readers.

Apart from the editorial board, the designing team has also invested a significant amount of their time in understanding the subject and creating the most relevant covers. They scrutinized every image to scout for the most suitable representation of the subject and create an appropriate cover for the book.

The publishing team has been an ardent support to the editorial, designing and production team. Their endless efforts to recruit the best for this project, has resulted in the accomplishment of this book. They are a veteran in the field of academics and their pool of knowledge is as vast as their experience in printing. Their expertise and guidance has proved useful at every step. Their uncompromising quality standards have made this book an exceptional effort. Their encouragement from time to time has been an inspiration for everyone.

The publisher and the editorial board hope that this book will prove to be a valuable piece of knowledge for researchers, students, practitioners and scholars across the globe.

List of Contributors

Haotian Wu and Xueping Gu
School of Electric and Electrical Engineering, North China Electric Power University, Baoding 071003, China

Hang Li,
CEPREI (Beijing) Industrial Technology Research Institute Co., Ltd., Beijing 100041, China

Xingli Zhai
Jinan Power Supply Company in Shandong Provincial Electric Power Company of State Grid, Jinan 250000, China

Ning Wang
College of Electrical Engineering, Zhejiang University, Hangzhou 310027, China

Yao Wang, HaiTao Yu, Zhiyuan Che, Yuchen Wang and Yulei Liu
School of Electrical Engineering, Southeast University, Nanjing 210096, China

Ammar Ali and Khurram Saleem Alimgeer
Department of Electrical and Computer Engineering, COMSATS University Islamabad, Islamabad 44000, Pakistan

Ghulam Hafeez
Department of Electrical and Computer Engineering, COMSATS University Islamabad, Islamabad 44000, Pakistan
Department of Electrical Engineering, University of Engineering and Technology, Mardan 23200, Pakistan

Muhammad Usman
Department of Electrical Engineering, University of Engineering and Technology, Mardan 23200, Pakistan

Noor Islam
Department of Electrical Engineering, CECOS University of IT & Emerging Sciences, Peshawar 25124, Pakistan

Salman Ahmad
Department of Electrical and Computer Engineering, COMSATS University Islamabad, Islamabad 44000, Pakistan
Department of Electrical Engineering, Wah Engineering College, University of Wah, Wah Cantt 47070, Pakistan

Athraa Ali Kadhem
Center for Advanced Power and Energy Research, Faculty of Engineering, University Putra Malaysia, Selangor 43400, Malaysia

Noor Izzri Abdul Wahab
Advanced Lightning, Power and Energy Research, Faculty of Engineering, University Putra Malaysia, Selangor 43400, Malaysia

Ahmed N. Abdalla
Faculty of Electronics Information Engineering, Huaiyin Institute of Technology, Huai'an 223003, China

Hongzhong Chen, Jun Tang, Xiaolei Wang and Yeying Mao
State Grid Suzhou Power Supply Company, Suzhou 215004, China

Lei Sun
School of Electrical and Automation Engineering, Hefei University of Technology, Hefei 230009, China

Jiawei Zhou
State Grid Suzhou Power Supply Company Suzhou Electric Power Design Institute Co., Ltd., Suzhou 215004, China

Yuanyuan Zhao, Xiangyu Si, Xiuli Wang, Rongsheng Zhu, Qiang Fu and Huazhou Zhong
National Research Center of Pumps, Jiangsu University, Zhenjiang 212013, China

Ning Gui
School of Comuputer Science and Engineering, Central South University, Changsha 410000, China

Jieli Lou
School of Mechanical Engineering and Automation, Zhejiang Sci-Tech. University, Hangzhou 310000, China

Zhifeng Qiu and Weihua Gui
School of Automation, Central South University, Changsha 410000, China

Hongqiang Guo, Shangye Du, Fengrui Zhao, Qinghu Cui and Weilong Ren
School of Mechanical & Automotive Engineering, Liaocheng University, Liaocheng 252059, China

Mohd. Wazir Mustafa and Madihah Md. Rasid
School of Electrical Engineering, Universiti Teknologi Malaysia, Johor Bahru 81310, Malaysia

Touqeer Ahmed Jumani
School of Electrical Engineering, Universiti Teknologi Malaysia, Johor Bahru 81310, Malaysia
Department of Electrical Engineering, Mehran University of Engineering and Technology SZAB Campus Khairpur Mirs, Khairpur 66020, Pakistan

Waqas Anjum
School of Electrical Engineering, Universiti Teknologi Malaysia, Johor Bahru 81310, Malaysia
Department of Electronic Engineering, The Islamia University of Bahawalpur, Bahawalpur 63100, Pakistan

Sara Ayub
School of Electrical Engineering, Universiti Teknologi Malaysia, Johor Bahru 81310, Malaysia
Department of Electrical Engineering, Balochistan University of Information Technology, Engineering and Management Sciences, Quetta 87300, Pakistan

Tao Zheng, Zemei Dai, Yufeng Yang and Jing Cao
NARI Group Corporation (State Grid Electric Power Research Institute), Nanjing 211106, China
NARI Technology Development Co. Ltd., Nanjing 211106, China
State Key Laboratory of Smart Grid Protection and Control, Nanjing 211106, China

Jiahao Yao
Key Laboratory of Measurement and Control of Complex Systems of Engineering, Ministry of Education, School of Automation, Southeast University, Nanjing 210096, China

Jan Hendrik Cloete and Schalk Cloete
Flow Technology Research Group, SINTEF Industry, 7465 Trondheim, Norway

Mohammed N. Khan
Department of Energy and Process Engineering, Norwegian University of Science and Technology, 7491 Trondheim, Norway

Shahriar Amini
Flow Technology Research Group, SINTEF Industry, 7465 Trondheim, Norway
Department of Energy and Process Engineering, Norwegian University of Science and Technology, 7491 Trondheim, Norway

Index